"As palm oil production continues to expand in Southeast Asia while also moving into Latin America and Africa, this timely volume challenges the notion that there can be only one single approach to making palm oil production sustainable. Instead, based on a great variety of case studies, the book highlights the diversity of national governance regimes that seek to find solutions to the longstanding challenge of reconciling environmental protection, social welfare and justice, and economic development. The book thus offers a novel perspective and new empirical insights on one of the world's most important, and most contested, agricultural commodities."

Karen M. Siegel, *Head of Research Group "Transformation and Sustainability Governance in South American Bioeconomies", University of Münster, Germany*

"This timely edited volume draws on the methodology of 'the palm oil assemblage' to reveal the deep complexities of this industry in different parts of the world. The book's comparative approach across two key palm oil-producing regions in the Global South, namely Southeast Asia and Latin America, coupled with a final chapter on the African setting, shows how the industry in different sites combine different mixes of human actors and non-human entities (landscapes, soils, pests, technology) that interact to shape diverse national and local governance complexes. Their insightful findings call into question the 'one-size fits all' approach to governing palm oil."

Helen E. S. Nesadurai, *Professor of International Political Economy, Monash University Malaysia*

"Amidst the seemingly relentless march towards the climate crisis, the desire for economic development, and the desperate calls for global justice, the palm oil industry continues to be one of the hottest—and most complex—sites of contestation between different forces with diverging interests. In this volume, a group of experts, gathered by the world's leading authorities on the topic, attempts to dissect these complexities by looking at the governance of the industry in key countries in the Global South. This is an indispensable volume for everyone concerned with global environmental politics, political economy, development, and—most importantly—our common future."

Shofwan Al Banna Choiruzzad, *Associate Professor, Faculty of Social and Political Sciences, Universitas Indonesia*

"Palm oil is a highly efficient and profitable system of production that has taken over forested landscapes across the tropical Global South, ruthlessly driving agrarian development that has not always benefited the local peoples. This book presents an excellent overview of governance of the palm oil industry, and provides an exciting approach to examining the complex assemblages of intertwined materialities, actors, interests, power, discourses, spatial dynamics, ecologies, and political-economic processes across different scales. This book is thus an important contribution to understanding the possibilities for if, and how, the palm oil 'complex' will be able to adequately respond to sustainability and justice demands."

Grace Y. Wong, *Associate Professor, Research Institute for Humanity and Nature and Stockholm Resilience Centre*

"The palm oil industry faces unprecedented challenges both in its main production regions and globally. This book highlights the policy, governance, management, and ethical challenges that will determine the future of oil palm as a global commodity, notwithstanding the existential threat posed by changing climates on productivity and yields. Those charged with charting a future for the palm oil sector are well advised to consider the questions that this timely, and indeed overdue, book raises. How these challenges are addressed will determine the future of oil palm landscapes as global public goods and/or increasingly stranded private assets."

Sayed Azam-Ali, *Professor and Chair (Emeritus) in Global Food Security, University of Nottingham and Member, UN High-Level Panel of Experts in Food Security and Nutrition*

GOVERNING THE PALM OIL INDUSTRY

This book examines how different countries across Southeast Asia and Latin America respond to the emergence and expansion of the lucrative, yet controversial palm oil industry, paying attention to how national policy and governance regimes are shaping this global industry.

With its historic roots in Southeast Asia, oil palm cultivation continues to expand beyond its historical centres. In Latin America, many countries are now developing their own policies to promote and govern oil palm cultivation. This book provides a unique examination of how different countries strive to strike a balance between developmental and environmental concerns, through case studies on Indonesia, Malaysia, the Philippines, Thailand, Colombia, Brazil, Ecuador, Honduras, and Mexico, and an outlook for the industry's prospects in Africa. This book applies an assemblage approach to draw out lessons on the global challenges posed by the industry and how differing national governance regimes and communities might respond to them. Rather than a single global industry, the book unveils a complex arrangement of national and even local palm oil assemblages, indicating that there is more than one way to do palm oil. In doing so, the book contributes to a better understanding of the drivers and processes that shape the governance of the industry, both in different nations and globally.

This book will be of great interest to students and scholars of the palm oil industry, as well as those interested in natural resource governance, sustainable agriculture, conservation, environmental justice, and environmental and development policy more broadly.

Patrick O'Reilly is a Teaching Fellow in Sociology at Liverpool John Moores University, UK. He has over 30 years of applied and research experience in rural development.

Helena Varkkey is an Associate Professor at Universiti Malaya, Malaysia. She works on global palm oil politics and transboundary haze governance in Southeast Asia.

Earthscan Studies in Natural Resource Management

For more information about this series, please visit: www.routledge.com/books/series/ECNRM/

GOVERNING THE PALM OIL INDUSTRY

Perspectives from Southeast Asia
and Latin America

Edited by
Patrick O'Reilly and Helena Varkkey

Routledge
Taylor & Francis Group
LONDON AND NEW YORK

earthscan
from Routledge

Designed cover image: Getty images

First published 2025
by Routledge
4 Park Square, Milton Park, Abingdon, Oxon OX14 4RN

and by Routledge
605 Third Avenue, New York, NY 10158

Routledge is an imprint of the Taylor & Francis Group, an informa business

British Library Cataloguing-in-Publication Data
A catalogue record for this book is available from the British Library

ISBN: 978-1-032-60555-5 (hbk)
ISBN: 978-1-032-60552-4 (pbk)
ISBN: 978-1-003-45960-6 (ebk)

DOI: 10.4324/9781003459606

Typeset in Times New Roman
by codeMantra

CONTENTS

SECTION 3
Outlook 253

CONTRIBUTORS

Rodolfo O. Abalus Jr. is a Director at the Palawan State University's Extension Services Office in Palawan, The Philippines. His research interest is in forest resources management, specialising in watershed management.

Sarah Ali is a PhD candidate at Universiti Malaya, Kuala Lumpur, Malaysia. As part of her thesis, she is currently studying the emotional responses and coping strategies of Malaysian men towards COVID-19, utilising digital ethnography.

Gusti Anshari is a Professor at Tanjungpura University, Pontianak, West Kalimantan, Indonesia. He has worked as a tropical peatland researcher for the last 20 years, and his expertise includes peatland ecology and conservation, and the role of peatlands in the global carbon cycle.

Emmy Antang was formerly a Senior Lecturer at the University of Palangka Raya, Kalimantan, Indonesia.

Corry Antang is a Lecturer at the University of Palangka Raya, Kalimantan, Indonesia.

Iñigo Arrazola Aranzabal belongs to the Critical Geography Collective of Ecuador. He also works at the Ecumenic Human Rights Commission of Ecuador. He is also a member of the Agrarian Reform Research Group (NERA) at the Universidade Federal da Bahia, Brazil. His research focuses on the agrarian question and rural transformations in both Ecuador and Brazil.

Diana Córdoba is an Assistant Professor at Queen's University, Canada. Her research focuses on the study of social and environmental impacts of new practices, technologies, and models of rural and territorial development, and the role played by the state, NGOs, social movements, and agrarian organisations in the implementation of these initiatives.

Francisca R. Dimaano is a Campus Director at the Palawan State University, San Rafael, The Philippines.

Stephanie Evers is a reader in Aquatic Ecology and Biogeochemistry at Liverpool John Moores University, as well as an Honorary Associate Professor in Environmental Science at the University of Nottingham, Malaysia campus. Her primary research focus is on wetlands ecology and biogeochemistry.

Mélanie Feurer is a Research Associate at the Bern University of Applied Sciences, Switzerland. Her areas of expertise include agroforestry, community forestry, ecosystem services, climate change mitigation, as well as international forest and climate policy and climate funding.

Ingrid Fromm is a Research Associate and Lecturer at the Bern University of Applied Sciences, Switzerland. She is an expert in value chain analysis in agriculture, the promotion of SMEs and entrepreneurship in developing countries, and socio-economic research methods for development.

Paul R. Furumo researched the Colombian oil palm sector as a Fulbright Fellow and National Geographic Explorer. His postdoctoral research at Stanford University focused on public-private strategies for zero-deforestation commodity production. Currently, he works on state climate policy as a science fellow with the California Council on Science & Technology in Sacramento, United States.

Julianne A. Hazlewood is an intercultural geographer and lecturer at Rachel Carson College at the University of California Santa Cruz, United States. She is also the co-founder and Executive Director of Roots & Routes IC, an organisation dedicated to facilitating the sharing of cultural ways of knowing and compassion between diverse cultures en route to responsibly stewarding a flourishing living world.

Adi Jaya has worked for the University of Palangka Raya, Kalimantan, Indonesia since 1988 and has over 20 years of experience researching tropical peatland ecosystems. He is currently the Director of the Centre for International Cooperation in the Sustainable Management of Tropical Peatland (UPT CIMTROP), University of Palangka Raya.

Antonio Jonay Jovani Sancho is a Greenhouse Gas Flux Field and Data Scientist at the UK Centre for Ecology & Hydrology. He has a PhD in Soil Science from the University of Limerick, Ireland.

Khor Yu Leng is the Director of Khor Reports and Segi Enam Advisors Pte Ltd, Singapore. She is a senior economist specialising in the environmental, social, and governance (ESG) framework, sustainable supply chains, and Southeast Asia.

May C. Lacao is Development Management Officer V for the Provincial Government of Palawan, Puerto Princesa City, in Palawan, The Philippines. Her areas of specialisation include natural capital accounting, conservation financing, enterprise development, and corporate finance.

Geovanna Lasso is an external lecturer at the Universidad Andina Simón Bolívar and the Universidad Politécnica Salesiana, Ecuador. A researcher and activist on issues of food sovereignty, political agroecology and sustainable territorial transformations, Geovanna forms part of the Agroecological Collective of Ecuador's coordination team.

Renata Moreno is a Professor at Universidad Autónoma de Occidente, Cali, Colombia. Her research is focused on environmental governance, hydro-social territories, as well as food and energy systems.

Sebastian Mengel is a Research Associate and Project Acquisition Manager at the Bern University of Applied Sciences, Switzerland. His expertise includes international and rural development, natural resource management, sustainable production systems, agricultural value chains, and market-systems approaches.

Olawale Emmanuel Olayide is a Senior Lecturer and Coordinator of the Sustainable Development Practice Programme at the University of Ibadan, Ibadan, Nigeria. He has expertise in agriculture and food systems, climate change, sustainable rural development, resource economics, circular economy, and industrial ecology.

Patrick O'Reilly currently works as a Teaching Fellow at Liverpool John Moores University. He has over 30 years of experience in Rural Development. During his time, he has been involved in multiple research and applied projects in over 30 countries in Europe, Asia, Africa and Latin America. His current work includes a focus on Southeast Asia, particularly Indonesia and Malaysia, focusing on the intersection between policy, livelihood and the environment. He also has research exploring peatland policy in Europe, comparative environmental policy, food insecurity, and transdisciplinary research.

Susan E. Page is a Professor of Physical Geography at the University of Leicester, United Kingdom. She studies the impacts of land use and fire on tropical peatland carbon dynamics, ecosystem restoration, ecosystem services and livelihoods.

María Moreno Parra is a research associate at FLACSO Ecuador and a lecturer at the University of Wisconsin-La Crosse in the United States. Her research is at the intersection of racism and antiracism, gender studies, and ethnic politics in the Andes.

Michael D. Pido is Director at the Centre for Strategic Policy and Governance, Palawan State University, Philippines. His professional interests include integrated coastal and river basin management, socioeconomic monitoring, fisheries co-management, rapid/participatory appraisals, and protected area management.

Erin C. Pischke is a Legislative Analyst for the Oregon Legislature, providing non-partisan, objective research, data and committee management for the members of the Oregon Legislature in the House Committees on Climate, Energy, and Environment; Environment and Natural Resources; and Water by leveraging her subject matter expertise in renewable energy policy, climate change, socioecological systems, and environmental policy.

John Francisco A. Pontillas is a Public Policy Analyst and Head of the ECAN Policy Research and Planning Division, Palawan Council for Sustainable Development, the Philippines. His professional interests include sustainable development, environment and natural resources economics, political economy of the environment, and natural resources management.

Sofie Sjogersten is affiliated with the School of Biosciences, University of Nottingham, Sutton Bonington, United Kingdom.

Daniel Sombra is a Professor at the Federal University of Pará, Belem, Brazil. He has research and extensive experience in the following areas: economic-ecological zoning, agroecological zoning, regionalisation, environmental analysis, environmental management and governance; socio-environmental conflicts, river basins and water resources, participatory cartography, and geotechnologies.

Nithiyah Tamilwanan is a Research Associate at Segi Enam Advisors Pte Ltd, Singapore. She is a law graduate from Queen's University Belfast, with experience in lecturing and digital marketing at Brickfields Asia College, a leading law school in Malaysia.

Caroline Upton is a Professor of Human Geography at the University of Leicester, United Kingdom. Most recently, she has worked with Susan E. Page and Patrick O'Reilly in interdisciplinary analyses of peatland livelihoods and conservation (e.g. Sustainpeat project, ongoing) and through attention to resilience, ecosystem service values, and governance challenges under climate change.

Helena Varkkey is an Associate Professor at Universiti Malaya in Kuala Lumpur, Malaysia. Her monograph "The Haze Problem in Southeast Asia: Palm Oil and Patronage" was published by Routledge Malaysia Studies Series in 2016.

Paul Wilson is a Professor of Agricultural Economics, at the University of Nottingham, United Kingdom. Paul's research interests are in the areas of agricultural economics and farm business management.

ACKNOWLEDGEMENTS

We acknowledge funding from the Equitable Society Research Cluster (ESRC), Universiti Malaya grant number GC003B-17SBS which supported a two-day International Workshop on Environmental Governance in the Palm Oil Sector in Southeast Asia and Latin America at the Centre for Latin American Studies in 2018. Here, scholars working on six countries came together to workshop early papers exploring policies towards the palm oil industry in both these regions. This workshop sparked the initial idea for this edited collection.

We would like to thank the International Review of Modern Sociology, particularly its editor-in-chief, Professor Sunil Kukreja from the University of Puget Sound, Washington, for inviting us to curate a special section on oil palm assemblages in Southeast Asia and Latin America in Volume 46, Issue 1 of the Review in 2020. This special section contained early versions of six chapters found in this collection:

- O'Reilly, P., & Varkkey, H. (2020). Palm oil governance in different locations: Using the assemblage approach to understand a "complex" sector. *International Review of Modern Sociology*, *46*(1), 1–17.
- Furumo, P. (2020). Assemblage of sustainability governance in the Colombian oil palm sector. *International Review of Modern Sociology*, *46*(1), 19–49.
- Varkkey, H. (2020). Palm oil, state autonomy, and assemblage of land use governance in Sarawak, Malaysia. *International Review of Modern Sociology*, *46*(1), 51–77.
- Fromm, I., Feurer, M., & Mengel, S. (2020). Sustainable palm oil production in Honduras: Myth or reality? *International Review of Modern Sociology*, *46*(1), 79–101.

- O'Reilly, P., Anshari, G., Sancho, J., Jaya, A., Antang, E., Antang, C., Evers, S., Evans, C., Wilson, P., Crout, N., Sjorgesten, S., Upton, C., & Page, S. (2020). Oil palm governance at the grassroots: How assemblage links oil palm, livelihoods, and local administration in an Indonesian village. *International Review of Modern Sociology*, *46*(1), 103–120.
- Pischke, E. (2020). Oil palm production regimes and resistance in Mexico's oil palm assemblage. *International Review of Modern Sociology*, *46*(1), 121–141.

The editors would like to express our sincerest appreciation to all the authors who contributed to this collection. Some of these authors were with us from the very beginning of this journey at the workshop in 2018, and some joined us later as we worked on bringing in representation from more countries from both of our focus regions, as well as an outlook chapter on Africa. We thank you for your willingness to share your subject matter knowledge and passion for your work with us.

Thank you to Sarah Ali, our editorial assistant, for her great attention to detail and dedication in preparing the manuscript for publication. We would like to also acknowledge Katie Stokes, Hannah Ferguson, and Rosie Anderson at Routledge, as well as all the anonymous reviewers, for believing in this project.

The authors of the chapter on Indonesia acknowledge the financial support provided by the Biotechnology and Biological Sciences Research Council (BBSRC), United Kingdom grant number BB/P023533/1 (SUSTAINPEAT). The authors would also like to dedicate their chapter to the memory of Emmy Antang who sadly passed away before this collection was published.

We would like to acknowledge all informants, interviewees, and respondents who kindly shared their insights and experiences to be integrated across the chapters. The authors of the chapter on the Philippines would like to particularly acknowledge the institutional support provided by the Palawan State University and Palawan Council for Sustainable Development, Philippines, and the contributions of the following individuals: Dr. Romeo Cabungcal of the Provincial Agriculture Office, Former Board Member Al Rama of the Legislative of the Provincial Government of Palawan, Chief Raul Aguilar, Ms. Marisol R. Ortiz, Mr. Ramon Rivera, and Ms. Noeme Kate Ben-Ek of the Philippine Coconut Authority, Provincial Office, and Dr. Romeo Lerom of Western Philippines University.

The opinions expressed in this collection are those solely of the authors. They do not represent the institutions where they belong.

ACRONYMS AND ABBREVIATIONS

A

ADM	Archer Daniel Midland
AEDP	Alternative Energy Development Plan
AF	*Amarelecimento fatal*; bud rot
AGPI	Agumil Philippines, Inc.
ALDAW	Ancestral Land/Domain Watch
ALRO	Agricultural Land Reform Office
ANP	*Agência Nacional do Petróleo, Gás Natural e Biocombustíveis*; National Petroleum Agency
APE	*Alianzas Productivas Estratégicas*; Strategic Productive Alliances

B

BAAC	Bank for Agriculture and Agricultural Cooperatives
BADANESA	National Development Bank
BBC	British Broadcasting Corporation
BRG	*Badan Restorasi Gambut*; Peat Restoration Agency

C

CAADP	Comprehensive Africa Agriculture Development Programme
CAVDEAL	Cavite Ideal International Construction and Development Corporation
CENRO	Community Environment and Natural Resources Offices
CDA	Cooperative Development Authority

CH_4	Methane
CIFOR	Centre for International Forestry Research
CO_2	Carbon dioxide
CONAICE	Confederation of Coastal Indigenous Nationalities of Ecuador
CPD	Cooperative Promotion Department
CPKO	Crude palm kernel oil
CPO	Crude palm oil
CPOPC	Council of Palm Oil Producing Countries
CSO	Civil society organisations
CSR	Corporate social responsibility

D

DA	Department of Agriculture
DAR	Department of Agrarian Reform
DENPASA	Dendê do Pará S/A
DENR	Department of Environment and Natural Resources
DIT	Department of Internal Trade
DOA	Department of Agriculture
DOAE	Department of Agricultural Extension
DOLE	Department of Labour and Employment

E

ELAC	Environmental Legal Assistance Centre, Inc.
Embrapa	*Empresa Brasileira de Pesquisa Agropecuária*; Brazilian Agricultural Research Corporation
EU	European Union
EUDR	European Union Deforestation Regulation
EV	Electric vehicle

F

FAO	Food and Agriculture Organisation
FARC	*Fuerzas Armadas Revolucionarias de Colombia*; Revolutionary Armed Forces of Colombia
FELDA	Federal Land Development Authority
FFB	Fresh fruit bunches
FFP	*Fondo de Fomento Palmero*; Fund for the Promotion of the Palm Sector
FMB	Forest Management Bureau
FPIC	Free Prior and Informed Consent

G

GDP	Gross Domestic Product
GIZ	*Deutsche Gesellschaft für Internationale Zusammenarbeit*; German Organisation for International Development

H

HCVRN	*Red de Recursos de Alto Valor de Conservación*; High Conservation Value Resource Network

I

IFA	*Instituto de Fomento Algodonero*; Cotton Development Institute
IMF	International Monetary Fund
INIFAP	*Instituto Nacional de Investigaciones Forestales, Agrícolas y Pecuarias*; National Institute for Forestry, Agriculture, and Fisheries Research
IPB	Independent Peat Basin
IRB	Inclusive Rural Business
ISPO	Indonesian Sustainable Palm Oil
ISSC	International Sustainability & Carbon Certification
IUCN	International Union for Conservation of Nature
IVPA	Indian Vegetable Oil Producers' Association

K

KLIA	Kuala Lumpur International Airport

L

LBP	Land Bank of the Philippines
LDD	Land Development Department
LGU	Local government units
LSC	Land Settlement Cooperatives

M

MA63	Malaysia Agreement 1963
MAATE	*Ministerio del Ambiente, Agua y Transición Ecológica*; Ministry of Environment, Water and Ecological Transition
MAE	Ecuadorian Ministry of Environment

MAGAP	*Ministerio de la Agricultura, Ganadería, Acuacultura y Pesca*; Ministry of Agriculture, Livestock, Aquaculture and Fisheries
MIDAS	More Investment for Sustainable Alternative Development
MMPL	Mt. Mantalingahan Protected Landscape
MNC	Multinational corporations
MOAC	Ministry of Agriculture and Cooperatives
MOC	Ministry of Commerce
MOU	Memorandum of understanding
MTPDP	Medium Term Philippine Development Plan

N

N_2O	Nitrous oxide
NAFTA	North American Free Trade Agreement
NCIP	National Commission on Indigenous Peoples
NCR	Native Customary Rights
NEP	National Energy Plan
NEPO	National Energy Policy Office
NES	Nucleus Estate and Smallholder Scheme
NGA	National government agency
NGO	Non-governmental organisation
NREB	Natural Resources and Environment Board
NTFP	Non-timber forest product
NYDF	New York Declaration on Forests

O

| OAE | Office of Agricultural Economics |
| OPPT | Oil Palm Production Technology |

P

PAMB	Protected Area Management Board
PCA	Philippine Coconut Authority
PCSD	Palawan Council for Sustainable Development
PENRO	Provincial Environment and Natural Resources Office
PIPOC	International Palm Oil Congress and Exhibition
PIR-Trans	Plasma Transmigration
PKO	Palm kernel meal
PKS	Palm kernel shells
PMC-PA	Oil Palm Competitive Improvement Plan
PNIS	*Plan Nacional Integral de Sustitución*; Comprehensive National Replacement Plan
PNNI	Palawan NGO Network, Inc.

PNPB	National Biodiesel Production and Usage Programme
PODES	*Potensial Desa*; Village Potential Statistics
PPKO	Processed palm kernel oil
PPO	Processed palm oil
PPODC	Philippine Palm Oil Development Council
PPOIDC	Palawan Palm Oil Industry Development Council
PPVOMI	Palawan Palm & Vegetable Oil Mills, Inc.
PRAS	Environmental and Social Remediation Programme
PROCEDE	*Programa de Certificación de Derechos Ejidales y Titulación de Solares*; Programme for Certification of Ejido Rights and Titling of Plots
PROFEPA	*Procuraduría Federal de Protección al Ambiente*; Federal Attorney for Environmental Protection
PRONAF	*Programa Nacional de Fortalecimento da Agricultura Familiar*; National Programme to Strengthen Family Farming
PSA	Philippine Statistics Authority
PT	*Partido dos Trabalhadores*; Workers' Party

R

RED II	Renewable Energy Directive Recast
REDD+	Reducing Emissions from Deforestation and Forest Degradation
REDP	Renewable Energy Development Plan
RSPO	Roundtable on Sustainable Palm Oil
RT	*Rukun Tetangga*; Neighbourhood Association
RW	*Rukun Warga*; Community Association

S

SAG	*Secretaría de Agricultura y Ganadería*; Ministry of Agriculture and Livestock
SAGARPA	*Secretaría de Agricultura, Ganadería, Desarrollo Rural, Pesca y Alimentación*; Ministry of Agriculture, Livestock, Rural Development, Fisheries and Food
SCORE	Sarawak Corridor of Renewable Energy
SEP	Strategic Environment Plan
SERNA	Ministry of Natural Resources
SIC	Ministry of Industry and Trade
SKT	*Surat Keterangan Tanah*; Land Certificate
SPA	State Planning Authority
SPFT	Southern Peasants' Federation of Thailand
SPOPP	Sustainable Oil Palm Production Programme
SPPN	Southern Poor People Network

T

TFA	Tropical Forest Alliance
TROPI	Tropical Peat Research Institute
TSPOA	Thailand Sustainable Palm Oil Alliance

U

UNDP	United Nations Development Programme
UNESCO	United Nations Educational, Scientific, and Cultural Organisation
UNODC	United Nations Office on Drugs and Crime
UPOIC	United Palm Oil Industry Company Limited
UPRA	National Rural Planning Office
USAID	United States Agency for International Development

W

WISSH	Wilmar Smallholders Support in Honduras Programme
WWF	World Wildlife Fund

Z

ZIDRES	*Zonas de Interés de Desarrollo Rural, Económico y Social*; Zones of Rural, Economic and Social Interest
ZAE-Dendê	*Zoneamento Agroecológico do Dendê*; Agroecological Zoning of Palm Oil

1

PALM OIL GOVERNANCE

A global industry assembled and reassembled by many people in many places

Patrick O'Reilly, Helena Varkkey and Sarah Ali

"You have to understand many small worlds to truly appreciate the big picture. And you have to be aware of the grand scale to know where all the pieces fit."

Introduction

Policymaking at multiple levels and across many sectors is faced with questions concerning the balance between economic and environmental priorities. In this context, the case of the debates around the cultivation of oil palm *(Elaeis guineensis)* in locations across the tropics stands out as a particularly controversial "battlefield" of policy knowledge and action. As the industry has grown and matured, so too has the range and number of studies, which presents a set of challenges for researchers, policymakers, and others with an interest in understanding and shaping the further development of the palm oil industry.

The story of the crop's commercialisation is worth noting. While known and cultivated for centuries near its centre of origin in West Africa, its commercialisation began in earnest in the 1960s when the newly-independent Malaysian government began a programme to develop the crop (Rajanaidu et al., 2013). Following this, the land area under palm oil cultivation rose spectacularly to 5.74 million hectares in 2016, with exports of 16.05 million tonnes. In neighbouring Indonesia, the figures are higher. In both countries, the presence of the crop is almost impossible to ignore. It is among the first sites visitors see upon arriving at the Kuala Lumpur International Airport (KLIA), Malaysia's main national airport; it lines the country's major roadways and adorns the Malaysian RM50 note. Unlike other plantation crops, oil palm is decidedly post-colonial, with knowledge and investment centres located in former colonial states. The place it occupies in the national

DOI: 10.4324/9781003459606-1

imagination as a driver of economic development is perhaps one of the defining features of Malaysia's postcolonial history.

In recent decades, the crop has expanded beyond its historical centres in Indonesia and Malaysia to encompass more countries in Southeast and Southern Asia, Africa, and Latin America. While the industry in Latin America is still considered small, it is experiencing substantial growth. Colombia is currently the fourth largest palm oil producer globally, with an annual production estimated at 1.35 million tonnes. Brazil, Mexico, Honduras, Peru, Ecuador, and Costa Rica also have significant land under oil palm. The expansion of the crop in this region has attracted the attention of multinational companies which dominate global production, including Southeast Asian companies based in Singapore, Malaysia, and Indonesia. Interest in the crop has also been initiated in Africa with what has, to date, seen mixed results (Olayide & O'Reilly, this volume).

The origin

Palm oil is widely used. Indeed, some might say it is an all-pervasive product; yet in many eyes, its production has come to represent the epitome of human's exploitative relationship with nature. Consequently, the pitch of debate that has emerged concerning its benefits and value on the one hand, and its negative environmental and social impacts on the other, is intense. In many respects, such concerns are not unprecedented. The question of how to balance the "benefits" of modern agricultural practice against the environmental and social costs associated with these activities is a perennial theme of contemporary agrarian and development studies as well as environmental science, one which plays an increasingly dominant role in almost all realms of policy concerning agricultural practices and calls for alternative approaches to agriculture (see for example Azam-Ali, 2021). However, the vehemence of this debate in relation to palm oil is striking, encompassing the work of multitudinous academics, industry players, environmental non-governmental organisations (NGOs), human rights activists, news outlets, and social media. Even the British Broadcasting Corporation (BBC) has got in on the act, featuring the crop in a documentary narrated by none other than David Attenborough. Indeed, such is the level of press interest in the crop and related environmental challenges that a sub-discipline has emerged, exploring media coverage and communication concerning palm oil and its related economic and environmental issues (Manzo et al., 2019).

It might be suggested that the scale and passion of these debates are a consequence of the accidents of history, geography, economics, and politics which have driven the pace of the oil palm industry's expansion, its spatial distribution, and socio-economic consequences. These are, in turn—at least in part—a consequence of the biophysical characteristics of the crop itself. Within a relatively broad range of parameters, oil palm grows well in most tropical environments. It is highly productive; trees fruit across the year, allowing up to four harvests per year with yields far exceeding those of any other oil crop (Jackson et al., 2019; Cramb & Curry, 2012). As a consequence—and as its advocates frequently point out—the crop produces far more vegetable oil on less land than alternatives, offering the prospect of reducing overall pressures to bring more land into production (Jackson et al., 2019). The product itself, chiefly palm oil, is regarded as being of high value with multiple uses, namely as a food ingredient, biofuel feedstock, and cosmetic additive; but with engineers and research scientists continually adding to the list of possible end uses both for the oil itself and of oil palm "by-products" promise an increasingly circular palm oil industry (see for example Hwang et al., 2022; Kasivisvanathan et al., 2012). When compared to other tropical cash crops, the oil palm is also widely perceived as offering better returns on labour—a factor which is tempting to both large- and small-scale producers (see Ogahara et al., 2022).

The species' capacity to produce exceptionally high yields of a valuable commodity on lands previously regarded as agriculturally marginal has contributed to a widely held belief in its potential as a tool of rural development in current and prospective producer countries (see for example Sibhatu, 2023; Qaim et al., 2020; Mingorria, 2014). Public discourse in the early adopting Southeast Asian countries, Malaysia and Indonesia, credit the crop with bringing millions of people out of poverty (Euler et al., 2017; Krishna et al., 2017), a fact that is also widely acknowledged even by researchers and commentators who have concerns regarding the negative impact of its expansion (Varkkey & O'Reilly, 2020; Bou Dib et al., 2018; Noor et al., 2017; Alam et al., 2016; Sayer et al., 2012). This, in turn, has informed wider perceptions of the crop as a means of delivering on a host of policy aims in the realms of the national economy, rural development, energy policy, and rural livelihood. Additional indirect benefits linked to the crop include the extension of amenities such as roads, electricity, and services such as schools and healthcare into remote, poorly developed areas (Gatto et al., 2017).

In Indonesia, palm oil has also been employed in policies supporting the movement of people out of overcrowded areas of Java to less populated locations (Widyatmoko & Dewi, 2019; Potter, 2012; McCarthy & Cramb, 2009). Whilst in Colombia, the crop has facilitated the country's peace process (Grajales, 2021; Genoud, 2020; Furumo, this volume) and efforts to reduce the narco-economy (Marin-Burgos & Clancy, 2017). Despite the reservations of many academics, these benefits are widely promoted by the industry itself, alongside supportive researchers and governments in producing countries (Jackson et al., 2019; Malaysian Palm Oil Council, 2010) and has contributed to the widespread interest in, and adoption of the crop outside of its original "core" producing centres in Malaysia and Indonesia (Potter, 2012). In this context, the geography of palm oil works in the industry's favour. Many of the countries in which suitable land exists face a combination of push factors, in the form of demands to deliver development via economic growth and supporting the wellbeing and income of rural dwellers; and pull factors, primarily in the form of the world's "anticipated increases in demand" for vegetable oils (Jackson et al., 2019), available investment, and the opportunities offered by cheap access to what is often characterised as underutilised lands that make the development of an oil palm sector a promising prospect in these states. Again, industry backers have been keen to promote this perception, coining the phrase "the golden crop" to describe oil palm's promise in these locations.

The crop's development is also associated with negative consequences and impacts. These result from the same biophysical characteristics that make the crop such a promising one for countries in the tropical belt. The lands on which palm oil can be cultivated include large tracts previously used for agriculture; however, it also includes vast areas which have not previously been used for large-scale agriculture at all. Indeed, questions of scale, titling, low population density, and the possibility of an additional income in the form of timber may make these areas better prospects for would-be palm oil producers. These lands include areas such as upland and lowland rainforest, tropical peat swamps, and wetlands which are deemed to be of "high ecological value". A key focus of criticism has thus addressed oil palm's implications for conservation, environmental well-being, and ecosystem function (Evers et al., 2016; Vijay et al., 2016; Page et al., 2011; Wilcove & Koh, 2010).

A particular and longstanding concern in this respect is that such areas are known to absorb and trap large quantities of carbon dioxide (CO_2) which are then stored in both living and dead biomass in the form of trees and organic peat soils (see Evers et al., 2016; Page et al., 2011). This has long underpinned a popular characterisation of such areas as a global asset, the "earth's lungs" (Centre for Ecology & Hydrology, 2013; Cox et al., 2013). This representation which has persisted for a considerable time, is one which is worth reflecting on and which we will return to later. Many of these locations are also known to be "biodiversity hotspots", hosting unique communities of flora and fauna. Furthermore, in many cases, such areas form the customary territories of forest-dwelling communities who often occupy ambiguous positions in relation to the states which claim jurisdiction over these locales and who live and manage these areas, employing customary forms of tenure and practice (Bennett et al., 2021; Varkkey & Ali, this volume).

Concerns have also been voiced concerning the implications of palm oil on the land and human rights of these communities and the consequences of their displacement for forest management, including those of various classes of smallholders. The range of dis-welfares linked to their experience of palm oil (Uda et al., 2017) include poor employment conditions (Barral, 2015), conflicts over land access rights (McCarthy et al., 2012b; Cramb & Sujang, 2011), the unequal distribution of the costs and benefits of the industry (McCarthy et al., 2012a), the extent to which they truly benefit from the crop (Ogahara et al., 2022; Castellanos-Navarette et al., 2021; Santika et al., 2019), and its impact on food security (Tabe-Ojong et al., 2023). The manner in which they are integrated into the industry and the consequence the different styles of integration have on the benefits and costs that accrue to them have also been raised (Bennett et al., 2019). These ideas have collectively contributed to the characterisation of these locations as resource or commodity frontiers (Sibhatu, 2023), whereby conflicts relating to the impact of human activities related to the cultivation of palm oil are regarded as particularly stark and the potential of those activities causing environmental damage is alarmingly high. It is worth bearing in mind that these problems have been identified both in early adopting countries and also in more recent industry entrants and even in situations where the latter claim to be implementing a sustainable model of oil palm development (Ogahara et al., 2022; Boron et al., 2016; Hazlewood et al., this volume).

The very visible, high profile negative environmental impacts associated with the crop are perhaps the primary reason why the crop has generated so much concern and debate. The impact of the crop on biodiversity conveyed to the public imagination via depictions of the threats posed to iconic species, its close association with infamous haze events in Southeast Asia which carries a significant loss of life and economic damage (Koplitz et al., 2017), the stark visual transformation that the crop brings to the areas which are cleared, and in particular the images of large-scale fires and their devastation are among the most evocative and stark depictions of environmental crises brought about by oil palm and have been fundamental in shaping the context and trajectory of debates over palm oil use and abuse. If it was not so closely associated with these environmental costs, it is scarcely conceivable that such a useful and productive crop would ever have been subject to such sustained opposition on the global stage.

Conversely, of course, it is scarcely conceivable that if oil palm was not such a useful and productive crop, it would be subject to such sustained and ardent support. In actual fact, regardless of concerns—including those listed above and their mobilisation in vigorous opposition—the scale of the commodity's production continues to grow. Since 2002, palm oil has been the world's most widely used vegetable oil. The area planted with oil palm now accounts for nearly one-tenth of the world's permanent cropland (Koh & Wilcove, 2008) and further land use conversion to oil palm is occurring (Carlson et al., 2012; Wilcove & Koh, 2010; Koh & Wilcove, 2007). In financial terms, the numbers still seem to add up. Demand is still rising and new markets are evolving with the increase in demand for biofuels. The relatively small populations on much of the land which is suitable for oil palm and the persistent perception (contrived or otherwise) of this land as underutilised, wasted, or idle persist (Lee et al., 2013; Majid-Cooke, 2006). Consequently, conversion continues. As much as public discourse may have highlighted the crop's environmental dark side, it remains for many "the golden crop" of the tropics. Across Southeast Asia, Latin America, and Africa, and in the face of a widespread recognition of the environmental and social costs involved, governments persist with policies to promote oil palm cultivation (for an exception, see Pido et al., this volume).

As a result, oil palm continues to present current and future countries with a puzzle—how to respond to the challenges presented by a crop which offers an almost irresistible economic allure while simultaneously posing almost immovable social and environmental hazards. Across the tropics, different countries have been forced to address this conundrum. In the process, adopting countries have been required to develop governance systems that address the question of the oil palm's role in different national contexts. The tensions and challenges posed by the crop, the opportunities and threats it presents in different locations to different groups, have been played out in different ways, resulting in the development of a range of diverse national governance regimes that impact multiple human actors—local communities, indigenous groups, landowners, industry players, and government officials—as well as non-human objects and entities; including the land itself, soil, plants, and animal species.

At the same time, however, we are also confronted by broader global questions concerning the crop and still expanding global industry of which it forms the basis; the organisation of that industry in a context where environmental dilemmas are deemed by many to be universally important appear to sit in direct opposition to some of the development needs and aspirations of many of those countries in which the crop is cultivated. For the global industry, key issues include efforts to establish global standards and a regulatory framework for the industry, the organisation of flows of investment, knowledge and know-how between producer countries, questions of control and environmental regulation within the global industry and its relationship with global environmental challenges, in addition to the management of relations between producer countries and those they supply with palm oil. Key issues at play in different countries are the balancing of environmental and developmental priorities, industry organisation, questions linked to land ownership and property rights, the status of indigenous communities within areas marked for palm oil expansion, social and environmental justice, the management of labour and protection of employment rights, as well as technology transfer and management of the palm oil value chain. Clearly, the global and local challenges are deeply interconnected and overlap with broader questions linked to environmental and development policy, climate justice, and equity.

Studies of local palm oil industries and the relationships between them (by which we mean national, provincial, and even local palm oil industries, their interrelationships and the links between them and the global palm oil industry), have an important part to play in understanding the dynamics and management of the global industry. Additionally, they have the capacity to generate useful insights concerning comparative governance approaches, models of industry best practice and value chain management, and also in testing the claims—both of the crop's advocates and critics—allowing us to explore the extent to which such claims are ubiquitous features of the oil palm crop or products of the accidents of history, climate, geography, society, and economy that have shaped the development of the industry in different places.

Recent years have seen a dramatic expansion in the scope and depth of the analysis of the policies adopted in producing countries in order to manage the development of the palm oil industry. However, even a brief survey of the palm oil literature indicates possible heavy bias, with much of the work focussed on practicalities of developing the industry (Paramananthan, 2013; Melling et al., 2011; World Bank, 2011; Mutert et al., 1999; Lim, n.d.) and the technical challenges associated with it. Such work is often published in highly specialised disciplinary journals or in-house journals which have links to the palm oil industry itself or the parastatals that support it. This phenomenon is not uncommon in the world of agricultural research (Pray & Echeverria, 2021; Bristow, 2011), and has been highlighted by some researchers who question the science and arguments of the pro-palm oil lobby (Liu et al., 2020; Goldstein, 2015).

More recently, research addressing technical challenges linked to making palm oil production more "sustainable" has become more prevalent. Subsequently, this work tends to produce "solutions" to the dilemmas of palm oil which are claimed to be objective, "evidence-based" and ultimately, apolitical. Yet they are also very much in keeping with popular political discourse concerning the development of the crop and the desire of the industry to be allowed to "clean up its own act" (Johnson, 2022; De Vos et al., 2021; Degli Innocenti & Oosterveer, 2020). This is scarcely surprising. In Malaysia, where the large-scale commercialisation of the crop was pioneered, oil palm has acquired strong ideological associations linking the crop to the country's long-held ambition of becoming a developed nation (Varkkey & O'Reilly, 2020). The country has invested substantial public and private resources and political capital in research via universities, parastatals, and industry bodies to generate and disseminate expertise concerning the crop's agronomy and uses, and to address environmental concerns associated with the oil palm business. Major efforts have been made to disseminate this knowledge to current and potential producing countries as well as consumers. In many respects, this has been successful, with a very wide acceptance of the need to promote sustainable palm oil.

As we have seen, this is juxtaposed by significant work concerning the adverse environmental impact of the crop. For example, the industry is linked to the destruction of rainforests via remote sensing studies, the release of carbon through fire events, and the drainage of large areas of tropical peat (Schrier-uijl et al., 2013; Miettinen et al., 2011; Wicke et al., 2011), resulting in peat oxidation during which greenhouse gasses are released (Evers et al., 2016). Oil palm expansion is also associated with forest fires and severe atmospheric pollution events that cross national boundaries (Varkkey, 2012). Again, a large proportion of this work focusses on the biophysical properties of these phenomena (Smith et al., 2020, 2016). A complicating factor is that much of this research is led by researchers based in or from the Global North (see, for example, Evers et al., 2016; Goldstein, 2015; Carlson et al., 2012). Varkkey (2018) has described how this fact is reflected in criticisms of what are depicted as biased Western scientists, resulting in what some regard as ineffectual and unfair European Union (EU) restrictions on palm oil imports.

It is by no means the case that the development-versus-environment dichotomy is exclusive, as we have already indicated above. There is a maturing literature drawing on a wide variety of social research methods, bringing greater nuance to debates over palm oil, helped in no small part by the increasing quantity and quality of research in the new producing regions resulting in the availability of national case studies drawing on different theoretical and analytical traditions. However, it remains the case that a particular development–environment dichotomy continues to exert an extraordinary influence in the academic debate around palm oil, having both a polarising and a "smoothing effect". Both poles of this dichotomy are united around, and draw on a substantial literature that supports two "truths". The first is that oil palm delivers short- and long-term economic benefits at both national and local levels and that many (though not all) communities and individuals in areas where it has been grown have derived measurable (though variously defined) benefits. Oil palm is thus constituted as a powerful tool for rural development (Euler et al., 2017; Cramb & Sujang 2013). The second is that the conversion of tropical forests to oil palm cultivation involves significant social, environmental, and health costs both locally and internationally, and little, if any, conclusive evidence has been produced to support the idea that the current "solutions to these problems" can work over the long term. Indeed, as the discussion above demonstrates, these problems have tended to accompany the industry on its global travels.

These two opposing truths pose a complex combination of challenges to researchers. On the one hand, they are engaged in efforts to gather, interpret and disseminate a range of scientific data that deals with the complexities of the biophysical, economic, and social processes involved in the development of oil palm. On the other, the knowledge produced has significant implications in normative discussions concerning policymaking and action on the ground. In effect, the opposing poles of conservation and development have a significant limiting effect on research trajectories. As illustrated in Figure 1.1, a substantial proportion of studies on oil palm can be located along a normative continuum ranging from those which support palm oil development (located at position A), through to those which stress the damage palm oil inflicts and argue against further palm oil expansion (located towards position B).

In practice, these two extremes are occupied by a (not insignificant) minority of contributors to the oil palm debate. However, many participants occupy a position between these two extremes, balancing environmental costs and development gains, under the rubric of sustainable development, with widely differing conceptualisations of what sustainability might look like (Bronkhorst et al., 2017; Nagiah & Azmi, 2012; Persey et al., 2011 Basiron et al., 2007). It might be argued that much of this work shares a common point of ontological departure which treats palm oil producing areas as "socio-ecological systems". Such approaches have been critiqued in other areas of environmental management. These approaches have been criticised because of their inability, and indeed, reluctance to integrate qualitative observations and insights, in addition to demonstrating a tendency to depoliticise environmental questions, employing approaches and solutions that "make the system work". The questions posed and even ideas about the types of methods involved, focus on quantitative "systems analysis", which comes at the expense of

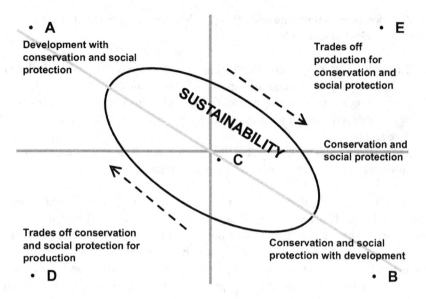

• **A**

Development with
conservation and social
protection

• **E**

Trades off
production for
conservation and
social protection

SUSTAINABILITY

• **C**

Conservation and
social protection

Trades off conservation
and social protection for
production

• **D**

Conservation and social
protection with development

• **B**

FIGURE 1.1   Normative framings in the literature on oil palm (O'Reilly & Varkkey, 2020).

more interpretive methods (see Stojanovic et al., 2016). However, research framed in these terms is in many ways quite restricted. These concerns are reflected in research on the performance of the Roundtable on Sustainable Palm Oil (RSPO) as an example of global multi-stakeholder governance (Schouten & Hospes, 2018; Hospes, 2014), the implications of oil palm expansion on land tenure (Rietberg & Hospes, 2018; McCarthy, 2010; McCarthy & Cramb, 2009), as well environmental and oil palm governance (Aubert et al., 2017).

Following Miller's (2014, 2019) useful critique of "sustainability", the framing of oil palm in terms of environmental, social and economic costs and benefits while embodying the "systems" problematic has significant consequences for research, reducing the level of attention paid to other issues and objects associated with the "palm oil controversy". Consider for example the ease in which diverse forest communities are subsumed under the term "indigenous", and diverse farming communities and households are converted into smallholders; how both of these groups are subsequently transformed into "land managers" alongside a plethora of other actors whose actions are assessed not in terms of their impacts on shaping assemblages of human and non-human actors in diverse and contested spaces but rather on their impacts on the "global" assets that Western commentators in particular have tended to describe as "the earth's lungs". A particular feature of this problem is the position of the nation-state, and governance frameworks more generally. Hospes (2014), for example, speaks of a gap in the literature concerning the "reactions of nation-states and producers in the South to the implementation or diffusion of global private partnership". He attributes this gap to a bias in the literature linked to several factors:

1 The widespread proposition that global governance has emerged to deal with problems which are beyond the capacity of the nation-state;
2 How concepts such as non-state (NGO and social movements) or market-based governance have drawn attention away from the responses of national governance regimes;
3 The overemphasis on how global governance affects actors in the South and how their responses affect global governance regimes.

We largely agree with the general point that Hospes makes. The dominance of some issues inevitably draws attention away from others, including the question of national palm oil governance. This point is particularly important given that a characteristic of the development-environment dichotomy is a tendency to strive for generalisation and scale. However, the extent to which results and recommendations of studies in specific countries, provinces, and communities are sufficiently portable to be applied to different national palm oil industries remains a key question, particularly as historically, many of these studies have occurred in Southeast Asia. Comparative work which makes greater use of the burgeoning palm oil literature in Latin America (including contributions in this volume) might reveal that the motivations underlying the industry's development, the ways in which oil palm is processed and consumed, impacts on land holding and smallholder income, and the management of supply chains differ radically between different countries in ways which are difficult to anticipate. Such work might test other generalisations; are the negatives linked to the crop—poor employment conditions, weakening of human rights, land grabbing, and environmental vandalism—more of a reflection of large-scale industrial agriculture than of the oil palm industry per se? Might such comparative work add to our knowledge of the relationship between formal governance, industry regulation and practices in the field (in some cases literally), and the balance of power between formal governance regimes and private capital?

A more nuanced approach to palm oil governance is emerging, proposing work that gives attention to historically and geographically situated national and local contexts (Bennett et al., 2019; Jelsma et al., 2017), where policymakers, industry actors, and local communities are engaged in the "working out" of the palm oil boom. Cramb and Curry (2012, p. 236) referred to the notion of a palm oil complex as a "whole series of differentiated actors (different types of plantation company, local community, landholder, migrant worker, government agency, local official, advocacy group and so on), each pursuing their own perceived interests and encountering unique sets of circumstances, interact in multiple ways to give rise to discernible, higher-order processes of far-reaching and often unintended or unplanned change." Work by Carmenta et al. (2017) speaks to the importance of recognising the multilayered nature of palm oil governance. Thorburn and Kull

(2014) highlight the complex interrelations between resource governance and environmental issues in palm oil-producing regions. Commentators have also identified the importance of scale (Hospes & Kentin, 2014) and the need to pay attention to livelihoods in palm oil governance (Deligiannis, 2012). These contributions are invaluable in calling attention to the diverse ways in which different actors pursuing a range of different interests collectively shape how the oil palm industry is embedded in different countries. While we by no means suggest that a focus on governance regimes within nations provides a definitive account of the industry's governance, understanding how national governance regimes are worked out forms an important part of the story, which needs to be understood if the governance of the sector is to be comprehensively mapped. It is our view that such an understanding can only be improved through national comparisons, which allow the exploration of how these dynamics are shaped by different national policy assemblages.

The path each producer country navigates between palm oil's developmental benefits (Cramb & Curry, 2012; Sayer et al., 2012; Rist et al., 2010; Agustira, 2008) and social and environmental costs (Barral, 2015) presents significant challenges for governance regimes in producing countries. These, in turn, have led to diverse governance responses at the national level as different states respond to these opportunities and threats differently. Conversely, the arrival of new players has implications for the global governance of the industry itself. This renders efforts to generalise the impact of oil palm highly problematic. In understanding the governance of the oil palm industry, we are at once dealing with global trends, diverse local responses to these trends, and the complex ways in which local and global governance arrangements interact (Cramb & McCarthy, 2016). While many of the issues and challenges encountered in these different national contexts are very similar, the ways these issues are worked out in different countries varies, in the process telling us something about how a highly problematic global entity, the international palm oil assemblage, influences and is influenced by its engagement with different national contexts, different legislative frameworks, different sets of power relations, as well as different political preferences and cultural understandings. Exploring these in-depth offers insights into whether and how nation-states can exert a degree of control over a seemingly homogenous global industry. Comparing these national industries open up intriguing possibilities in generating new questions and reappraising others which have occupied a significant place in studies of the oil palm assemblage.

The initial impetus for this book arose from a series of discussions between several colleagues regarding policy research related to oil palm. Among the issues raised was the methodological challenge of evaluating policy making and policy practice in relation to the industry, in particular, the question of how to generalise insights concerning a global industry and how to think about the diverse ways in which different countries are articulated into the global oil palm industry. Clearly, national-level responses are not uniform. The history of the countries involved differs greatly. They possess widely varying landscapes, agricultural systems, and styles of farming, and occupy different positions in relation to global circuits of capital, knowledge, and commerce. The areas in these countries in which palm oil development takes place include the customary lands of different forest peoples with their own knowledge, interests, practices, needs and wants. Views and aspirations for development differ as do views with regard to the role of new crops in old landscapes.

The experience of palm oil internationally reveals multiple industry configurations reflecting both the different histories and institutional frameworks in different countries and potentially at least, offering the prospects of radically different models for the industry and different outcomes in terms of the distribution of benefits and costs. Equally, much emphasis in the literature has tended to be placed on the conversion of rainforests to palm oil cultivation. Yet, we know from other countries that plantings are occurring in areas that have previously been employed for other agricultural purposes (Khor & Tamilwanan, this volume; Pischke, this volume). Different communities enjoy widely varying styles and levels of engagement in different countries, which again has an impact on the way that palm oil is incorporated into different rural landscapes. Furthermore, the scale and desired role apportioned to palm oil in different countries differ, with implications for the scale and ownership of the business.

Exploring these variations offers not only a useful way of learning about different national palm oil industries and their variations but also may provide a window into the way the global palm oil industry itself works, along with insights into the extent to which national governance can influence the development of an industry in ways which safeguard its peoples and environments. Bringing together different approaches to address a range of questions related to different aspects of the oil palm industry in different countries may offer us ways to enrich the research conducted and the learnings achieved. However, capturing the full complexity of these arrangements in one volume is not possible. Here, our intention is to outline some ideas about the value of cross-country comparisons and the merits of a multi-perspectival, rather than integrative approach to multi-disciplinary research.

In the remainder of this introductory chapter, we consider whether and what such a comparative framework might include and what it might be able to contribute to our understanding of the relationships between the global palm oil industry and national governance systems and, in so doing, contribute to our broader understanding of how the national governance of an environmentally problematic global industry functions. In the next section, we explore the influence that existing approaches to looking at the palm oil industry have on our understanding of the debate and pose the question as to whether these approaches adequately capture the wide variety of diverse activities in different arenas through which palm oil is produced. We briefly consider the implications that current trends in oil palm research have on current understandings of the crop and proposed interventions. We suggest that current research trends pay limited attention to the day-to-day processes through which policies are made and implemented at the grassroots level.

We suggest an alternative framing that conceptualises the oil palm industry as an assemblage. Following this, we outline how an approach drawing on the assemblage theory employed by environmental researchers such as Murray Li (2007)—and further developed by (Müller, 2015)—can be utilised to better understand the dynamics of oil palm governance.

Conceptualising oil palm using an assemblage approach

An exploration of national and sub-national governance involves a range of questions; how the industry is currently controlled at the national and local level, who controls it and how, who benefits from the industry and to what extent, how its introduction affects existing social, economic, and land tenure arrangements, and how the costs associated with the adoption of such arrangements should be shared. While these issues are not unknown, current framings of research into oil palm governance employ approaches which "background" these issues.

As we have argued, the way in which governance arrangements and policy regimes have evolved in different national contexts may have varying environmental, social, and economic drivers, influences, and implications. This can include implications over land use in producing locations and on the rights and interests of a wide variety of actors, including, but not restricted to, local communities, large-scale producers, and market intermediaries. Cramb and McCarthy (2016) describe the industry as a "complex" in which multiple actors, located in different sites and circumstances and pursuing different goals in a range of ways collectively, contribute to higher-order processes that give the industry its shape.

The way that locations become involved in emerging industries such as palm oil cannot be simply represented *a priori* as a form of "factory floor agriculture" or resource extraction driven by technology, land/resource availability, capital inputs and output costs, nor can the spaces in which this occurs be simplistically conceptualised as frontiers or two-dimensional spaces (Goldstein, 2019). It also cannot be assumed that power and decision-making within such industries operate in a linear fashion along the lines suggested by organisational charts and policy documents (O'Reilly et al., this volume). When capitalist and technological processes of accumulation are emphasised at the expense of acknowledging other entities, such approaches detach human actors and non-human biological and physical entities from the specific "conjunctures of circumstances, events, and relationships that are integral to regional change" (Blanco et al., 2015, p. 179). In understanding industries such as oil palm, we are simultaneously dealing with global trends, discourses, and entities, and diverse local responses to the challenges, opportunities, and imperatives the global phenomena create.

Assemblage has been employed to conceptualise and to facilitate an exploration of issues of scale, livelihood, power, and ordering. It embarks from a broad ontological position which conceptualises the socio-material phenomena such as the oil palm industry as consisting of a series of human and non-human actors and objects which are brought into relationships with each other, thereby forming an assemblage organised around specific reasons and purposes for a certain period of time. A particular feature of this approach is that it centres on the interaction of human and non-human entities in such practices and the resulting assemblages without presuming that the non-human entities are incapable of agency (Thornton et al., 2020). Assemblage describes both the process and the result of processes through which heterogeneous entities are brought together to serve certain functions for a certain time. Within the approach, a key role is given to how these entities are linked together through "relations of exteriority". Multiple sets of such relationships link different entities and determine the shape and effect of an assemblage at any given time. Assemblage-based approaches thus seek to avoid *a priori* assumptions about who holds power and how social structure is constituted; rather treating these as emergent properties of the processes through which assemblage itself is constructed by human and non-human actors (Müller, 2015). In considering the oil palm industry, a phenomenon that brings together a wide range of human actors, products, infrastructure such as industrial mills, oil palm trees, and pathogens, the attraction of an assemblage approach in considering how all these elements are integrated into the approach should be obvious.

Assemblage-inspired work explores how entities pursuing differing projects are brought together and linked via a series of relations in ways that have power and structuring effects (Umans & Arce, 2014). A critical benefit of this approach is that it does not privilege one site or set of power relations. Rather than suggesting that organisations or industries are the results of defined institutional practices underpinned by discreet driving forces such as capital or technology, the approach emphasises contingency, material transversal associations, and events (Deleuze & Guattari, 1987). Power concerns the capacity of entities to "fix" relationships between different components of the assemblage in ways that support their projects or interests, a process termed as "territorialisation". Such territorialisations are prone to mutation, transformation, and "break up". Power and agency are thus contingent and emerge from dynamic processes (see Fromm et al., this volume).

Assemblage rejects notions of linear arrangements between different "levels" of governance and *a priori* assumptions about which the most important decisions are. Instead, these questions are opened as objects of inquiry. In this approach, governance is understood as the efforts of some of these actors to enact specific relations and maintain these, locking and fixing actors and objects into arrangements maintained over time (Murray Li, 2007). Assemblage theory also suggests that assemblages are themselves made up of assemblages and that relationships between these can shift over time. In the case of palm oil, this may help enable the simultaneous analysis of governance in specific sites and its articulation with the wider palm oil industry.

Employing this approach allows us to redefine oil palm governance as an open-ended assemblage of assemblages linking a range of human and non-human actors in ways that are continually being renegotiated by different entities which pursue separate objectives. From this perspective, we can view oil palm governance as a multidimensional process occurring in multiple places at once, involving a variety of actors and objects whose articulation is contingent. This allows us to explore how the operation of power in different arenas influences the overall shape of the assemblage. Such an approach helps to bring local actors into view and exposes the potential for new and different trajectories of national, regional, and local action to shape the governance of the oil palm industry, drawing attention to how certain actors have sought to arrange the components of (territorialise) this assemblage in ways that reflect their interests.

The chapters in this collection offer an overview of governance issues that confront the oil palm phenomenon in Southeast Asia, Latin America, and Africa. In doing so, it is not our intention to provide a definitive account of the industry in different countries but rather to explore how different research on the oil palm industry in different countries can shed light on key processes and practices of assemblage in national palm oil industries, combining insights from different countries about how national regimes have sought to address some of the challenges that confront the industry. In addition, this will help us to consider how these local assemblages, in turn, are integrated into the global palm oil industry, how they shape and are shaped by it. By applying elements of assemblage theory to dissect governance issues in different regions and countries, we hope that lessons can be extracted concerning the global challenges the industry faces and how differing national governance regimes and affected communities respond to these challenges. Hence, these chapters will contribute towards a better understanding of the drivers and processes that shape the governance of the industry, both in different nations and globally. Such knowledge can hopefully support a more grounded appraisal of the palm oil sector and contribute to a comprehensive analysis of the forces shaping oil palm governance.

When we look at palm oil in different parts of the world, we are confronted with a diverse range of situations. Rather than a unified and rationally organised system or network, we can understand the palm oil industry in each location as being comprised of heterogeneous elements. Non-human actors like the oil palm tree itself, pests, technologies, and landscapes interact with human actors at the national, regional, and local levels, contributing to the shaping of the oil palm assemblage in each area. Within these national assemblages, certain actors have sought to shape the palm oil industry in different ways, which reflect their different understandings of how the palm oil industry may impact on their economic, social, environmental, and political interests and their efforts to arrange the components of the oil palm assemblage in ways that reflect these interests. The chapters demonstrate and explore the heterogeneous nature of oil palm governance and the industry itself. They illustrate that, despite some superficial similarities, the palm oil industry in different countries follows (often radically) varying trajectories. Thus, a critical question concerns how we conceptualise the process through which oil palm comes to be adopted and incorporated into socio-economic-environmental practice in particular locations and the role that governance plays in these processes. The approach is decidedly not to apply uniform approaches to address uniform issues to produce uniform sets of data, but rather to adopt a multi-perspectival approach, bringing together multiple viewpoints to generate a range of associated insights (Noe et al., 2008).

Thus, governance of this sector involves choices between livelihood and conservation, between the interests of large-scale producers and small-scale producers, and between conflicting national environmental and economic policy goals. Achieving a balance between these conflicting interests has proven extremely difficult. Governance has not facilitated the equitable development of the industry in ways that support the interests of all the affected entities to the same degree. Rather, the outcomes have tended to be heavily skewed in favour of certain interests over others. Resource-poor indigenous villagers and transmigrants have often borne the brunt of government-sponsored "lose-lose" and "win-lose" initiatives prioritising conservation over livelihood goals (Jewitt et al., 2014), while at the same time, large-scale producers have often benefitted generously from governments applying a light-touch approach to their regulation. These outcomes draw attention to the wide range of challenges that oil palm poses for governance and the presence of dominant tensions between different entities that shape governance arrangements.

This collection is timely as Latin American countries are currently embarking on policies to support expanded palm oil production, even though assessments of the crop's positive and negative impact are less well-rehearsed here. However, the Southeast Asian experience suggests any expansion of oil palm in Latin America is likely to have implications for communities living in affected areas and involve environmental costs, such as putting the biodiversity of the Amazon basin (see Cordóba et al., this volume) and other areas of high ecological value at risk and threatening globally important carbon sinks. Bearing this in mind, the chapters in this collection seek to pose and answer several key questions when considering both the global and local oil palm assemblage:

1 How do different governance arrangements enable and legitimise expansion?
2 How do power differentials affect oil palm governance?
3 How does governance enable the accumulation of wealth?
4 How is conflict governed and moderated in the oil palm sector?

References

Agustira, M. A., Rañola, R. F., Sajise, A. J. U., & Florece, L. M. (2015). Economic impacts of smallholder oil palm (Elaeis guineensis jacq.) plantations on peatlands in Indonesia. *Journal of Economics, Management & Agricultural Development*, *1*(2), 105–123. https://doi.org/10.22004/ag.econ.309271

Alam, A. S. A. F., Er, A. C., Begum, H., & Siwar, C. (2016). Smallholders the prominent contributor towards sustainable oil palm sector. *International Journal of Advanced and Applied Sciences*, *3*(2), 20–24.

Aubert, P. M., Chakib, A., & Laurans, Y. (2017). *Implementation and effectiveness of sustainability initiatives in the palm oil sector: A review* (pp. 1–55). Institut du Développement Durable et des Relations Internationales.

Azam-Ali, S. N. (2021). *The ninth revolution: Transforming food systems for good.* World Scientific Publishing Company.

Barral, S. (2015). Labour issues in Indonesian plantations, from indenture to enterpreneurship. *Global Labour Column, 177,* 1–2.

Basiron, Y. (2007). Palm oil production through sustainable plantations. *European Journal of Lipid Science and Technology, 109*(4), 289–295. https://doi.org/10.1002/ejlt.200600223

Bennett, A., Ravikumar, A., McDermott, C., & Malhi, Y. (2019). Smallholder oil palm production in the Peruvian Amazon: Rethinking the promise of associations and partnerships for economically sustainable livelihoods. *Frontiers in Forests and Global Change, 2.* https://doi.org/10.3389/ffgc.2019.00014

Bennett, A. B., Larson, A. M., Zamorra Rios, A., Gamarra Agama, S., & Monterroso, I. (2021). Forests regenerate on titled indigenous territories: A multiscale interdisciplinary analysis of 25 indigenous communities over 40 years in the Peruvian Amazon. *Center for International Forestry Research.* https://doi.org/10.17528/cifor/008387

Blanco, G., Arce, A., & Fisher, E. (2015). Becoming a region, becoming global, becoming imperceptible: Territorialising salmon in Chilean Patagonia. *Journal of Rural Studies, 42,* 179–190. https://doi.org/10.1016/j.jrurstud.2015.10.007

Boron, V., Payán, E., MacMillan, D., & Tzanopoulos, J. (2016). Achieving sustainable development in rural areas in Colombia: Future scenarios for biodiversity conservation under land use change. *Land Use Policy, 59,* 27–37. https://doi.org/10.1016/j.landusepol.2016.08.017

Bou Dib, J., Krishna, V. V., Alamsyah, Z., & Qaim, M. (2018). Land-use change and livelihoods of non-farm households: The role of income from employment in oil palm and rubber in rural Indonesia. *Land Use Policy, 76,* 828–838. https://doi.org/10.1016/j.landusepol.2018.03.020

Bristow, E. (2011). Global climate change and the industrial animal agriculture link: The construction of risk. *Society & Animals, 19*(3), 205–224. https://doi.org/10.1163/156853011x578893

Bronkhorst, E., Cavallo, E., van Dorth tot Medler, M., Klinghammer, S., Smit, H. H., Gijsenbergh, A., & van der Laan, C. (2017). Current practices and innovations in smallholder palm oil finance in Indonesia and Malaysia: Long-term financing solutions to promote sustainable supply chains. *Centre for International Forestry Research.* https://doi.org/10.17528/cifor/006585

Carlson, K. M., Curran, L. M., Asner, G. P., Pittman, A. M., Trigg, S. N., & Marion Adeney, J. (2012). Carbon emissions from forest conversion by Kalimantan oil palm plantations. *Nature Climate Change, 3*(3), 283–287. https://doi.org/10.1038/nclimate1702

Carmenta, R., Zabala, A., Daeli, W., & Phelps, J. (2017). Perceptions across scales of governance and the Indonesian peatland fires. *Global Environmental Change, 46,* 50–59. https://doi.org/10.1016/j.gloenvcha.2017.08.001

Castellanos-Navarrete, A., de Castro, F., & Pacheco, P. (2021). The impact of oil palm on rural livelihoods and tropical forest landscapes in Latin America. *Journal of Rural Studies, 81,* 294–304. https://doi.org/10.1016/j.jrurstud.2020.10.047

Centre for Ecology & Hydrology. (2013, February 7). *Tropical rainforests, "lungs" of the planet, reveal true sensitivity to global warming.* https://www.ceh.ac.uk/news-and-media/news/tropical-rainforests-lungs-planet-reveal-true-sensitivity-global-warming

Cox, P. M., Pearson, D., Booth, B. B., Friedlingstein, P., Huntingford, C., Jones, C. D., & Luke, C. M. (2013). Sensitivity of tropical carbon to climate change constrained by carbon dioxide variability. *Nature*, *494*(7437), 341–344. https://doi.org/10.1038/nature11882

Cramb, R. A., & Curry, G. N. (2012). Oil palm and rural livelihoods in the Asia-Pacific region: An overview. *Asia Pacific Viewpoint*, *53*(3), 223–239. https://doi.org/10.1111/j.1467-8373.2012.01495.x

Cramb, R. A., & McCarthy, J. F. (2016). *The oil palm complex: Smallholders, agribusiness and the state in Indonesia and Malaysia*. NUS Press.

Cramb, R. A., & Sujang, P. S. (2011). "Shifting ground": Renegotiating land rights and rural livelihoods in Sarawak, Malaysia. *Asia Pacific Viewpoint*, *52*(2), 136–147. https://doi.org/10.1111/j.1467-8373.2011.01446.x

Cramb, R. A., & Sujang, P. S. (2013). The mouse deer and the crocodile: Oil palm small-holders and livelihood strategies in Sarawak, Malaysia. *Journal of Peasant Studies*, *40*(1), 129–154. https://doi.org/10.1080/03066150.2012.750241

De Vos, L., Biemans, H., Doelman, J. C., Stehfest, E., & van Vuuren, D. P. (2021). Trade-offs between water needs for food, utilities, and the environment—a nexus quantification at different scales. *Environmental Research Letters*, *16*(11), 115003. https://doi.org/10.1088/1748-9326/ac2b5e

Degli Innocenti, E., & Oosterveer, P. (2020). Opportunities and bottlenecks for upstream learning within RSPO certified palm oil value chains: A comparative analysis between Indonesia and Thailand. *Journal of Rural Studies*, *78*, 426–437. https://doi.org/10.1016/j.jrurstud.2020.07.004

Deleuze, G., & Guattari, F. (1987). *A thousand plateaus: Capitalism and schizophrenia*. Bloomsbury.

Deligiannis, T. (2012). The evolution of environment-conflict research: Toward a livelihood framework. *Global Environmental Politics*, *12*(1), 78–100. https://doi.org/10.1162/glep_a_00098

Euler, M., Krishna, V., Schwarze, S., Siregar, H., & Qaim, M. (2017). Oil palm adoption, household welfare, and nutrition among smallholder farmers in Indonesia. *World Development*, *93*, 219–235. https://doi.org/10.1016/j.worlddev.2016.12.019

Evers, S., Yule, C. M., Padfield, R., O'Reilly, P., & Varkkey, H. (2016). Keep wetlands wet: The myth of sustainable development of tropical peatlands—implications for policies and management. *Global Change Biology*, *23*(2), 534–549. https://doi.org/10.1111/gcb.13422

Gatto, M., Wollni, M., Asnawi, R., & Qaim, M. (2017). Oil palm boom, contract farming, and rural economic development: Village-level evidence from Indonesia. *World Development*, *95*, 127–140. https://doi.org/10.1016/j.worlddev.2017.02.013

Genoud, C. (2020). Access to land and the round table on sustainable palm oil in Colombia. *Globalisations*, *18*(3), 1–18. https://doi.org/10.1080/14747731.2020.1716480

Goldstein, J. E. (2015). Knowing the subterranean: Land grabbing, oil palm, and divergent expertise in Indonesia's peat soil. *Environment and Planning A: Economy and Space*, *48*(4), 754–770. https://doi.org/10.1177/0308518x15599787

Goldstein, J. E. (2019). The volumetric political forest: Territory, satellite fire mapping, and Indonesia's burning peatland. *Antipode*. https://doi.org/10.1111/anti.12576

Grajales, J. (2021). *Agrarian capitalism, war and peace in Colombia: Beyond dispossession*. Routledge.

Hospes, O. (2014). Marking the success or end of global multi-stakeholder governance? The rise of national sustainability standards in Indonesia and Brazil for palm oil and soy. *Agriculture and Human Values*, *31*(3), 425–437. https://doi.org/10.1007/ s10460-014-9511-9

Hospes, O., & Kentin, A. (2014). Tensions between global-scale and national-scale governance: The strategic use of scale frames to promote sustainable palm oil production in Indonesia. In F. Padt, P. Opdam, N. Polman, & C. Termeer (Eds.), *Scale-sensitive Governance of the Environment* (pp. 203–219). John Wiley & Sons, Ltd.

Hwang, J. Z. H., Andiappan, V., & Ng, D. K. S. (2022). Promoting circular economy between palm oil sector with multiple industries. *IOP Conference Series: Materials Science and Engineering*, *1257*(1), 012005. https://doi.org/10.1088/1757-899x/1257/1/012005

Jackson, T. A., Crawford, J. W., Traeholt, C., & Sanders, T. A. B. (2019). Learning to love the world's most hated crop. *Journal of Oil Palm Research*, *31*(3). https://doi.org/10.21894/ jopr.2019.0046

Jelsma, I., Schoneveld, G. C., Zoomers, A., & van Westen, A. C. M. (2017). Unpacking Indonesia's independent oil palm smallholders: An actor-disaggregated approach to identifying environmental and social performance challenges. *Land Use Policy*, *69*, 281–297. https://doi.org/10.1016/j.landusepol.2017.08.012

Jewitt, S. L., Nasir, D., Page, S. E., Rieley, J. O., & Khanal, K. (2014). Indonesia's contested domains. Deforestation, rehabilitation and conservation-with-development in Central Kalimantan's tropical peatlands. *The International Forestry Review*, *16*(4), 405–420. https://www.jstor.org/stable/24310693

Johnson, A. (2022). The Roundtable on Sustainable Palm Oil (RSPO) and transnational hybrid governance in Ecuador's palm oil industry. *World Development*, *149*, 105710. https://doi.org/10.1016/j.worlddev.2021.105710

Kasivisvanathan, H., Ng, R. T. L., Tay, D. H. S., & Ng, D. K. S. (2012). Fuzzy optimisation for retrofitting a palm oil mill into a sustainable palm oil-based integrated biorefinery. *Chemical Engineering Journal*, *200–202*, 694–709. https://doi.org/10.1016/j. cej.2012.05.113

Koh, L. P., & Wilcove, D. S. (2007). Cashing in palm oil for conservation. *Nature*, *448*(7157), 993–994. https://doi.org/10.1038/448993a

Koh, L. P., & Wilcove, D. S. (2008). Is oil palm agriculture really destroying tropical biodiversity? *Conservation Letters*, *1*(2), 60–64. https://doi.org/10.1111/j.1755-263x.2008. 00011.x

Koplitz, S. N., Jacob, D. J., Sulprizio, M. P., Myllyvirta, L., & Reid, C. (2017). Burden of disease from rising coal-fired power plant emissions in Southeast Asia. *Environmental Science & Technology*, *51*(3), 1467–1476. https://doi.org/10.1021/acs.est.6b03731

Krishna, V., Euler, M., Siregar, H., & Qaim, M. (2017). Differential livelihood impacts of oil palm expansion in Indonesia. *Agricultural Economics*, *48*(5), 639–653. https://doi. org/10.1111/agec.12363

Lee, J. S. H., Abood, S., Ghazoul, J., Barus, B., Obidzinski, K., & Koh, L. P. (2013). Environmental impacts of large-scale oil palm enterprises exceed that of smallholdings in Indonesia. *Conservation Letters*, *7*(1), 25–33. https://doi.org/10.1111/conl.12039

Lim, K. H. (n.d.). *Key areas for improving oil palm productivity on peat soils with special reference to fertiliser and water management* (pp. 1–23). Retrieved September 25, 2023, from http://www.oneoilpalm.com/wp-content/uploads/2015/11/Improving-Oil-Palm-Productivity-on-Peat-Soils.pdf

Liu, F. H. M., Ganesan, V., & Smith, T. E. L. (2020). Contrasting communications of sustainability science in the media coverage of palm oil agriculture on tropical peatlands in Indonesia, Malaysia and Singapore. *Environmental Science & Policy*, *114*, 162–169. https://doi.org/10.1016/j.envsci.2020.07.004

Majid Cooke, F. (2006). *State, communities and forests in contemporary Borneo*. ANU Press.

Malaysian Palm Oil Council. (2010, February 21). *Palm oil: A crop of peace & prosperity*. Palm Oil Today. https://palmoiltoday.net/palm-oil-a-crop-of-peace-properity/

Manzo, K., Padfield, R., & Varkkey, H. (2019). Envisioning tropical environments: Representations of peatlands in Malaysian media. *Environment and Planning E: Nature and Space*, *3*(3), 251484861988089. https://doi.org/10.1177/2514848619880895

Marin-Burgos, V., & Clancy, J. S. (2017). Understanding the expansion of energy crops beyond the global biofuel boom: Evidence from oil palm expansion in Colombia. *Energy, Sustainability and Society*, *7*(1). https://doi.org/10.1186/s13705-017-0123-2

McCarthy, J. F. (2010). Processes of inclusion and adverse incorporation: Oil palm and agrarian change in Sumatra, Indonesia. *The Journal of Peasant Studies*, *37*(4), 821–850. https://doi.org/10.1080/03066150.2010.512460

McCarthy, J. F., & Cramb, R. A. (2009). Policy narratives, landholder engagement, and oil palm expansion on the Malaysian and Indonesian frontiers. *Geographical Journal*, *175*(2), 112–123. https://doi.org/10.1111/j.1475-4959.2009.00322.x

McCarthy, J. F., Gillespie, P., & Zen, Z. (2012a). Swimming upstream: Local Indonesian production networks in "globalised" palm oil production. *World Development*, *40*(3), 555–569. https://doi.org/10.1016/j.worlddev.2011.07.012

McCarthy, J. F., Vel, J. A. C., & Afiff, S. (2012b). Trajectories of land acquisition and enclosure: Development schemes, virtual land grabs, and green acquisitions in Indonesia's outer islands. *Journal of Peasant Studies*, *39*(2), 521–549. https://doi.org/10.1080/03066150.2012.671768

Melling, L., Chua, K. H., & Lim, K. H. (2011). Managing peat soils under oil palm. In K. J. Goh, S. Chiu, & S. Paramananthan (Eds.), *Agronomic Principles and Practices of Oil Palm Cultivation* (pp. 695–728). Petaling Jaya: Agricultural Crop Trust.

Miettinen, J., Shi, C., & Liew, S. C. (2011). Deforestation rates in insular Southeast Asia between 2000 and 2010. *Global Change Biology*, *17*(7), 2261–2270. https://doi.org/10.1111/j.1365-2486.2011.02398.x

Miller, E. (2014). Economisation and beyond: (Re)Composing livelihoods in Maine, USA. *Environment and Planning A: Economy and Space*, *46*(11), 2735–2751. https://doi.org/10.1068/a130172p

Miller, E. (2019). *Reimagining livelihoods: Life beyond economy, society, and environment*. University of Minnesota Press.

Mingorría, S., Gamboa, G., Martín-López, B., & Corbera, E. (2014). The oil palm boom: Socio-economic implications for Q'eqchi' households in the Polochic valley, Guatemala. *Environment, Development and Sustainability*, *16*(4), 841–871. https://doi.org/10.1007/s10668-014-9530-0

Müller, M. (2015). Assemblages and actor-networks: Rethinking socio-material power, politics and space. *Geography Compass*, *9*(1), 27–41. https://doi.org/10.1111/gec3.12192

Murray Li, T. (2007). Practices of assemblage and community forest management. *Economy and Society*, *36*(2), 263–293. https://doi.org/10.1080/03085140701254308

Mutert, E., Fairhurst, T. H., & von Uexküll, H. R. (1999). Agronomic management of oil palms on deep peat. *Better Crops International*, *13*(1), 22–27.

Nagiah, C., & Azmi, R. (2012). A review of smallholder oil palm production: Challenges and opportunities for enhancing sustainability—A Malaysian perspective. *Journal of Oil Palm &the Environment*, *3*(12), 114–120. https://doi.org/10.5366/jope.2012.12

Noe, E., Alrøe, H. F., & Langvad, A. M. S. (2008). A polyocular framework for research on multifunctional farming and rural development. *Sociologia Ruralis*, *48*(1), 1–15. https://doi.org/10.1111/j.1467-9523.2008.00451.x

Noor, F. M. M., Gassner, A., Terheggen, A., & Dobie, P. (2017). Beyond sustainability criteria and principles in palm oil production: Addressing consumer concerns through insetting. *Ecology and Society*, *22*(2). https://www.jstor.org/stable/26270132

O'Reilly, P., & Varkkey, H. (2020). Palm oil governance in different locations: Using the assemblage approach to understand a "complex" sector. *International Review of Modern Sociology*, *46*(1–2), 1–17.

Ogahara, Z., Jespersen, K., Theilade, I., & Nielsen, M. R. (2022). Review of smallholder palm oil sustainability reveals limited positive impacts and identifies key implementation and knowledge gaps. *Land Use Policy*, *120*, 106258. https://doi.org/10.1016/j.landusepol.2022.106258

Page, S. E., Rieley, J. O., & Banks, C. J. (2011). Global and regional importance of the tropical peatland carbon pool. *Global Change Biology*, *17*(2), 798–818. https://doi.org/10.1111/j.1365-2486.2010.02279.x

Paramananthan, S. (2013). Managing marginal soils for sustainable growth of oil palms in the tropics. *Journal of Oil Palm and the Environment*, *4*(1), 1–16. https://doi.org/10.5366/jope.2013.1

Persey, S., Nussbaum, R., Hatchwell, M., Christie, S., & Crowley, H. (2011). *Towards sustainable palm oil: A framework for action* (pp. 1–37). Zoological Society of London.

Potter, L. (2012). New transmigration "paradigm" in Indonesia: Examples from Kalimantan. *Asia Pacific Viewpoint*, *53*(3), 272–287. https://doi.org/10.1111/j.1467-8373.2012.01492.x

Pray, C., & Echeverria, R. (2021). Private sector agricultural research and technology transfer links in developing countries. In D. Kaimowitz (Ed.), *Making the Link Agricultural Research and Technology Transfer in Developing Countries*. Routledge.

Qaim, M., Sibhatu, K. T., Siregar, H., & Grass, I. (2020). Environmental, economic, and social consequences of the oil palm boom. *Annual Review of Resource Economics*, *12*(1). https://doi.org/10.1146/annurev-resource-110119-024922

Rajanaidu, N., Kushairi, A., Marhalil, M., Din, A., Fadilla, A., Noh, A., Isa, Z. A., Chan, K. W., & Raviga, S. (2013). *Breeding of oil palm for the strategic requirement of the industry*. International Palm Oil Congress and Exhibition.

Rietberg, P. I., & Hospes, O. (2018). Unpacking land acquisition at the oil palm frontier: Obscuring customary rights and local authority in West Kalimantan, Indonesia. *Asia Pacific Viewpoint*, *59*(3), 338–348. https://doi.org/10.1111/apv.12206

Rist, L., Feintrenie, L., & Levang, P. (2010). The livelihood impacts of oil palm: Smallholders in Indonesia. *Biodiversity and Conservation*, *19*(4), 1009–1024. https://doi.org/10.1007/s10531-010-9815-z

Santika, T., Wilson, K. A., Budiharta, S., Law, E. A., Poh, T. M., Ancrenaz, M., Struebig, M. J., & Meijaard, E. (2019). Does oil palm agriculture help alleviate poverty? A multidimensional counterfactual assessment of oil palm development in Indonesia. *World Development*, *120*, 105–117. https://doi.org/10.1016/j.worlddev.2019.04.012

Sayer, J., Ghazoul, J., Nelson, P., & Klintuni Boedhihartono, A. (2012). Oil palm expansion transforms tropical landscapes and livelihoods. *Global Food Security*, *1*(2), 114–119. https://doi.org/10.1016/j.gfs.2012.10.003

Schouten, G., & Hospes, O. (2018). Public and private governance in interaction: Changing interpretations of sovereignty in the field of sustainable palm oil. *Sustainability*, *10*(12), 4811. https://doi.org/10.3390/su10124811

Schrier-uijl, A., Silvius, M., Parish, F., Lim, K. H., Rosediana, S., & Anshari, G. (2013). *Environmental and social impacts of oil palm cultivation on tropical peat—a scientific review* (pp. 1–40). Roundtable on Sustainable Palm Oil.

Sibhatu, K. T. (2023). Oil palm boom: Its socioeconomic use and abuse. *Frontiers in Sustainable Food Systems*, *7*. https://doi.org/10.3389/fsufs.2023.1083022

Smith, T., Evers, S., Lupascu, M., & Chiu, H. (2020). How do tropical peatland greenhouse gas emissions respond in the immediate aftermath of a fire? *EGU General Assembly 2020*, Online, May 4–8, 2020. https://doi.org/10.5194/egusphere-egu2020-12567

Smith, T. E. L., Yule, C., Evers, S., Paton-Walsh, C., & Jing, Y. G. (2016). First in situ measurements of tropical peatland fire emissions: New emission factors for greenhouse gas reporting and haze forecasting. *15th International Peat Congress*, 353–354.

Stojanovic, T., McNae, H. M., Tett, P., Potts, T. W., Reis, J., Smith, H. D., & Dillingham, I. (2016). The "social" aspect of social-ecological systems: A critique of analytical frameworks and findings from a multisite study of coastal sustainability. *Ecology and Society, 21*(3). https://doi.org/10.5751/es-08633-210315

Tabe-Ojong, M. P. Jr., Alamsyah, Z., & Sibhatu, K. T. (2023). Oil palm expansion, food security and diets: Comparative evidence from Cameroon and Indonesia. *Journal of Cleaner Production, 418*, 138085. https://doi.org/10.1016/j.jclepro.2023.138085

Thorburn, C. C., & Kull, C. A. (2014). Peatlands and plantations in Sumatra, Indonesia: Complex realities for resource governance, rural development and climate change mitigation. *Asia Pacific Viewpoint, 56*(1), 153–168. https://doi.org/10.1111/apv.12045

Thornton, S. A., Setiana, E., Yoyo, K., Dudin, Yulintine, Harrison, M. E., Page, S. E., & Upton, C. (2020). Towards biocultural approaches to peatland conservation: The case for fish and livelihoods in Indonesia. *Environmental Science & Policy, 114*, 341–351. https://doi.org/10.1016/j.envsci.2020.08.018

Uda, S. K., Hein, L., & Sumarga, E. (2017). Towards sustainable management of Indonesian tropical peatlands. *Wetlands Ecology and Management, 25*(6), 683–701. https://doi.org/10.1007/s11273-017-9544-0

Umans, L., & Arce, A. (2014). Fixing rural development cooperation? Not in situations involving blurring and fluidity. *Journal of Rural Studies, 34*, 337–344. https://doi.org/10.1016/j.jrurstud.2014.03.004

Varkkey, H. (2012). Patronage politics and natural resources: A historical case study of Southeast Asia and Indonesia. *Asian Politics, 40*(5), 438–448.

Varkkey, H. (2018). EU's anti-palm oil measures do not help the environment. *ASEAN Focus, 20*(1), 8–9.

Varkkey, H., & O'Reilly, P. (2020). Sociopolitical responses toward transboundary haze: The oil palm in Malaysia's discourse. In S. Kukreja (Ed.), *Southeast Asia and Environmental Sustainability in Context* (pp. 65–88). Lexington Books.

Vijay, V., Pimm, S. L., Jenkins, C. N., & Smith, S. J. (2016). The impacts of oil palm on recent deforestation and biodiversity loss. *PLOS One, 11*(7), e0159668. https://doi.org/10.1371/journal.pone.0159668

Wicke, B., Sikkema, R., Dornburg, V., & Faaij, A. (2011). Exploring land use changes and the role of palm oil production in Indonesia and Malaysia. *Land Use Policy, 28*(1), 193–206. https://doi.org/10.1016/j.landusepol.2010.06.001

Widyatmoko, B., & Dewi, R. (2019). Dynamics of transmigration policy as supporting policy of palm oil plantation development in Indonesia. *Journal of Indonesian Social Sciences and Humanities, 9*(1), 35–55. https://doi.org/10.14203/jissh.v9i1.139

Wilcove, D. S., & Koh, L. P. (2010). Addressing the threats to biodiversity from oil-palm agriculture. *Biodiversity and Conservation, 19*(4), 999–1007. https://doi.org/10.1007/s10531-009-9760-x

World Bank. (2011). *The World Bank Group framework and IFC strategy for engagement in the palm oil sector* (pp. 1–88). The World Bank and International Finance Corporation.

SECTION 1
Southeast Asia

2

GRASSROOTS GOVERNANCE

How assemblage links oil palm, livelihoods, and local administration in an Indonesian village

Patrick O'Reilly, Gusti Anshari, Antonio Jonay Jovani Sancho, Adi Jaya, Emmy Antang, Corry Antang, Stephanie Evers, Paul Wilson, Sofie Sjogersten, Caroline Upton and Susan E. Page

Introduction

The Indonesian state plans further expansion of oil palm cultivation, justified based on its ability to deliver developmental benefits (Susanti & Maryudi, 2016). Substantial scientific literature exists concerning the Indonesian oil palm sector. Much of this literature recognises the benefits of oil palm (Euler et al., 2017; Agustira et al., 2015) and measures to further promote it. Another significant body of work examines its social and environmental costs (Vijay et al., 2016). The desire to strike a balance between development and conservation has given rise to numerous proposals for the adoption of alternative practices within and outside the industry (Padfield et al., 2016; Hansen et al., 2015; Paramananthan, 2013; Nagiah & Azmi, 2012). However, the actual uptake of such practices is patchy; only a proportion of global palm oil is produced in compliance with sustainability standards, and the effectiveness of these standards is questioned (Gassler & Spiller, 2018; Carlson et al., 2017; Cattau et al., 2016; Ruysschaert & Salles, 2014), while evidence that less damaging forms of agriculture are capable of replacing the cultivation of oil palm on a significant scale is limited (see for example Tan et al., 2021; Giesen & Sari, 2018). As a consequence, and despite intense scientific endeavours, environmental problems associated with the industry persist.

The case of oil palm highlights broader challenges in managing science-to-policy relationships in environmental governance. While the oil palm boom has generated an enormous degree of research activity, the extent to which this work contributes to policy varies. Inevitably, this problem has itself become the subject of research to "bridge the gap" between research and policy (see for example Ivancic & Koh, 2016; Hansen et al., 2015). Valuable as this work is, the interface between environmental research and oil palm governance remains challenging,

DOI: 10.4324/9781003459606-3

reflected in work supporting the need for "evidence-based" policy approaches (Ruysschaert & Salles, 2018), as well as in the analysis of the reasons why the warnings implicit in so much research appears to go unheeded (Liu et al., 2020). While environmental scientists and the advocates of more sustainable, or even less, palm oil may rightly suggest that this failure relates to aspects of the policymaking process that are beyond their control, it may at least be worth considering the extent to which current approaches to studying the oil palm industry and its governance framework provide the basis for designing effective policy interventions and engagement strategies.

Over the past decade, the role of smallholders in oil palm cultivation has been increasingly recognised. Estimates in numerous publications suggest that the extent of smallholder involvement in global palm oil production stands at around 40 percent (Lai et al., 2022); a similar figure has been cited with respect to their role in Indonesian palm oil production (Andrianto et al., 2019). Work by researchers based in the Centre for International Forestry Research (CIFOR) and elsewhere has focussed in particular on the growing role of an independent smallholder sector. Jelsma et al. (2017) and subsequently Andrianto et al. (2019) have created typologies of smallholders with a view to addressing questions as to how these producers can be encouraged to adopt more sustainable cultivation practices. Such questions have generated further work and reflection (see for example Fleming et al., 2021; Kinseng et al., 2019). This increased research activity coincides with a period of increased efforts to integrate such smallholders into policy. Of note in this respect was the introduction of Indonesia's *Badan Restorasi Gambut* (Peatland Restoration Agency/BRG), as well as efforts to involve smallholders in Roundtable on Sustainable Palm Oil (RSPO) certification (De Vos et al., 2021). While the former intervention does not specifically and singly concern oil palm, the issues it addresses—a desire to restore degraded peatlands and subsequently mangroves—are indelibly associated with oil palm, as many of the locations and communities in which the BRG works are deeply integrated into the Indonesian oil palm industry. While these efforts are widely applauded, there is some evidence that implementing such policies remains challenging. Ogahara et al.'s (2022) review of sustainability interventions notes only limited positive impacts from sustainability measures, while Alfajri et al. (2023) provide evidence to suggest that the positive impacts of BRG interventions may be short-lived. Common across these critiques is a concern with the extent to which sustainability-related policies understand and address the priorities and needs of local communities. In brief, while policies and research may have correctly turned their attention to smallholder communities, the question remains: do current approaches to research and policymaking have the conceptual and applied tools needed to work with communities?

It is worth considering the extent to which these approaches are capable of addressing some of the key questions concerning palm oil governance. For example, how do these various governance arrangements enable and determine the shape of oil palm expansion, and how do these arrangements influence the accumulation of wealth among the actors in the assemblage? Perhaps most critically, do current approaches provide us with the most useful perspectives on how power operates and is distributed within the oil palm assemblage? In this chapter, we consider these questions in particular, focussing on the governance of the Indonesian oil palm industry. We suggest that any analysis of the Indonesian oil palm industry that proceeds from *a priori* assumptions about where power lies and a focus on the formal elements of oil palm governance is likely to fail. Instead, we explore the palm oil industry in Indonesia as an assemblage that is constantly being (re)produced through the practices of the entities of which it is composed. In this context, we suggest that alongside a consideration of how actors are incorporated into the palm oil industry, it is equally important to consider how the industry is incorporated into the lifeworlds and livelihood strategies of those who come into contact with it. We illustrate this through a case study of the practices engaged in by smallholders in one village, exploring how they participate in "grassroots governance" of the oil palm industry. The local practices of assemblage in which they are involved (re)produce the oil palm assemblage in that space, a contingent and open-ended socio-materiality that transcends the formal components of the oil palm industry and its governance. We believe that this framing allows for a richer analysis of how different actors' decisions and actions at different points in the industry contribute to the shape that its governance takes. By presenting an empirical description of how a complex oil palm assemblage emerges in one village in Indonesia, we illustrate the significance of locally-based studies and understandings in the development of effective policies to engage with smallholders, which informs the implications of this perspective for research into the governance of the industry.

Oil palm in Indonesia

The Indonesian oil palm industry initially lagged behind that of neighbouring Malaysia. However, the country's oil palm expansion in the 1990s and early 2000s accelerated, resulting in the country becoming the world's largest palm oil producer in the early twenty-first century. It now controls 46 percent of the world market. 12.3 million hectares (6 percent) of Indonesian soil is planted with oil palm, contributing to around 7 percent of Indonesia's Gross Domestic Product (GDP) (Varkkey et al., 2018).

As in Malaysia, oil palm has been present in Indonesia for many decades, having first been brought into the country as an ornamental plant in the nineteenth century. However, before the 1960s, the crop's commercial value was viewed as limited. It was overshadowed by the production of other, more established crops in colonial commodity supply chains (rubber being particularly significant in this system). During the early years of independence, a period of "guided democracy" emerged in Indonesia. A form of neo-consensus operated within the country, with the influence of economic and political interest groups exerting a significant influence on national policies. Attitudes towards agriculture reflected a broader decolonisation agenda, shifting from commodity production as a focus for rural policies and an emphasis on measures that supported the food and income security of rural populations, agrarian and land reform, as well as economic self-sufficiency. These policies came to be linked with a period of economic and political instability, poor agricultural performance and a turning away from global markets (which were changing dramatically at this time).

Indonesian economic priorities experienced another significant shift beginning in 1964. Economic and political problems contributed to the fall of the regime of its first president, Sukarno in 1965 and—following a period of significant upheaval and political instability during which opposition voices were purged—in 1967, Suharto assumed power (Yazid, 2014). Suharto's regime was underpinned by the support of the military and was informed intellectually by the New Order movement, which supported an opening up of trade and foreign policy. This regime vested significant power in an executive headed by Suharto and was supported electorally via the Golkar organisation (which later became a political party in 1999). In economic terms, Suharto and the New Order regime were credited with turning the country's fortunes around in the late 1960s and 1970s by adopting more outward-looking, pro-Western economic and foreign policies (Yazid, 2014).

Under Suharto, Indonesia has been described as having evolved into a unitary developmental state (Thee, 2012). Under this model, the state retained a strong role in steering economic development. However, rather than using this power to placate the demands of national interest groups, this power was mobilised to support the development of a trading economy that could grow GDP via a series of five-year economic plans. Natural resource extraction was an important component of this approach in the early 1970s, and forestry was identified as a mechanism for accelerating economic development (Susanti & Maryudi, 2016), enacted through the first five-year national development programme (Thee, 2012; Rudner, 1976). At the same time, the lack of a strong opposition meant that during the New Order era, Indonesian political and economic life revolved around the ruling regime and its supporting organisations, notably Golkar members and the military (Thee, 2012). One result of this is was that while the state successfully grew the national economy, a business and policy culture developed in which patronage relationships between commercial actors and certain local figures played an important role (Tyson et al., 2018).

The Indonesian take up of the oil palm itself began in earnest only during the 1970s under the New Order. Initial oil palm development took place in areas of logged-over rainforest and subsequently involved the conversion of primary forest into both large- and small-scale plantations. The original centres of the industry were in areas of mineral soil in Sumatra. However, as demand and production technologies evolved, oil palm cultivation expanded in the peatlands of Borneo and West Papua.

Like Malaysia, Indonesian discourse concerning oil palm supported the idea that the crop could perform a dual role: as an engine for increasing GDP and as a mechanism for supporting rural development and anti-poverty measures. From a relatively small base in the 1970s, production expanded. The 1980s and 1990s saw large areas of former forest and rubber cultivation being given over to oil palm production, accompanied by a range of policies aimed at supporting an industry that involved both large companies and small producers. A key driver of the expansionary strategy was the idea that the demand for palm oil would continue to grow. Simultaneously, the limited extent of mechanisation in the industry permitted participation by large and small operators to continue in parallel. Many large operators benefited from the support of key local stakeholders via patronage relationships (Varkkey, 2012).

Concurrently, policies were developed on a range of partnership arrangements through which smallholders could also participate in the palm oil boom (see McCarthy, 2010; McCarthy & Zen, 2009). The primary form such policies took during the New Order period were variations of so-called nucleus estate and smallholder (NES) schemes (World Bank, 2011). The principal idea underpinning these schemes was that a large company would enter into a relationship with smaller-scale producers. In the initial versions of the scheme, the large company would develop a parent estate and possibly a mill, while providing technical support for smallholders who agreed to develop plots and sell fresh fruit bunches (FFBs) to the companies. During this stage, the plantations and smallholdings comprised 80 percent and 20 percent, respectively, of the total scheme. In some cases, this approach also linked the issue of overpopulation and land scarcity by marrying the NES concept to transmigration policies. This was made explicit in the Plasma Transmigration (PIR-Trans) programme, which supported the creation of NES schemes in which the smallholders were transmigrants (McCarthy & Cramb, 2009).

Following varied levels of success, NES schemes were effectively discontinued in 2001 (McCarthy & Cramb, 2009). This followed the end of the Suharto regime and the appointment of a new government, which adopted a more neo-liberal approach to economic policy. This approach implied a smaller direct role for the state in economic development and a greater emphasis on corporate performance. Consequently, the development of new palm oil schemes assumed a more expressly commercial focus. This coincided with the pledge under the *Reformasi* (reformation policy) to devolve more powers to the provincial, district, and even village governments, under which these layers of local government assumed a greater degree of control over land. Under these arrangements, the local government was given some involvement in the allocation of land to new oil palm developments by companies. These companies were obliged to enter into negotiations with communities located where they wished to establish oil palm plantations and form partnerships with them (*Kemitraan*), under which local communities would receive ownership of up to 20 percent of the planted area. According to McCarthy and Zen (2009), this new arrangement has proven to be somewhat problematic as it allocates considerable authority to provincial and district governments in reinterpreting national guidelines. Critically, questions emerge concerning how village governments manage governance at the local level.

Oil palm and village life – Teluk Empening

In this section, we address the question of how local governance of the palm oil assemblage is worked out in the case of one village in West Kalimantan. The village of Teluk Empening is located in the Kubu Raya Regency, 48 km south of Pontianak. The village is situated on a bend of the Kapuas River (0°23'S, 109°36'E) and possesses a mixture of peat and mineral soils. The area has seen the large-scale conversion of tropical forests. A large oil palm plantation is located to the southwest of the village, while land to the north of the village proper has been developed for rice cultivation. Within and around the village, there is, therefore, a range of areas that have been cleared, drained, and converted into different agricultural land use classes at different points in time. A small area of the village land (300 hectares) is part of a large oil palm plantation, and besides this area, the bulk of land in the village is divided among locals and is used for a wider variety of long- and short-term crops and agricultural activities (PODES data). Teluk Empening is not a transmigration community; however, few of the inhabitants come from indigenous ethnic groups. Most of those living in the area have been drawn there for multiple reasons. In interviews, collaborators cited the opportunities offered by the presence of significant areas of land for rice production, a perception that land was available to those willing to clear it, and employment in the oil palm plantation. The piecemeal nature of its settlement has resulted in an area with a diverse ethnic base, dominated by Buginese and Madurese people and including others from a range of ethnic backgrounds.

Despite its riverside location, the village retains a largely agricultural character. Fishing undertaken there is extremely small scale; the only fishing activity observed by researchers during this study was trapping or fishing for snakehead in drainage ditches. While of negligible commercial value, these fish do constitute an extra source of protein in the village diet. Interviews with locals and observations made during the research confirm that while rice cultivation dominates in the areas of mineral soil, many of the area's households cultivate crops primarily or exclusively on peatland. While limited evidence of conflict relating to land use was detected during the research, unclaimed land was being actively cleared and brought into production in the area, which raises issues concerning land titles. During the period the researchers spent in the village, it was evident that issues with the fertility of peat soil were of major importance to the local community. In interviews, about 50 percent of local farmers stated that they used burning to treat these soils. Bans on land burning were in effect during the time the researchers spent in the location, and concerns were expressed about both land burning itself and restrictions on burning. While a small number of people mentioned environmental issues and the fact that burning attracted the attention of enforcement agents, the main focus of these concerns was related to the implications of burning bans on the ability of locals to clear and farm land.

Methods

The study was designed around grounded theory, in which the researchers tried to avoid making prior assumptions about local livelihood practices and the operation of policies. Instead, the research team came to Teluk Empening in the hope of exploring how local livelihood and cropping practices intersected with other aspects of daily life (including formal governance entities) and the natural environment. The approach drew on assemblage theory: livelihood was conceptualised as a set of practices during which individuals and households assembled a range of entities and objects in ways that make possible the reproduction of a meaningful daily life. Besides the production of physical elements of livelihood, attention was also paid to the sociocultural elements, including power and identity. In this context, oil palm was not singled out as the focus of the investigation. Rather, observations concerning the role of oil palm were allowed to emerge from a holistic account of how people assembled livelihoods in the specific biophysical, economic, and sociocultural context of Teluk Empening.

Data was gathered via a mixed-methods approach combining a quantitative survey of household livelihood practices and qualitative work. The quantitative survey was targeted towards members of local households who engaged in agriculture on peatland as part of their livelihoods. The questionnaire consisted of a detailed series of closed-ended questions exploring multiple themes relating to livelihood. This included questions on household composition, agricultural and non-agricultural income, land ownership and transfers, crop choices and outcomes, agricultural and land management practices, as well as market access and support structures. In addition to the quantitative survey, in-depth interviews were undertaken with multiple farmers, traders (shopkeepers and agricultural traders), members of local non-governmental organisations (NGOs), figures in the local administration, including the current and former heads of village, and figures in the provincial administration and the BRG. Participants were identified via a reputational sampling technique. This involved asking collaborators to suggest who else the researchers should talk to until a degree of saturation was achieved (i.e. until no new names were being mentioned by additional interviewees).

A final element of the research involved the staging of several participatory action research exercises. These took the form of workshops where locals and researchers discussed local administrative and economic issues, key problems and opportunities the area faces, in addition to aspirations for the future.

Palm oil and livelihood assemblage

The survey data largely confirm the findings of work undertaken by Jelsma et al. (2017) and suggestions made by McCarthy (2010), highlighting the variability of livelihood among smallholders in palm oil-producing areas. In particular, it was clear that within the village, oil palm constituted one of a number of income streams. Neither participation in the oil palm industry nor non-participation could be said to define individual households since hardly any households relied solely on the crop. Rather, households in the area produced a wide range of crops. Furthermore, besides their own farming, agricultural households drew on multiple income sources, including employment in the village government, plantation work, trading, and labouring. Except for rice, most agricultural produce would typically go to the market. Therefore, by and large, this is a community engaged in petty commodity production rather than subsistence agriculture. The survey data found only weak relationships between crop choice and levels of income available to a household. Among the factors that did seem to have a bearing on income were the level of education of different households, their relationship with the village government, ethnicity, and size of the holding.

The survey, interviews, and observations showed that livelihood practices in Teluk Empening were not homogeneous. Within the broad category described as smallholders, the researchers identified a wide range of strategies employed by different individuals and households, resulting in significantly different outcomes. While most had a low income, the village also had a small group of people who appeared to be doing significantly better.

In-depth interviews with members of this group revealed that even among them, there was a high degree of variation in the strategies they employed. One collaborator attributed his success to his trading business, which had allowed him to generate additional money to increase his land holding. Another worked exclusively on peat soil and employed a strategy whereby he would grow a short-term cash crop for some years until the land "was no longer good", at which point he would plant oil palm or rubber on these areas and clear more land, building up the size of his landholding in this way. A third attributed his success to the fact that he had developed a substantial commercial oil palm holding combined with rice in other sites. What was notable about each of these collaborators is that they enjoyed relatively good relationships with the village government, they belonged to the same ethnic group as the village leadership, and in two cases, also held minor official positions within the village government itself. They also revealed a considerable degree of knowledge of additional support that could be accessed via the government and skills in negotiating additional resources through these and other links. In particular, they had successfully made use of the somewhat ambiguous land titling powers of the different layers of government to increase the amount of land under their control.

In assemblage terms, we see the different individuals seeking to bring together a variety of different human and non-human objects in ways that support their livelihood. The degree to which they can fix, or territorialise these different human and non-human components in ways that support their interests is reflected in the outcomes they achieve. It is important to stress that, in some cases, the maximisation of income may not be the actors' goal. Therefore, it is not necessarily the case that those who generate the largest income are the most successful in fulfilling their livelihood goals. However, those who do pursue strategies that lead to the generation of the most income exhibit several similarities. These include a wider social network, connections to the village government, and strategies that result in increasing landholding. Crucially, their relationship with the palm oil assemblage was instrumental, tangential, and variable. One of those mentioned above had effectively established himself as a small-scale plantation owner while combining this business with rice cultivation. A second of the more successful farmers discussed above used oil palm along with rubber as a means of securing and maintaining the title to the land. By contrast, others in the village merely used the oil palm industry as a source of employment or cultivated small plots, which yielded limited benefits.

In the case of this village, palm oil is used by local people in different ways, not simply to boost income, but also to secure and maintain control of the land, as savings and as a response to government incentives. Therefore, the realities of oil palm cultivation at the grassroots level are shaped by how locals incorporate it into their lifeworlds and livelihoods. At the same time, others choose not to engage with the crop at all. In this case, the so-called palm oil frontier mosaic associated with smallholders is not so much a palm oil landscape, rather, it is a landscape in which palm oil is included.

Palm oil assemblage and local institutions

Figure 2.1 is based on a description of local institutional arrangements provided by a senior member of the local government in Central Kalimantan. However, it is essentially that which can be found in any basic textbook on Indonesian governance. This distinguishes five layers of government, ranging from the nation to the village. As it is currently constituted, the Indonesian system of government is, to some degree, a legacy of previous government policies. The idea of five strictly nested forms of government was initially linked to the desire of the early postcolonial regime to avoid any suggestion of federalism that might invite ideas about separatism across the archipelago. Under the New Order, the layers of government became the mechanisms for ensuring compliance with the regime's rules and regulations via a strict and rigidly controlled hierarchy of reporting arrangements. Since the end of that period, the different layers of government within Indonesia have evolved into the basis of a less centralised system of government with extensive devolved powers.

However, as mentioned above, the data from the extended and group interviews illustrates that local government and governance in Indonesian communities are complex. Besides the village government itself, a range of "subordinate" layers exist in the form of sub-villages, community organisations (*Rukun Warga* or RW), and neighbourhood organisations (*Rukun Tetangga* or RT). In addition, a range of quasi-autonomous organisations has evolved at the village and sub-village levels. Sometimes, these entities are formally subordinate to the village government. In

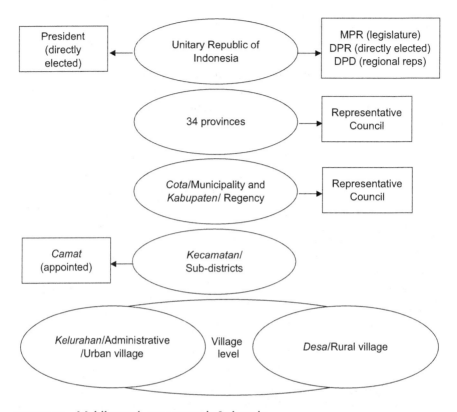

FIGURE 2.1 Multilayered government in Indonesia.

the case of others, while they may collaborate with village leadership, they are funded and administered under different national, provincial, and district arrangements. Interviews and observations in Teluk Empening revealed a dense network of local organisations and entities connected to the village government in different ways (see Figure 2.2).

Kemitraan negotiations

The in-depth interviews gave some insight into how these complex arrangements are negotiated, and also concerning the impact of *Reformasi* measures on the responsibilities and operation of the village government. As a rural village (*desa*), the village government of Teluk Empening is subject to a degree of democratic control. In theory, the village head, as well as the heads of the RW and RT, are elected. However, in practice, the elected village head of Teluk Empening has generally come from within one ethnic community. Furthermore, the village head has a very strong influence on the composition of the village government, appointing its executive officers. They may also invite certain people to stand as RT and RW heads, as these elections are often unopposed.

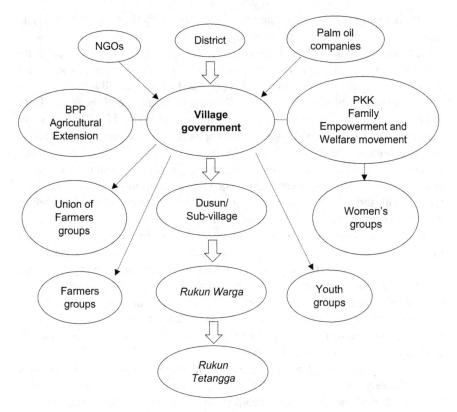

FIGURE 2.2   Local administration and influences.

Appointment to these positions is not a minor issue. Many of the positions come with payments, and in some cases, provide access to significant resources and the levers of local power. Under *Reformasi*, most of those interviewed felt that the village government had improved with additional resources and training, and the devolution of more control over local processes of land acquisition and land use planning. This was consolidated under Law No. 6 (2014) and Regulation No. 60 (2014), which extended additional power and budget to village governments. In theory, this extends to land use: villages retain the power to issue letters to signify that individuals have rights to land (*Surat Keterangan Tanah* or SKT). Villages have responsibility for local environmental issues, and palm oil concession holders must negotiate the terms under which they can plant oil palm.

Village governments thus retain considerable control over civic groups, land access, and titling. In addition, however, they are now key actors in the new and more commercialised "partnership" arrangements in relation to the development of new oil palm concessions. Under such arrangements, a large palm oil company was issued permission to plant on specific tracts of land close to Teluk Empening. Consequently, the company was required to negotiate a partnership arrangement with

Teluk Empening and neighbouring villages, under which the benefits of the plantations were to be shared with the villages. While in theory, this process extends some autonomy to the village, in practice, the village only has scope to discuss the terms under which the concession is operated. The interviews revealed a sense that the village had little real control over this process. Instead, the village leaders were essentially presented with a *fait accompli*: in the words of one of the respondents, "The district government told the palm oil company which land they had to use for their plantation and which villages the land belonged to. The company then had to negotiate with the different villages". This arrangement places the onus on village governments and companies to reach acceptable terms under which the company could use land in the village catchment. The village government entered into negotiations with the palm oil company concerning 300 hectares of local land, which had been granted a concession to plant oil palm.

While under the partnership arrangements, some benefit is expected to accrue to the village, in this case, those interviewed indicated that the extent of the benefits to Teluk Empening was extremely limited. Indeed, those currently serving the village government expressed some dissatisfaction with the results of the negotiation and were at pains to stress that the deal had been done before their appointment. Under the agreement, the village did not receive any share of income from the estate. Rather, a small donation amounting to a contribution to the wage costs of the village administration was agreed upon.

While they did not claim to be pleased with the outcome, those interviewed were relatively understanding of the position of the then village head, stressing the difficult position he was in. According to two of those interviewed, this was because the company had more experience and resources in dealing with these types of negotiations: "The company negotiators were highly skilled and were able to offer incentives to the village government". This suggests that while in theory, *Kemitraan* arrangements offer the prospect for villages to develop mutually beneficial partnerships with oil palm companies, such negotiations are undertaken in conditions in which power relations are asymmetrical and can lead to arrangements that heavily favour plantation companies over local communities.

Again, drawing on assemblage theory, we can understand that the company's relationship with other branches of government, access to social and human capital (in the form of its relationship with entities and actors with significant negotiating skills), experience, and its influence with people in other tiers of government place it in a good position to "fix" relationships with the village in ways that support its interests.

Discussion

Exploring the relationship between oil palm and assemblage in Teluk Empening via an assemblage approach draws attention to the existence of diverse arrangements through which the industry is articulated in local livelihoods and local administration. Indeed, it is possible to discern dual trajectories relating to the laws, processes and practices governing the ways in which palm oil companies "negotiate" with villages and the less formal ways in which particular local farmers engage with the industry. The latter appear in the interviews to devise highly individualised responses. We see responses ranging from non-engagement with the oil palm assemblage to those who take advantage of the presence of the oil palm assemblage, features of local land governance, and land titling in order to incorporate palm oil into a portfolio of livelihood options. For this group, the crop is seen as both a source of income and as a means for expanding their capital base and claims to land ownership. An examination of the practices of the small farmers who live there thus reveals highly differentiated strategies that result in a diverse range of outcomes. An examination of their interaction with the palm oil assemblage confirms that while it is the presence of the oil palm industry that provides the opportunity, these small-scale "independent" producers are not slavishly incorporated into the assemblage in ways that are dictated by the industry. Rather, they are knowledgeable and capable, and they respond to the introduction of the crop in multiple ways. Critically, these responses may or may not correspond to how the industry envisages such relationships to evolve, or to those forms of behaviour anticipated by policymakers or researchers. In effect, in Teluk Empening, we see some local actors adopting practices of assemblage which align entities from the palm oil assemblage in their livelihoods. Governance is not simply handed down to these groups. Rather, they have the capacity to interpret, resist, modify, and mutate the palm oil assemblage, in effect doing governance at the farm level.

The position of local governments is much more difficult. Situated in a space in which they have only limited scope to overturn or influence agreements made at the district level, they are also hampered by being placed into uneven negotiations with people who have access to forms of knowledge, social capital, and political influence that they lack. Ironically, therefore, dividing the various governance roles between different layers of government in relation to palm oil and "empowering" village governments to participate in partnership negotiations have potentially left local communities in a weaker position.

However, it would be misleading to suggest that the devolution of power to the village government solely plays into the interests of large-scale plantation companies. As noted above, the livelihood strategies of this smallholder community are diverse. Different local actors operating in similar agro-ecological conditions exercise a wide degree of agency and assemble various combinations of activities in household livelihood portfolios, which include both agricultural and non-agricultural components. Some local actors have demonstrated considerable skill in their dealings with the village government, employing formal and informal relationships with figures within the village government, and knowledge of its procedures and positions within it to pursue strategies that have enabled them to accumulate significant holdings and capital through the cultivation of a combination of different crops, land clearance, and allied activities. The village government also provides a direct route to prosperity for senior figures within the leadership and facilitates the wider dissemination of benefits in the area. As we have seen, the extent to which different locals were able to do so was unclear. There was some evidence that, as Bebbington et al. (2006) suggest, variations in the social capital of different locals have a strong bearing on the extent to which they were able to make use of the opportunities offered by decentralisation and the local government. This perhaps suggests that some investment in measures to grow social capital within existing local governance structures can play an important role in steering the future development of the oil palm assemblage in Indonesia.

These observations suggest that palm oil governance is more nuanced and complex than many policymakers suggest. In particular, it draws attention to the role of grassroots governance in the (re)production of the palm oil assemblage. Rather than simply seeking to resolve the large questions concerning the role of oil palm in national conservation and development strategies, it is also important to consider local factors that may influence responses to, and shape the outcomes of policies. Crucially, in Teluk Empening, we are not dealing with people who fall into specific oil palm smallholder typologies at all. But rather with actors whose broader livelihood strategies may or may not have led them to adopt oil palm as an option within their portfolio of economic practices. Similarly, it is reasonable to suggest that interventions aimed at making oil palm more sustainable are likely to be subject to similar strategic appraisal by individual farmers, with responses reflecting the livelihood interests and spatio-temporal framings of the smallholders themselves, rather than those that are close to the hearts of those who frame and make national policies and programmes. Similarly, at the level of the village, the decisions of a community's leadership are almost certainly not going to be simply driven by the priorities of palm oil or environmental policymakers, a consideration that should inform and temper calls for environmental conditionality for village governments' USD60,000 annual *"Dana Desa"* village development payments (see Naylor et al., 2019). Rather, an important question for policy design might concern the reality of diverse local responses to their proposed interventions and the development of ways to align these with the interests of other policy actiors.

Conclusion

Indonesia is the world's largest producer of oil palm. The industry absorbs an area of over 6 million hectares of land, with plans afoot to further expand on this total. Simultaneously, the development of the oil palm industry has attracted considerable criticism from scientists and others who have raised significant concerns regarding its environmental impact. However, despite a wealth of evidence concerning the adverse environmental impacts of current oil palm industry practices, the extent to which these concerns impact oil palm governance remains open to question. We argue that this is due to the way in which scientists and policymakers perceive the palm oil problem. Too often, scientific approaches to the question of oil palm governance rely on a relatively simplistic view of the relationship between science and policymaking, which employs a "fix the problem" approach, in which the role of research is clearly defined as involving the provision of discreet solutions to isolated technical problems. As a counterpoint, we suggest that an approach to exploring the palm oil industry as an assemblage (re)produced through the active practices of entities within the assemblage located in different global, national, and even local sites may provide useful additional insights.

Employing such an approach in the case of Teluk Empening reveals the challenges of attempting to design interventions that impact oil palm expansion. Understanding how palm oil expansion is being undertaken is imperative; it cannot simply be treated as a form of economic development driven by technology, land availability, and capital, or even as an industry-specific challenge. Indeed, such narratives carry the danger of detaching the oil palm, associated actors, and the biological and physical environment from the "particular conjunctures of circumstances, events, and relationships that are integral to regional change" (Blanco et al., 2015, p. 179), the reality of which is as much a consequence of grassroots governance as it is of the decisions and choices made in national and international arenas.

References

Agustira, M. A., Rañola Jr., R. F., Sajise, A. J. U., & Florece, L. M. (2015). Economic impacts of smallholder oil palm (Elaeis guineensis Jacq.) plantations on peatlands in Indonesia. *Journal of Economics, Management & Agricultural Development*, *1*(2), 105–123.

Alfajri, A., Varkkey, H., O'Reilly, P., & Hamzah, T. A. (2023). *Peatland multilevel governance: Exploring the policy impact among local actors in shaping peatland policy intervention in Indonesia*. MBF, Universiti Malaya.

Andrianto, A., Fauzi, A., & Falatehan, A. (2019). The typologies and the sustainability in oil palm plantation controlled by independent smallholders in Central Kalimantan. In R. Kinseng, A. Dharmawan, D. Lubis, & A. Seminar (Eds.), *Rural socio-economic transformation: Agrarian, ecology, communication and community, development perspectives* (pp. 3–14). CRC Press.

Bebbington, A., Dharmawan, L., Fahmi, E., & Guggenheim, S. (2006). Local capacity, village governance, and the political economy of rural development in Indonesia. *World Development*, *34*(11), 1958–1976. https://doi.org/10.1016/j.worlddev.2005.11.025

Blanco, G., Arce, A., & Fisher, E. (2015). Becoming a region, becoming global, becoming imperceptible: Territorialising salmon in Chilean Patagonia. *Journal of Rural Studies*, *42*, 179–190. https://doi.org/10.1016/j.jrurstud.2015.10.007

Carlson, K. M., Heilmayr, R., Gibbs, H. K., Noojipady, P., Burns, D. N., Morton, D. C., Walker, N. F., Paoli, G. D., & Kremen, C. (2017). Effect of oil palm sustainability certification on deforestation and fire in Indonesia. *Proceedings of the National Academy of Sciences*, *115*(1), 121–126. https://doi.org/10.1073/pnas.1704728114

Cattau, M. E., Marlier, M. E., & DeFries, R. (2016). Effectiveness of Roundtable on Sustainable Palm Oil (RSPO) for reducing fires on oil palm concessions in Indonesia from 2012 to 2015. *Environmental Research Letters*, *11*(10), 1–11. https://doi.org/10.1088/1748-9326/11/10/105007

De Vos, R. E., Suwarno, A., Slingerland, M., Van Der Meer, P. J., & Lucey, J. M. (2021). Independent oil palm smallholder management practices and yields: can RSPO certification make a difference? *Environmental Research Letters*, *16*(6), 065015.

Euler, M., Krishna, V., Schwarze, S., Siregar, H., & Qaim, M. (2017). Oil palm adoption, household welfare, and nutrition among smallholder farmers in Indonesia. *World Development, 93,* 219–235. https://doi.org/10.1016/j.worlddev.2016.12.019

Fleming, A., Agrawal, S., Dinomika, Fransisca, Y., Graham, L., Lestari, S., Mendham, D., O'Connell, D., Paul, B., Po, M., Rawluk, A., Sakuntaladewi, N., Winarno, B., & Yuwati, T. W. (2021). Reflections on integrated research from community engagement in peatland restoration. *Humanities and Social Sciences Communications, 8*(1). https://doi.org/10.1057/s41599-021-00878-8

Gassler, B., & Spiller, A. (2018). Is it all in the MIX? Consumer preferences for segregated and mass balance certified sustainable palm oil. *Journal of Cleaner Production, 195,* 21–31. https://doi.org/10.1016/j.jclepro.2018.05.039

Giesen, W., & Sari, E. N. N. (2018). *Tropical peatland restoration report: The Indonesian case* (pp. 1–82). Mott MacDonald.

Hansen, S. B., Padfield, R., Syayuti, K., Evers, S., Zakariah, Z., & Mastura, S. (2015). Trends in global palm oil sustainability research. *Journal of Cleaner Production, 100,* 140–149. https://doi.org/10.1016/j.jclepro.2015.03.051

Ivancic, H., & Koh, L. P. (2016). Evolution of sustainable palm oil policy in Southeast Asia. *Cogent Environmental Science, 2*(1). https://doi.org/10.1080/23311843.2016.1195032

Jelsma, I., Schoneveld, G. C., Zoomers, A., & van Westen, A. C. M. (2017). Unpacking Indonesia's independent oil palm smallholders: An actor-disaggregated approach to identifying environmental and social performance challenges. *Land Use Policy, 69,* 281–297. https://doi.org/10.1016/j.landusepol.2017.08.012

Kinseng, R. A., Dharmawan, A. H., Lubis, D., & Seminar, A. U. (Eds.). (2019). *Rural socio-economic transformation: Agrarian, ecology, communication and community, development perspectives.* CRC Press.

Lai, J. Y., Mardiyaningsih, D. I., Rahmadian, F., & Hamzah, N. (2022). What evidence exists on the impact of sustainability initiatives on smallholder engagement in sustainable palm oil practices in Southeast Asia: A systematic map protocol. *Environmental Evidence, 11*(1). https://doi.org/10.1186/s13750-022-00283-x

Liu, F. H. M., Ganesan, V., & Smith, T. E. L. (2020). Contrasting communications of sustainability science in the media coverage of palm oil agriculture on tropical peatlands in Indonesia, Malaysia and Singapore. *Environmental Science & Policy, 114,* 162–169. https://doi.org/10.1016/j.envsci.2020.07.004

McCarthy, J. (2010). Processes of inclusion and adverse incorporation: Oil palm and agrarian change in Sumatra, Indonesia. *The Journal of Peasant Studies, 37*(4), 821–850. https://doi.org/10.1080/03066150.2010.512460

McCarthy, J., & Cramb, R. A. (2009). Policy narratives, landholder engagement, and oil palm expansion on the Malaysian and Indonesian frontiers. *Geographical Journal, 175*(2), 112–123. https://doi.org/10.1111/j.1475-4959.2009.00322.x

McCarthy, J., & Zen, Z. (2009). Regulating the oil palm boom: Assessing the effectiveness of environmental governance approaches to agro-industrial pollution in Indonesia. *Law & Policy, 32*(1), 153–179. https://doi.org/10.1111/j.1467-9930.2009.00312.x

Nagiah, C., & Azmi, R. (2012). A review of smallholder oil palm production: Challenges and opportunities for enhancing sustainability—A Malaysian perspective. *Journal of Oil Palm & the Environment, 3*(12), 114–120. https://doi.org/10.5366/jope.2012.12

Naylor, R. L., Higgins, M. M., Edwards, R. B., & Falcon, W. P. (2019). Decentralisation and the environment: Assessing smallholder oil palm development in Indonesia. *Ambio, 48*(10), 1195–1208. https://doi.org/10.1007/s13280-018-1135-7

Ogahara, Z., Jespersen, K., Theilade, I., & Nielsen, M. R. (2022). Review of smallholder palm oil sustainability reveals limited positive impacts and identifies key implementation and knowledge gaps. *Land Use Policy*, *120*, 106258. https://doi.org/10.1016/j.landusepol.2022.106258

Padfield, R., Drew, S., Syayuti, K., Page, S., Evers, S., Campos-Arceiz, A., Kangayatkarasu, N., Sayok, A., Hansen, S., Schouten, G., Maulidia, M., Papargyropoulou, E., & Tham, M. H. (2016). Landscapes in transition: An analysis of sustainable policy initiatives and emerging corporate commitments in the palm oil industry. *Landscape Research*, *41*(7), 744–756. https://doi.org/10.1080/01426397.2016.1173660

Paramananthan, S. (2013). Managing marginal soils for sustainable growth of oil palms in the tropics. *Journal of Oil Palm and the Environment*, *4*(1), 1–16. https://doi.org/10.5366/jope.2013.1

Rudner, M. (1976). The Indonesian military and economic policy: The goals and performance of the first five-year development plan, 1969–1974. *Modern Asian Studies*, *10*(2), 249–284. https://www.jstor.org/stable/311808

Ruysschaert, D., & Salles, D. (2014). Towards global voluntary standards: Questioning the effectiveness in attaining conservation goals. *Ecological Economics*, *107*(107), 438–446. https://doi.org/10.1016/j.ecolecon.2014.09.016

Ruysschaert, D., & Salles, D. (2018). The strategies and effectiveness of conservation NGOs in the global voluntary standards: The case of the Roundtable on Sustainable Palm Oil. In P. B. Larsen & D. Brockington (Eds.), *The anthropology of conservation NGOs: Rethinking the boundaries* (pp. 121–149). Palgrave.

Susanti, A., & Maryudi, A. (2016). Development narratives, notions of forest crisis, and boom of oil palm plantations in Indonesia. *Forest Policy and Economics*, *73*, 130–139. https://doi.org/10.1016/j.forpol.2016.09.009

Tan, Z. D., Lupascu, M., & Wijedasa, L. S. (2021). Paludiculture as a sustainable land use alternative for tropical peatlands: A review. *Science of the Total Environment*, *753*, 1–14. https://doi.org/10.1016/j.scitotenv.2020.142111

Thee, K. W. (2012). *Indonesia's economy since independence*. Institute of Southeast Asian Studies.

Tyson, A., Varkkey, H., & Choiruzzad, S. A. B. (2018). Deconstructing the palm oil industry narrative in Indonesia: Evidence from Riau province. *Contemporary Southeast Asia*, *40*(3), 422–448. https://doi.org/10.1355/cs40-3d

Varkkey, H. (2012). Patronage politics and natural resources: A historical case study of Southeast Asia and Indonesia. *Asian Profile*, *40*(5), 438–448.

Varkkey, H., Tyson, A., & Choiruzzad, S. A. B. (2018). Palm oil intensification and expansion in Indonesia and Malaysia: Environmental and socio-political factors influencing policy. *Forest Policy and Economics*, *92*, 148–159. https://doi.org/10.1016/j.forpol.2018.05.002

Vijay, V., Pimm, S. L., Jenkins, C. N., & Smith, S. J. (2016). The impacts of oil palm on recent deforestation and biodiversity loss. *PLOS One*, *11*(7), 1–19. https://doi.org/10.1371/journal.pone.0159668

World Bank. (2011). *The World Bank Group framework and IFC strategy for engagement in the palm oil sector* (pp. 1–88). The World Bank and International Finance Corporation.

Yazid, M. N. M. (2014). The Indonesian economic development after 1965: Developmental state, radical politics & regional cooperation. *Transactions on Economic Research*, *1*(3), 1–14. https://doi.org/10.15764/er.2014.03001

3

PALM OIL, STATE AUTONOMY, AND ASSEMBLAGE OF LAND USE GOVERNANCE IN SARAWAK, MALAYSIA

Helena Varkkey and Sarah Ali

Introduction

Out of the total 13 states in Malaysia, Peninsular Malaysia on mainland Southeast Asia has 11 states, and East Malaysia on the island of Borneo is made up of the two large states of Sarawak and Sabah. Malaysia was an early pioneer of intensive oil palm production, with the crop first being cultivated in 1917 for commercial production in Tennamaram Estate in Peninsular Malaysia. When synthetic rubber caused rubber prices to drop in the 1960s and -70s, oil palm production became more profitable than natural rubber. This led to large swathes of once-lucrative rubber plantations across the Peninsular and Sabah being swiftly converted into oil palm plantations. Malaysia quickly became the world's largest producer of palm oil and maintained that position until Indonesia surpassed its output capacity in 2008. Today, Malaysia is the world's second-largest producer of palm oil behind Indonesia, contributing approximately 24 percent of global production (US Department of Agriculture, 2023).

Sarawak, the biggest and most heavily forested state in Malaysia, was initially not considered very suitable for oil palm cultivation because of its challenging terrain (Varkkey et al., 2018). However, by 2005, oil palm became the most widely grown crop in Sarawak based on hectarage (Kamlun et al., 2012). In 2015, Sarawak's oil palm planted area stood at around 1.4 million hectares (Varkkey et al., 2018). In 2022, this figure rose to 1.62 million hectares, making up 28.6 percent of Malaysia's oil palm planted area (Malaysian Palm Oil Board, 2022). The cultivation of palm oil in Sarawak has not been without controversy. Much of the extension has involved the large-scale conversion of agriculturally-poor but biodiversity-rich rainforests into plantations, with significant environmental costs. Furthermore, much of this land had previously been held under customary forms of tenure.

DOI: 10.4324/9781003459606-4

The palm oil industry has caused massive landscape, environmental, and socio-economic changes across the state. In this chapter, an assemblage lens is employed to explore how, despite the constraints mentioned above, powerful entities have used governance to enable the exploitation of, firstly, Native Customary Rights (NCR) lands and, later, peatlands in Sarawak for conversion into oil palm plantations to suit their interests.

Assemblages are defined as relational constructs, in which several human and non-human entities are brought together towards certain strategic ends, in particular spaces and times (Blanco et al., 2015). Assemblage processes represent continuous activities through which these relational constructs are being constantly remade over time. During these processes, actors may seek to fix other elements of an assemblage in patterns of relationship which support their projects, a process described in the literature as territorialisation. In this context, governance can be understood as a process through which different actors engage in efforts to steer change, which involves fixing relationships between entities to support their aims. Such activities lead to the emergence of governance assemblages. These are provisional, situated, and unique compositions, continuously emerging from the processes through which these entities come together to achieve particular, issue-related goals. In the course of such assemblage-making, powerful entities may seek to establish specific sets of relationships between entities that define or enrol them in ways that support their perceived interests (Murray Li, 2007).

This chapter addresses the first and second research questions set out in this book, highlighting how power differentials affect palm oil governance in Sarawak, through a discussion of how the different governance arrangements enable and legitimise the expansion of plantations. It pays close attention to the role played by the Sarawak State Government by exploring how it has maintained a position of significant power in the state's oil palm assemblage, and critically, how the role of its autonomous control over land use governance within Sarawak has contributed to this success. This autonomy was particularly useful for the Sarawak State Government in solving the problem posed by the quickly dwindling availability of arable land for conversion into oil palm plantations in the late 1980s. Significant local autonomy over land use stems from the history of the state, its distinct traditions of governance and its relationship with the Peninsular. All these elements have placed the Sarawak State Government in a unique position in shaping relationships between different entities in the state's palm oil assemblage.

The chapter further examines the measures undertaken to moderate conflicts which occur in the Sarawak palm oil sector, as outlined in the book's fourth research question. A feature of this process has been the capacity of the Sarawak State Government to fix relationships between land and other components of the assemblage in ways that constitute land as an object of economic potential. Through the process of territorialisation and reterritorialisation, the State Government was able to ensure the availability of an "unlimited" supply of lands (Cramb, 2011b) to plantation investors through the formulation and adaptation of the state-level forest and land use policy to expedite the conversion of these lands into oil palm plantations.

The chapter first identifies the key non-human and human entities in the Sarawak palm oil governance assemblage. It then explains how the Sarawak State Government carried out territorialisation and reterritorialisation by assembling other entities in ways that support its interests, firstly with NCR lands and later with peatlands. In the same section, the chapter also acknowledges the dynamic nature of assemblages by delving into the efforts of other entities attempting to reterritorialise palm oil governance in the state, although with varied success. This chapter concludes with reflections on power and agency in the palm oil sector in Sarawak and Malaysia, more broadly.

Key entities in the Sarawak palm oil assemblage

The assemblage approach pays close attention to how non-human and human entities are combined in the generation of assemblages. Maintaining these entities in specific territorialised relationships requires the active agency of these non-human and human actors (Blanco et al., 2015). In this case, key non-human entities include the oil palm tree itself, the NCR lands and the peatlands of Sarawak. In terms of human entities, key actors include indigenous groups, the Federal Government, the Sarawak State Government, commercial interests, and scientific communities.

Non-human entities

The oil palm

The oil palm tree (*Elaeis guineensis*) has West African origins and now flourishes throughout the topics, growing commercially in almost 20 countries worldwide. The fruit of this tree produces palm oil, the consumption and use of which has a recorded history of 5,000 years. It is an important global source of food and biofuel and has various industrial uses. Among the 13 major vegetable oils produced worldwide, palm oil holds the largest single share of the market, accounting for about 35 percent of world production. It also has the lowest production cost and is the most efficiently grown vegetable oil in land use per unit of output (Varkkey, 2016).

NCR lands

Before joining the Federation of Malaysia, Sarawak had been under colonial rule for more than a century. This began in 1841 when James Brooke was installed as the first Rajah of Sarawak. The Brooke administration determined two types of land tenure systems. NCR lands were allocated based on native customary law or *adat* and perpetuated traditional land use and farming systems among Sarawak's indigenous groups. The other was a codified land system, which legalised private land ownership and supported the commercialisation of agriculture. However, native rights to land were considered a threat to the Brooke administration because they interfered with the regime's political and economic goals to control natural resources in Sarawak. Subsequently, the Brooke administration proclaimed that all lands which were not codified and/or privately owned belonged to the state. The succeeding colonial government adopted a similar principle ruling that all lands belonged to the Crown when Sarawak was ceded to the British in 1946 (Ngidang, 2005). This approach was also adopted by the new administration when Sarawak joined the Malaysian Federation in 1963. Consequently, in Sarawak, formal titles to much of the land are vested in the state, concentrating power over land use and ownership within the local state.

Between the 1960s and 1980s, the State Government's efforts were mostly focused on improving the productivity of smallholder agriculture on NCR lands. As a result of these efforts, there was a gradual shift from traditional land use to semi-intensive land utilisation. Cash cultivation became a prominent feature of farming systems in rural Sarawak. Initially, promoting large-scale commercial plantations on NCR lands was problematic because of the complex landownership arrangements based on the *adat* system. However, land policies and practices in the 1990s were changed and redesigned in favour of large-scale utilisation of ancestral lands for commercial plantations, citing the need to eradicate rural poverty and to integrate the periphery to the centre (Ngidang, 2005).

Peatlands

The majority of Malaysia's 2.4 million hectares of peatland are found in Sarawak, totalling about 1.7 million hectares (Melling et al., 2011). Here, peatlands are found along the coasts and in the Rajang Delta and Baram River areas, with some peat deposits being as deep as 20 metres (Phillips, 1998). Waterlogged tropical lowland peat swamps perform vital ecological and environmental functions (Evers et al., 2016). These include maintaining the intricate hydrological balance of the lower regions of large river basins, serving as a buffer between salt- and freshwater and providing a habitat for animal life that cannot exist in other ecosystems (Phillips, 1998). It also plays an important role in the global carbon balance (Page et al., 2011). Carbon is stored in the peat formed from organic material such as tree and vegetation litter, sediments and organic soil accumulated over thousands of years. The anaerobic conditions and low availability of nutrients in these peat swamps constrain the decomposition of these organic materials, allowing carbon stocks to continue growing (Parish & Looi, 1999).

When peatlands are drained for agricultural use, decomposition occurs quickly, and carbon dioxide (CO_2) is released into the atmosphere, contributing to climate change. Drained peatlands also dry out fast and become highly vulnerable to fire.

Accidental or intentional—often started as a cheap and quick way to clear the land before planting—peat fires can go deep underground and burn up even more carbon. The smoke from these fires also produces serious localised and transboundary air pollution (Varkkey, 2016). Traditionally, peatlands have been regarded as a "difficult" soil with low fertility due to its impenetrable physiognomy and waterlogged conditions (Phillips, 1998). In their natural state, peat swamps are poorly suited to agriculture as most plantation trees and other crops are not adapted to a semi-aquatic environment, and if they grow at all, are unstable and can easily succumb to windthrow (Varkkey, 2016). Difficulties also emerge with working such land and providing access routes by which raw materials and inputs can be brought into and out of these areas.

Before recent times, peatlands have seen limited local exploitation by indigenous communities for this reason. To convert and prepare peat swamps to plant crops on commercial scales, these areas have to be cleared of vegetation, drained and dried so that the water table drops. This is achieved through an expensive process that involves constructing ditches to allow water to drain out of the area. It involves the use of machinery and drainage technologies, which have only become available relatively recently.

Even once it has been converted, deep peat is still considered problematic. Peat soil is acidic, low in oxygen and inorganic ions, high in carbon and has high concentrations of humic acid. These factors create conditions that are suboptimal for most crops, necessitating significant soil management interventions to bring this land into production (O'Reilly et al., this volume). While the techniques used to adapt these areas for agriculture may deliver productivity in the shorter term, they cause further problems in the longer term, including surface subsidence due to soil shrinkage and compacting, decomposition, leaching, loss of peat material during reclamation, irreversible drying, heavy metal poisoning, and increased incidences of burning.

Human entities

Indigenous groups

Around two-thirds of the people of Sarawak are made up of indigenous groups. The indigenous inhabitants of Sarawak can be divided into two main groups, the coastal peoples (the Malay and the Melanau) and the inland or interior peoples, the Dayak. Broadly, the Dayak peoples include the Bidayuh (Land Dayak), Iban (Sea Dayak), Kenyah, Kayan, Kedayan, Murut, Penan, Bisayah, Kelabit, and other groups. Historically, these groups developed and practised a way of life based on shifting agriculture complemented by hunting and gathering (Kaur, 1998), and some of these practices continue. However, there has also been a steady trend of rural-urban migration since the 1980s in Sarawak (Cramb, 2011a), changing the population makeup of urban areas previously dominated by the minority Chinese.

TABLE 3.1 Distribution of powers in the Federal Constitution (Abdul Rahman et al., 2019)

Federal list	State list	Concurrent list
Mineral resources	Land	Wildlife
Marine and estuarine	Agriculture	Town and country planning
fisheries	Forestry	National parks
Pest control	Infrastructure activities for	Rehabilitation
Shipping and navigation	state works	Eroded and mined land
Water supplies	Water	Drainage
Tourism	Riverine fisheries	Irrigation
Infrastructure activities for		Housing
federal works		

Source: Adopted from the Ninth Schedule of the Federal Constitution.

Federal Government

Malaysia observes a two-tier government system, with both the Federal Government and State Governments having constitutions and legislative powers (Yong, 2006). The Ninth Schedule to the Malaysian Constitution provides for the general distribution of legislative powers between the federal and state Governments (Hansen, 2003). The Federal Government has jurisdiction over matters such as foreign affairs, international trade, defence, internal security, finance, communications, transport, and education. The individual states retain the power to formulate their own policies relating to natural resources such as land, forest, wildlife, agriculture, water resources, fisheries, minerals, and mining (Hezri & Hasan 2006; Yong, 2006; Memon, 2000).

According to the Food and Agriculture Organisation (FAO) of the United Nations, most forests are public lands administered by the state, except for some (about 2 percent) alienated land where forest clearance is permitted for private use (FAO, 2010). Each state has its own Forestry Department and related institutions to manage these public forest resources and implement forestry policies at the state, district, and local levels (Yong, 2006). The Federal Government does not have any powers to enforce forestry laws in the states. The role of the Federal Government towards state forestry matters is confined to some administrative matters such as research, training and advisory roles, without the power of enforcement (Ariffin, 2015, see Table 3.1).

Sarawak State Government

Sarawak and Sabah have more "independence" than Peninsular states due to their later entrance into the Malaysian Federation under the 1963 Malaysia Agreement (MA63), following the departure of the British from Borneo (Cooke, 1997). This Agreement included conditions meant to allay the fears of both states that they would be overwhelmed economically and politically by the more developed

Peninsular states when they joined Malaysia. Hence, the Ninth Schedule of the Malaysian Federal Constitution includes List 2A (Supplement to State List for Sabah and Sarawak) and List 3A (Supplement to Concurrent List for the States of Sabah and Sarawak), which accorded even greater control to the two East Malaysian states over various areas of governance and particularly natural resources (Memon, 2000).

Therefore, Sarawak has exclusive legal jurisdiction to make laws affecting land use, forestry (including the removal of timber and biomass), impounding of inland water and the diversion of rivers, electricity, and the production of electricity generated by water (Memon, 2000). Furthermore, upon joining the Federation, Sarawak negotiated special provisions concerning land-related legislation that provides greater state control over land utilisation policy (Hansen, 2003). Hence, it is empowered to gazette[1] reserves, issue forestry or agriculture permits, collect royalties and premiums, decide on land use as well as allocation of the forest and its development, and so on (Yong, 2006).

This was further underpinned by the new developmentalist ideology adopted by recent leadership, which espoused the need for the state to "catch up" with the rest of Malaysia. Sarawak leaders often admonish the central government for not paying adequate attention to Sarawak, resulting in a large development gap between Sarawak and Peninsular Malaysia. The rapid expansion into palm oil is part of the state's insistence that "Sarawak should not be left behind" (Ling, 2016). Because the economy of Sarawak heavily depends on the exploitation and export of natural resources, the state has, over the years, zealously guarded its constitutional rights against interference from the Federal Government (Hansen, 2003; Memon, 2000). For example, the Chief Minister traditionally holds the role of Minister for Urban Development and Natural Resources in Sarawak. Hence, they oversee the Department of Lands and Surveys, the Forest Department and the Sarawak Natural Resources and Environment Board (NREB—see below).

Furthermore, the allocation of state land is the jurisdiction of the State Planning Authority (SPA) under the Chief Minister's Office (Hansen, 2003; Taylor et al., 1994). The SPA allocates land and forests according to the two principal pieces of written land legislation: the 1953 Forest Ordinance and the 1958 Land Code (Taylor et al., 1994). The formulation of policies on forestry and land development in Sarawak is thus concentrated at the highest level of authority (Yong, 2006).

Commercial interests

The first commercial palm oil plantation in Sarawak was established relatively late and on quite a small scale; on 3,000 hectares of state land in Miri in 1968 under a joint venture between the Sarawak Government and the Commonwealth Development Corporation (Lau, 2016). However, despite its huge land size of 12.4 million hectares, only 28 percent of Sarawak's land has historically been considered suitable for commercial crops, with the remaining land made up of 58 percent steep land, 13 percent peatland, and 1 percent infertile land (Varkkey et al., 2018). Thus, Sarawak was largely "left behind" during the early phase of commercial palm oil expansion, which saw vast areas of forests in the rest of Malaysia converted into palm oil plantations in the 1960s and 1970s.

As arable land for palm oil plantations dwindled in the rest of Malaysia in the 1980s, commercial plantation investors began to move to neighbouring land-rich countries such as Indonesia, the Philippines, and Papua New Guinea to establish new oil palm plantations. In response to this, Sarawak's then-newly sworn-in Chief Minister, Taib Mahmud, gave an open invitation to local investors: "Why not come to Sarawak to invest? You can operate from Kuala Lumpur. Sarawak welcomes you" (Cramb, 2011b). This was a welcome development to these plantation corporations who were beginning to come up against foreign investment barriers abroad (Varkkey, 2016) and marked the beginning of large-scale palm oil investments in Sarawak. Sarawak earned its nickname as Malaysia's "final frontier" for commercial oil palm cultivation. However, expansion was still limited to the small areas of available arable land in the state. Until the late 1990s, Sarawak continued to lag behind smaller states in total oil palm planted area.

Scientific communities

These are networks of knowledge-based experts who help decision-makers define the problems they face, identify various policy solutions, and assess the policy outcomes (Haas, 1989). In recent years, there has been an added interest in peatland science and policy within scientific communities due to the important role of tropical peatlands in the global carbon balance and the increasingly significant use of these lands for important agricultural commodities such as palm oil. To this end, the extent to which cultivation of oil palm on peat can be sustainable is the subject of considerable debate within these networks.

On the one hand, many scholars consider undisturbed peat swamps as more important to humans than if drained and developed (Wijedasa et al., 2017; Evers et al., 2016; Phillips, 1998). Many peatland experts suggest that large areas of currently drained coastal peatlands will be progressively subject to longer periods of inundation by the river and ultimately seawater, making agriculture in these areas increasingly untenable (Wijedasa et al., 2017). In short, they are generally of the view that the long-term potential of drained peatlands continuing to sustain production-to-cost ratios at commercially viable levels is limited. On the other hand, there also exist certain state-sponsored scientific groups that argue for the viable and productive use of peatlands for such commercial agriculture. Most notably, the Sarawak State Government established the Tropical Peat Research Laboratory (since renamed Tropical Peat Research Institute, or TROPI in 2016), which provides scientific support for developing oil palm plantations on peat (Cheng, 2016) to complement Sarawak's agricultural goals.

Assembling palm oil governance in Sarawak

Within assemblages, various actors devise and employ strategies to establish specific sets of relationships between entities that enrol them in ways that support their perceived interests. The extent to which they have the resources and capacity to do so successfully constitutes the degree of power these entities have. Murray Li (2007) describes these processes as taking the form of six assemblage practices, including *forging alignments, rendering technical, authorising knowledge, managing failures and contradictions, anti-politics, and reassembling*. The most powerful entities in any assemblage at a time are successful in defining these relationships in ways that suit their goals. This section explores the practices through which the Sarawak State Government sought to (re)arrange entities in assemblages that support its interests; in this case, in continuing the appropriation and conversion of lands in Sarawak for the production of palm oil. In this process, the State Government first attempted to territorialise NCR lands and later peatlands.

Lacking the presence of the type of traditional governance systems which were inveigled in the colonial project on the Peninsular, Sarawak was governed by a unique colonial regime for much of the twentieth century, which involved the highly centralised control of the formal economy, limited democratic accountability to certain subsets of the local population, and at best, the neglect of native rights and interests. While independence may have removed many of the features of a colonial regime and empower new actors, the new administration rarely starts from scratch. New governance actors seldom entirely sweep away all aspects of previous governance assemblages. Rather, objects and sets of arrangements that formed part of that previous assemblage may be reconfigured and incorporated into new governance assemblages. In this way, some aspects of the deep structure of colonial governance assemblages may be retained. Often, this includes legislative hangovers and existing administrative practices that are perceived to be mundane or lacking in ideological significance.

In the case of Sarawak, features of previous administrative assemblages, including highly centralised control over commercial activities and an ambiguous relationship with indigenous groups, appear to have been continued into the post-colonial era. Arguably, this results in a governance system that reflects the interests of a commercial and administrative elite, which can exercise considerable control over land-use policies affecting large swathes of land due to pre-independence legislation and policy habits. The following sections attempt to answer the first, second, and fourth research questions set out in the book by exploring the different governance arrangements used in Sarawak, alongside its power differentials which enable and justify the expansion of oil palm plantations within the state, as well as the steps taken to mitigate potential conflicts that arise.

(Re)territorialisation of NCR Lands

In 2000, Sarawak set itself an agricultural development target to become the new leader in palm oil production by 2010, hoping to increase planted areas from 300,000 hectares to 1 million hectares by the end of the decade (Hon, 2011; Hansen, 2003). However, until the 1990s, the movement of commercial plantations into Sarawak was limited due to land constraints. This did not mean that there was a lack of interest; for example, a senior official from the Federal Land Development Agency (FELDA) indicated in 1982 that the agency was keen to utilise the "idle" lands of Sarawak, but needed clear titles to the land, unencumbered by bothersome NCR claims (Cramb, 2011b). Hence, to maintain the interest of investors and the steady flow of profits into state coffers, the then-Chief Minister, Taib, quickly provided a solution: NCR lands would be made available for conversion into oil palm plantations. Such is an example of *forging alignments* within the assemblage—the work of linking together the objectives of various parties (Murray Li, 2007), in this case, the State Government that governs conduct and the commercial plantations whose conduct are to be governed.

NCR lands were assumed "idle" land simply because the people who occupy it were regarded as "idle" (Cramb, 2011b; Doolittle, 2007). Of course, identifying these lands as idle and underutilised is problematic (Carlson et al., 2012). These lands are often used for community farming or as areas for hunting and gathering. Some groups, such as the Dayak and Malays, took up rubber planting on these customary lands from early in the twentieth century. Dayak customary smallholders went through an agricultural revolution by cultivating the technically-, financially-, and managerially-intensive pepper crop in the 1970s (Cramb, 2011b). Oil palm is currently also one of the crops widely cultivated by indigenous smallholders on customary lands in Sarawak. Hence, customary land tenure has not been an obstacle to the adoption and expansion of smallholder cash crops among the Sarawak indigenous groups (Cramb & Sujang, 2013).

However, since colonial times, traditional land use in Sarawak has been in tension with the more profitable commercial plantation operations. The desire of the latter was often allied with the State Government—leading to negative representations of local small-scale agriculture as less efficient and poorly organised, and of local populations as idle or backward. The representations of these entities in these terms supported the creation of new mutually reinforcing relationships between the State Government and large-scale commercial actors. They made for the reterritorialisation and utilisation of these ancestral lands for commercial purposes justified in terms of a "civilising process", to help eliminate poverty among indigenous groups (Cramb, 2011b) and integrate them from the periphery of the state's vision of civil society into its centre. As a result, the various political and administrative entities in Sarawak have, throughout its history, used legislative means to reconstruct land ownership, land values, and present specific ideas as to how best to develop huge tracts of ancestral land (Ngidang, 2005).

This narrative has been modernised and formalised through the Sarawak Ministry for Agriculture Modernisation and Rural Economy, which justifies its focus on NCR land as a move to transform large tracts of unproductive and underutilised lands into viable economic units (Goh, 2016) to help boost the rural economy (Chia & Ten, 2016). This was the State Government's attempt to *manage failures and contradictions* within the assemblage by presenting failure (indigenous groups being "backward" and "less civilised") as the outcome of rectifiable deficiencies (transforming "idle" customary land into viable economic units).

As land matters are under state jurisdiction, the state can circumscribe the rules governing NCR or use its powers to oppose native titles and rights. Land and forest-related legislation has consistently and progressively been amended to introduce increasingly aggressive clauses to limit and compromise NCR over the years (Yong, 2006). NCR to land was given limited recognition in the 1958 Land Code, just before the state's independence. This Code created a racially based relationship between different groups of people and land in Sarawak via a system of zoning land, by generating different categories of land which could be owned by different groups of people: (1) Mixed Zone Land, to which anyone can hold a title; (2) Native Area Land, to which only Sarawak "natives" can hold a title; (3) NCR, which is not under title but subject to NCR; (4) Reserved Land or land held by the government principally as forest reserves; and (5) Interior Area Land, which is the residual bulk of the state's land. The introduction of a distinct category of NCR alongside written titles adds further complexity to these arrangements. NCR land status can be superimposed over the other land classes (Taylor et al., 1994), except Reserved Land. Untitled land within a region classified as Mixed Zone or Native Area Land is typically "state land subject to NCR". Also, the proportion of Interior Area Land that is subject to customary rights before 1958 continues to be legally recognised as NCR (Cramb, 2011b).

The Code defines that these rights could be acquired by proving that before 1958, there occurred (1) the felling of virgin jungle and the occupation of the land thereby cleared; (2) the planting of land with fruit trees; (3) the occupation or cultivation of land; (4) the use of land for a burial ground or shrine; (5) the use of land of any class for rights of way; or (6) any other lawful method. Once "legitimate" customary land has been established, the holder of customary rights could be issued with a grant in perpetuity free of rent, implying recognition that NCR amounts to a form of ownership (Cramb, 2011b). Until then, the land was held "by license from the state" (Cramb, 2016).

This potentially broad approach to NCR recognition reflected the de-facto situation in Sarawak before 1958, whereby indigenous groups habitually used large areas of land, which were perceived as having only limited economic value. New technologies and new species (in particular palm oil) have changed that situation, resulting in the emergence of an assemblage in which the commercial potential of land affected by NCR increased. This is linked to new strategies being adopted by the Sarawak government to enrol this land in the emerging palm oil assemblage. Policies and laws are developed and amended accordingly to mitigate conflict between the parties involved in the palm oil sector. The Land Code (Land Code— Chapter 81, 2022) was amended by the Sarawak Government in 1994 to enable the government to extinguish NCR to land, making it accessible for large-scale private land developers. It was amended again in 1996, with a new section stating that

> whenever any dispute shall arise as to whether any NCR exists or subsists over any state land, it shall be presumed until the contrary is proved, that such state land is free of and not encumbered by any such rights.

A 1997 amendment allowed the government to amalgamate NCR within a "development area" into a single parcel of land, and to grant a lease of up to 60 years over the land to a body corporate approved by the Minister and "deemed to be a native" by the cabinet. An amendment in 1998 streamlined the extinguishment of NCR and minimised compensation pay-outs. Finally, in 2000, an amendment removed the sixth method listed for creating NCR, namely "any other lawful method", which excluded forest land for community use within a longhouse territory (*menoa*) (Cramb, 2011b).

The Forest Ordinance has been amended for this purpose as well. In 1984, a government directive restricted the rights to make NCR claims, citing the fear that native forest dwellers may destroy commercially viable species through swidden agriculture or for private consumption, for example, in boat-making or building community longhouses (Cooke, 1997). A 1987 amendment made it a criminal offence to barricade logging roads (Leigh, 1998). Furthermore, the establishment of a Permanent Forest Estate now specifically requires the termination of NCR. Also, new laws such as the Land Surveyors Ordinance 2001 have been enacted to criminalise activities related to land rights defence, such as community mapping activities, and remove from the courts the power to decide on the admissibility of community maps as evidence at the courts (Yong, 2006).

Moreover, the areas estimated as NCR by the state often differ drastically from figures used by claimants, customary courts, and non-state actors (Cramb, 2011b, 2016). For example, in 1988, Cramb and Dixon (1988) estimated the total area of NCR as 3.1 million hectares, about 25 percent of Sarawak's land area. However, in 1997, Sarawak officially asserted that there are only 1.5 million hectares of NCR in the state (Cramb, 2011b). Official figures were based on aerial photographic evidence of cultivation before 1958 only and excluded forested land reserved within community territories (Cramb, 2016). The lower figure could reflect the government's desire to exclude as much land as possible from officially recognised customary claims (Cramb, 2011b).

Resistance and attempts at destabilisation

Assemblages and the territories they create are dynamic and contingent. At any one time, other entities which negatively perceive current arrangements may destabilise existing arrangements, forming new patterns of relationships within an assemblage (Murray Li, 2007). In the reterritorialisation process of NCR lands, the Sarawak State Government ran into such destabilising obstacles from indigenous groups and their supporters. Consequently, plantation investors still faced the problem that extensive areas of suitable land remained "encumbered" with *claims* of NCR by various indigenous groups. Even ostensibly unencumbered state land earmarked for land development has often turned out to be subject to NCR claims, especially in Central and Northern Sarawak, where settlement is more recent. For example, some longhouse communities that cleared forests after 1958 for shifting cultivation still consider the remaining old-growth forests within their territory subject to NCR claims (Cramb, 2011b).

As a result, companies that receive concessions from the State Government, including to NCR lands, were almost certain to be involved in conflicts with local communities. Indeed, a list by Danish forest consultants, Pro Regenwald (2010), identified at least 57 land conflicts from 1995 to 2010 related to oil palm plantations on NCR lands. In today's hyper-connected world, companies can ill-afford negative publicity relating to land disputes, which have often been couched in terms of big business bullies against marginalised natives. A particularly high-profile case was that of Tabung Haji, one of Malaysia's biggest oil palm plantation companies, clashing with over 100 Iban families near Serian as they blocked the company from harvesting oil palm on 3,000 hectares of their NCR land (Papau, 2014). Amid negative publicity, Tabung Haji was compelled to abandon its plans.

Hence, these indigenous groups, often working closely together with interest groups and international media, have managed to destabilise the Sarawak State Government's efforts to advantageously subvert the position of former NCR lands according to its interests. Despite the legislative measures described above, the conversion of NCR lands into oil palm plantations was relatively slow. As of 2016, only 328,000 hectares of NCR land have been converted into oil palm plantations (22 percent of the official and 11 percent of the unofficial NCR area) (Goh, 2016). This situation presented a challenge to the Sarawak State Government, which had placed significant political capital into the development of oil palm. Consequently, it sought other ways to open up and streamline access to land for private estate development (Cramb, 2011b).

Reterritorialisation of peatlands

The Sarawak State Government thus sought other avenues to find land for palm oil. Sarawak proposed its 1.7 million hectares (Lau, 2016) of peatlands as a solution to this problem, a viable source of land to meet the demand of commercial oil palm plantations in the state. While less productive than plantations on mineral soil, oil palm can be grown in the extreme conditions found on drained peatland (Tan et al., 2009). It must be noted, however, that such cost calculations rarely extend to considerations of long-term environmental, public health and rehabilitation costs associated with these activities.

This is facilitated by the biological characteristics of the oil palm itself and the economics of the industry. Biologically, oil palm has a high tolerance for areas with fluctuating water tables (Liew, 2010). There is anecdotal evidence that oil palm grown on reclaimed peat soil can produce comparably high yields (van Noordwijk et al., 2008). Economically, the sheer productivity and profitability of the oil palm industry are such that even though the constraints discussed above make oil palm development on peat soil more expensive (with set-up costs on peatlands almost double those on regular mineral soil) (Liew, 2010), high sustained demand and trading prices for palm oil continue to make this economically viable. Secondly, peatlands are attractive because of the valuable types of commercial timber growing in these areas, such as the Alan (*Shorea albida*), Ramin (*Gonystylus bancanus*) and Terentang (*Campnosperma spp.*) (Phillips, 1998; Parish, 1997). Concessionaires can log these areas during land clearing, with the profit from the sale of this timber used as revenue to fund plantation start-up costs (Stone, 2007). Finally, the usually secluded nature of peatlands (far away from towns and cities) also means that plantation companies can conduct their practices relatively free from scrutiny by environmental authorities (Varkkey, 2016).

Hence, significant actors in the Sarawak governance undertook a new process of fixing peatlands in the oil palm assemblage. Similar to the NCR narrative, Sarawak presented its peatlands as "idle" or wastelands (see Manzo et al., 2019). Thus, it was a viable (and cheap) source of land to meet oil palm cultivation goals in the state. In a further attempt to *manage failures and contradictions*, the State Government argued that given the scarcity of agricultural land on mineral soils in Sarawak, it is necessary to develop peatlands in plantations to improve the livelihood of local communities (Singapore Institute of International Affairs, 2017).

Similar to the case of the NCR lands detailed above, the Sarawak State Government has used its autonomous jurisdictional rights to amend forest-related legislation to bring about a transformation of the relationship between peatland and the oil palm assemblage. This began in earnest in the 1990s, when it became obvious that the complications related to NCR lands were reducing the attractiveness of these areas to investors. In Sarawak, most of the peat swamp forests are gazetted as "permanent" (managed to maintain forest cover for multiple purposes) but not "totally protected" (managed to preserve biodiversity and natural ecosystems in situ with no commercial activity allowed other than tourism) (Phillips, 1998). This means that most peatlands would fall either under the category of Permanent Forest Estates (including Forest Reserves) or State Land Forests (not reserved permanently and can be alienated for agriculture) under the Forest Ordinance. With the amendments to the Land Code and Forest detailed above, much of the state's peatlands would fall outside what could be considered to be NCR.

In 1996, the Forest Ordinance was further amended, and a mechanism was created for oil palm estates to be established within Forest Reserves (Cramb, 2016). In 1997, new Forests (Planted Forests) Rules were introduced, providing for the issuance of "licenses for planted forest" within land zoned as Forest Reserves and allowing licensees to plant oil palm on 20 percent of the plantable area for one cycle of 25 years (Nature Economy and People Connected, 2017; Cramb, 2016).

In the early 2000s, Sarawak classified agricultural plantations as a "public good", which implied that the SPA can expropriate land for oil palm plantations (Hansen, 2003). Since then, the Sarawak government has classified large tracts of lands claimed as belonging to the state as State Land Forests, enabling it to be alienated for other land uses such as agricultural plantations (Nature Economy and People Connected, 2017). As a result, palm oil plantation activities have accelerated land-use change in the lesser regulated government-owned State Land Forests (Hon & Shibata, 2013).

A state-wide Independent Peat Basin (IPB) study was also conducted in 1992, which classified 91 out of the 109 IPBs (one-third of the total 825,156 hectares identified) as having potential for agricultural development. To quickly tap this potential, road development in the peat-rich coastal plains of Sarawak was increased in the late 1990s to improve access to peatlands. As a result, peat swamp forests in Sarawak suffered a tremendous loss of almost 50 percent between 1990 and 2009, especially in coastal areas due to intensified oil palm development (Kamlun et al., 2012). Following this, Sarawak launched its Sarawak Corridor of Renewable Energy (SCORE) in 2006, its long-term development strategy for the central region. SCORE reaffirmed the state's commitment to further oil palm expansion in peatland areas. The three key growth industry areas identified—Tanjung Manis, Mukah, and Similanjau—were dominated by peat swamp forests (Hon, 2011).

Other assemblage practices such as *authorising knowledge* (by specifying the requisite body of knowledge, confirming enabling assumptions, and containing critiques) and *anti-politics* (reposing political questions as matters of technique, closing down debate about how and what to govern, and the distributive effects of particular arrangements by reference to expertise, and encouraging citizens from engaging in debate while limiting the agenda) (Murray Li, 2007) can also be observed within the Sarawak State Government's efforts at reterritorialising peatlands. In this context, the capacity of the State Government to successfully fix the peatlands in an oil palm assemblage also depended on it being able to establish that the cultivation of oil palm in these areas was technically viable. To do so, new actors drawn from the scientific community were brought into play. Sarawak invested significant resources in supporting research aimed at enhancing peatland conversion and cultivation for oil palm. Most notably, the state-sponsored TROPI remains the sole research unit under the Chief Minister's Department, reporting directly to the Chief Minister, and formally recognised by the State Government as the *knowledge authority* of peatlands in Sarawak. This had the effect of bringing new entities in the assemblage in the form of researchers and research techniques, which played an important role in establishing that such areas could be used to cultivate oil palm.

TROPI maintains that oil palm is currently the most economical perennial crop for planting on peat soils as it gives the best return of investment when properly managed (Melling et al., 2011). It points out that oil palm has been successfully grown on peat in Malaysia for two to three generations now, and there are oil palm plantations on peatlands that have matched the productivity of those on mineral soils (Singapore Institute of International Affairs, 2017). TROPI argues that it is possible to cultivate almost all peat areas while mitigating its possible negative impacts using specialised agricultural techniques. This includes artificial soil compaction using excavators, a controlled drainage system to maintain stable water table levels, planting oil palm trees on raised mounds to prevent leaning, and using excavators to push leaning trees upright. According to TROPI research, artificial soil compaction can improve yield and reduce the need for regular fertilising, as compacted peat retains fertiliser better. It also increases peat density and capillary rise, which reduces the rate of carbon emissions and reduces the amount of oxygen in the soil, reducing the risk of fire (Singapore Institute of International Affairs, 2017). This could be a practice of *rendering technical* and *anti-politics;* reposing political questions (the sustainability and long-term tenability of palm oil on peat) as matters of technique (compaction, controlled drainage, etc.).

Within the academic field, there is some criticism of the technical solutions proposed by TROPI. Compaction has been found to have negative effects (Singapore Institute of International Affairs, 2017), as compacted soil impedes root growth and penetration, reducing the uptake of water and nutrients, and possibly resulting in stunted, drought-stressed plants and lower yields. It also reduces the ability of the soil to store water and regulate water flows, possibly increasing the severity of seasonal droughts and floods. Furthermore, it does not address the issue of long-term peat subsidence, as peat above the water table level will continue to decompose and subside until the area becomes permanently flooded and thus unsuitable for planting. Despite this, TROPI retains a key role as the official source of scientific knowledge on peatland cultivation. Its "best practices" for oil palm development on peat are widely promoted, enabling State Government officials to maintain that oil palm growers in Sarawak are employing measures that minimise fire and environmental damage.

The introduction of legislative changes and knowledge authorities to the assemblage allowed new alignments forged between the oil palm tree, governance entities, plantation companies, and peatlands. This mapped well onto the growing desire of palm oil companies (responding to growing pressure from human rights interest groups) to avoid conflicts with local communities by exploiting apparently "idle" peat swamps where community conflicts were less in evidence (though not wholly absent). As a result of these various assemblage practices, oil palm cultivation on peatland in Sarawak has expanded dramatically to become the most widely grown crop in the state based on hectarage in 2005. In 2006, Sarawak achieved its 1-million-hectare goal (Kamlun et al., 2012), surpassing the planted areas of the Peninsular states of Johor and Pahang, which were formerly the top two states in terms of oil palm hectarage in Malaysia (Sarawak Report, 2014; Hon, 2011). Over the next two years, 70 percent of the increase in the national total planted area took place in Sarawak (Nature Economy and People Connected, 2017).

When Taib stepped down in 2014, his successor, Adenan Satem, continued Taib's trajectory when he announced that Sarawak would continue to open up coastal lowland areas (peatlands) to encourage the expanding oil palm industry (Cheng & Sibon, 2016). Even though it was declared that only logged-over peat swamp forests could be developed into oil palm plantations in Sarawak (Melling et al., 2011), this hurdle was easily overcome as planted peat areas have been logged since the 1950s. Sarawak's planted area continued to expand, amounting to 1.4 million hectares in 2015, or 25.5 percent of all palm oil planted land in Malaysia. It contributed about USD2.03 billion (9 percent) of the state's total exports in that year, from which the state can collect substantial tax (Chia & Ten, 2016). Today, out of the 1.7 million hectares of peatland in Sarawak, about 44 percent, have been planted with oil palm (Awang et al., 2021).

Resistance and attempts at destabilisation

Many experts in this field have generated evidence which resolutely contests Sarawak's position. For example, more than 100 local and international scientists (including this chapter's first author) wrote a strongly worded Letter to the Editor of the journal, Global Change Biology, declaring that peatland development in Sarawak for oil palm would have dire immediate and future consequences (Wijedasa et al., 2017). The letter compiled extensive findings that show how contemporary agriculture techniques on peatlands have heavily impacted the peatland ecosystem through land clearance, drainage, fertilisation and, often, fire (Page et al., 2002). No current techniques have been shown to prevent the loss and subsidence of peat following drainage. These studies thus call into question the very notion of "long-term sustainability of tropical peatland agriculture" (Evers et al., 2016). They argue that the Sarawak government's approach prioritises short-term profit production that irreversibly damages the peat swamp ecosystem (Wijedasa et al., 2017).

Notable bodies, such as the RSPO, the world's leading palm oil certification system, have taken note of these scientific developments. As part of the RSPO's latest Principles and Criteria 2018, RSPO certifiable members should no longer carry out new oil palm plantings on peat after November 2018 (RSPO, 2019). However, Malaysia's Federal Government-facilitated Malaysian Sustainable Palm Oil (MSPO) certification system acknowledges that state law allows for planting on peatlands and has developed "best practice" guidelines for these peatlands (Malaysian Palm Oil Board, 2020).

It is still unclear if these destabilisation attempts by scientific communities (again, supported by interest groups and international media) will succeed. Sarawak still maintains that peatlands can and should be developed sustainably for agriculture, especially palm oil (Nurbianto, 2016). However, recently in 2019, the current Chief Minister, Abang Abdul Rahman Zohari, announced a moratorium on new licenses for timber and oil palm except for communal and NCR lands (Bernama, 2019). Simultaneously, the Sarawak State Government maintains its Sustainable Land Use Policy, which targeted 3 million hectares of oil palm plantations in the state by 2020 (Lau, 2016). There is a lack of clarity over whether peatlands which have already been licensed out before the moratorium can still be cleared for plantations. On another note, the refocus on NCR lands may be the precursors to yet another attempt at reterritorialisation to stay on track for state development goals.

Conclusion

This chapter has detailed how evolving governance arrangements and power differentials can enable and legitimise palm oil expansion, in addition to the strategies adopted in order to mitigate the conflicts that arise in the state of Sarawak, Malaysia, effectively shedding light on the inquiries posed by the first, second, and fourth research questions outlined in the book. In Sarawak, the State Government was able to retain substantial autonomous control over its land use policy due to its unique governance traditions which stem from historical factors related to its former colonial status and current membership of the Malaysian Federation. Hence, the State Government currently appears to be the most powerful actor in this assemblage as it has the strongest capacity to enact other actors and objects in ways that support its interests (Nuijten, 2005). This has placed the Sarawak State Government in a unique position to shape relationships between different elements in the palm oil assemblage in the state, and over how land is constituted as an object of economic potential. When faced with resistance and obstacles, the Sarawak State Government has acted in purposive and thoroughly rational ways to shape and reterritorialise elements of the assemblage in ways which support its perceived interests.

New forms of resistance, however, may be coming from within the industry itself, as some countries have taken steps to pressure the industry to adopt more environmentally sustainable practices, and at the same time, palm oil-producing countries and companies have taken individual and collective steps at both national and international levels to respond to global concerns regarding its environmental impact. Examples include the international industry response of the RSPO and private sector pledges such as "No Deforestation, No Peat, No Exploitation"; the European Union's (EU) revised Renewable Energy Directive Recast (RED II), which aims to phase out biofuels linked to high indirect land-use change by 2030 (EU, 2019); and the national response of Indonesia in declaring a more well-defined moratorium on peatland conversion.

However, the extent to which all this is likely to impact on the governance of the palm oil industry in Sarawak is open to question. The Sarawak government's autonomy over its land-use policies has been remarkably durable, as it enjoys a significant degree of autonomy from the Federal Government while simultaneously retaining a considerable degree of centralised control at the state level. Despite recent shakeups in Malaysia politics, Gabungan Parti Sarawak, which has joined the new Pakatan Harapan coalition to form the central government, still includes many former stalwarts of the old guard, making it unlikely to result in NCR and peatland politics in Sarawak coming under closer scrutiny. Hence, it remains unclear how much effect central scrutiny (if any) and indirect controls will have on state-level land use policies.

Note

1 The "gazetting" of land under Malaysian law is the process of the State Authority publishing its decision to compulsorily acquire the land in question for either public purpose (e.g. forest reserve) or private purpose (e.g. commercial agriculture) in the Government Gazette, and giving public notice of the same. The Sarawak Land Code and subsequent amendments allow the State Authority to gazette or degazette land as native areas and NCR (TRAFFIC International, 2004).

References

Abdul Rahman, M. A., Ghazali, F., Mohd Rusli, M. H., Aziz, N., & Wan Talaat, W. I. A. (2019). Marine protected areas in Peninsular Malaysia: Shifting from political process to co-management. *Journal of Politics and Law*, *12*(4), 22. https://doi.org/10.5539/jpl.v12n4p22

Ariffin, R. (2015). Environmental legislation for forestry protection. *Forum on Environmental Legislation Forestry Department of Peninsular Malaysia*, 1–39. Forestry Department Peninsular Malaysia.

Awang, A. H., Rela, I. Z., Abas, A., Johari, M. A., Marzuki, M. E., Mohd Faudzi, M. N. R., & Musa, A. (2021). Peat land oil palm farmers' direct and indirect benefits from good agriculture practices. *Sustainability*, *13*(14). https://doi.org/10.3390/su13147843

Bernama. (2019, August 23). Dr M dismisses claims linking palm oil to deforestation. *New Straits Times*. https://www.nst.com.my/news/nation/2019/08/515272/dr-m-dismisses-claims-linking-palm-oil-deforestation

Blanco, G., Arce, A., & Fisher, E. (2015). Becoming a region, becoming global, becoming imperceptible: Territorialising salmon in Chilean Patagonia. *Journal of Rural Studies*, *42*, 179–190. https://doi.org/10.1016/j.jrurstud.2015.10.007

Carlson, K. M., Curran, L. M., Ratnasari, D., Pittman, A. M., Soares-Filho, B. S., Asner, G. P., Trigg, S. N., Gaveau, D. A., Lawrence, D., & Rodrigues, H. O. (2012). Committed carbon emissions, deforestation, and community land conversion from oil palm plantation expansion in West Kalimantan, Indonesia. *Proceedings of the National Academy of Sciences*, *109*(19), 7559–7564. https://doi.org/10.1073/pnas.1200452109

Cheng, L. (2016, August 4). *Peatland—The last frontier of oil palm industry*. Borneo Post Online. https://www.theborneopost.com/2016/08/04/peatland-the-last-frontier-of-oil-palm-industry/

Cheng, L., & Sibon, P. (2016, August 17). *Sarawak opening up coastal lowland areas for agriculture, plantation devt—Adenan*. Borneo Post Online. https://www.theborneopost.com/2016/08/17/sarawak-opening-up-coastal-lowland-areas-for-agriculture-plantation-devt-adenan/

Chia, J., & Ten, M. (2016, July 19). *Master plan in the works to boost state's palm oil industry*. Borneo Post Online. https://www.theborneopost.com/2016/07/19/master-plan-in-the-works-to-boost-states-palm-oil-industry/

Cooke, F. M. (1997). The politics of "sustainability" in Sarawak. *Journal of Contemporary Asia*, *27*(2), 217–241. https://doi.org/10.1080/00472339780000141

Cramb, R. A. (2011a). Agrarian transitions in Sarawak: Intensification and expansion recon-sidered. In R. de Koninck, S. Bernard, & J.-F. Bissonnette (Eds.), *Borneo transformed: Agricultural expansion on the Southeast Asian frontier* (pp. 44–93). NUS Press.

Cramb, R. A. (2011b). Re-inventing dualism: Policy narratives and modes of oil palm expansion in Sarawak, Malaysian. *The Journal of Development Studies, 47*(2), 274–293. https://doi.org/10.1080/00220380903428381

Cramb, R. A. (2016). The political economy of large-scale oil palm development in Sarawak. In R. A. Cramb & J. F. McCarthy (Eds.), *The Oil Palm Complex: Smallholders, agribusiness and the State in Indonesia and Malaysia* (pp. 189–246). NUS Press.

Cramb, R. A., & Dixon, G. (1988). Development in Sarawak: An overview. In R. A. Cramb & R. H. W. Reece (Eds.), *Development in Sarawak: Historical and Contemporary Perspectives* (pp. 349–350). Centre of Southeast Asian Studies, Monash University.

Cramb, R. A., & Sujang, P. S. (2013). The mouse deer and the crocodile: Oil palm small-holders and livelihood strategies in Sarawak, Malaysia. *Journal of Peasant Studies, 40*(1), 129–154. https://doi.org/10.1080/03066150.2012.750241

Doolittle, A. A. (2007). Native land tenure, conservation, and development in a pseudo-democracy: Sabah, Malaysia. *The Journal of Peasant Studies, 34*(3–4), 474–497. https://doi.org/10.1080/03066150701802793

EU. (2019). *Renewable energy—Recast to 2030 (RED II)*. EU Science Hub; Euro-pean Commission. https://joint-research-centre.ec.europa.eu/welcome-jec-website/reference-regulatory-framework/renewable-energy-recast-2030-red-ii_en

Evers, S., Yule, C. M., Padfield, R., O'Reilly, P., & Varkkey, H. (2016). Keep wetlands wet: The myth of sustainable development of tropical peatlands—implications for poli-cies and management. *Global Change Biology, 23*(2), 534–549. https://doi.org/10.1111/gcb.13422

Food and Agriculture Organisation of the United Nations. (2010). *Global forest resources assessment 2010* (pp. 1–340). FAO.

Goh, P. P. (2016, November 30). Sarawak NCR land: Only 328,000 out of 1.5 mil-lion hectares developed, planted with oil palm. *New Straits Times*. https://www.nst.com.my/news/2016/11/193118/sarawak-ncr-land-only-328000-out-15-million-hectares-developed-planted-oil-palm?d=1

Haas, P. M. (1989). Do regimes matter? Epistemic communities and Mediterranean pollu-tion control. *International Organization, 43*(3), 377–403. https://www.jstor.org/stable/2706652

Hansen, T. S. (2003). A step in the right direction: Towards integrated natural resource management in Sarawak, Malaysia. *Water Resources Systems—Water Availability and Global Change, 280*, 175–183.

Hezri, A. A., & Hasan, Mohd. N. (2006). Towards sustainable development? The evolution of environmental policy in Malaysia. *Natural Resources Forum, 30*(1), 37–50. https://doi.org/10.1111/j.1477-8947.2006.00156.x

Hon, J. (2011). SOS: Save our swamps for peat's sake. *SANSAI: An Environmental Journal for the Global Community, 5*, 51–65. http://hdl.handle.net/2433/143608

Hon, J., & Shibata, S. (2013). A review on land use in the Malaysian state of Sarawak, Borneo and recommendations for wildlife conservation inside production forest envi-ronment. *Borneo Journal of Resource Science and Technology, 3*(2), 22–35. https://doi.org/10.33736/bjrst.244.2013

Kamlun, K. U., Goh, M. H., Teo, S., Tsuyuki, S., & Phua, M.-H. (2012). Monitoring of defor-estation and fragmentation in Sarawak, Malaysia between 1990 and 2009 using landsat and SPOT images. *Journal of Forest and Environmental Science*, *28*(3), 152–157. https://doi.org/10.7747/jfs.2012.28.3.152

Kaur, A. (1998). A history of forestry in Sarawak. *Modern Asian Studies*, *32*(1), 117–147. https://www.jstor.org/stable/312971

Lau, H. (2016). Plantation: Friend or foe from Sarawak perspective. *15th International Peat Congress (IPC)*, 2. Kuching, Sarawak.

Leigh, M. (1998). Political economy of logging in Sarawak, Malaysia. In P. Hirsch & C. Warren (Eds.), *The Politics of Environment in Southeast Asia* (pp. 93–106). Routledge.

Liew, S. F. (2010). *A fine balance: Stories from peatland communities in Malaysia*. Malaysian Palm Oil Council.

Ling, S. (2016, November 21). *Adenan: Sarawak should not be left behind*. The Star. https://www.thestar.com.my/news/nation/2016/11/21/adenan-sarawak-should-not-be-left-behind

Malaysian Palm Oil Board. (2020). *Certification of Malaysian sustainable palm oil (MSPO)—Frequently asked questions*. MPOB. https://mspo.mpob.gov.my/?#section-faq

Malaysian Palm Oil Board. (2022). *Overview of the Malaysian palm oil industry 2022* (pp. 1–6). MPOB.

Manzo, K., Padfield, R., & Varkkey, H. (2019). Envisioning tropical environments: Representations of peatlands in Malaysian media. *Environment and Planning E: Nature and Space*, *3*(3), 251484861988089. https://doi.org/10.1177/2514848619880895

Melling, L., Chua, K. H., & Lim, K. H. (2011). Managing peat soils under oil palm. In K. J. Goh, S. B. Chiu, & S. Paramananthan (Eds.), *Agronomic principles and practices of oil palm cultivation* (pp. 695–728). Petaling Jaya: Agricultural Crop Trust.

Memon, P. A. (2000). Devolution of environmental regulation: Environmental impact assessment in Malaysia. *Impact Assessment and Project Appraisal*, *18*(4), 283–293. https://doi.org/10.3152/147154600781767295

Murray Li, T. (2007). Practices of assemblage and community forest management. *Economy and Society*, *36*(2), 263–293. https://doi.org/10.1080/03085140701254308

Nature Economy and People Connected. (2017). *Palm oil risk assessment: Malaysia—Sarawak* (pp. 1–100). NEPCon.

Ngidang, D. (2005). Deconstruction and reconstruction of native customary land tenure in Sarawak. *Southeast Asian Studies*, *43*(1), 47–75.

Nuijten, M. (2005). Power in practice: A force field approach to power in natural resource management. *The Journal of Transdisciplinary Environmental Studies*, *4*(2), 3–14.

Nurbianto, B. (2016, August 24). *Malaysia challenges the world over palm oil on peatland*. The Jakarta Post. https://www.thejakartapost.com/news/2016/08/24/malaysia-challenges-the-world-over-palm-oil-on-peatland.html

Page, S. E., Rieley, J. O., & Banks, C. J. (2011). Global and regional importance of the tropical peatland carbon pool. *Global Change Biology*, *17*(2), 798–818. https://doi.org/10.1111/j.1365-2486.2010.02279.x

Page, S. E., Siegert, F., Rieley, J. O., Boehm, H.-D. V., Jaya, A., & Limin, S. (2002). The amount of carbon released from peat and forest fires in Indonesia during 1997. *Nature*, *420*, 61–65. https://doi.org/10.1038/nature01131

Papau, D. (2014, April 16). *Angry Iban block Tabung Haji access to NCR land*. Malaysiakini. https://www.malaysiakini.com/news/260184

Parish, F. (1997). The Asian region: An overview of Asian wetlands. In A. J. Hails (Ed.), *Wetlands, Biodiversity and the Ramsar convention: The role of the convention on Wetlands in the conservation and wise use of biodiversity* (pp. 54–80). Ramsar Convention Bureau, Ministry of Environment and Forests India.

Parish, F., & Looi, C. C. (1999). *Options and needs for enhanced linkage between the Ramsar Convention on Wetlands, Convention on biological diversity and UN framework convention on climate change* (pp. 1–17). Global Environment Network.

Phillips, V. D. (1998). Peatswamp ecology and sustainable development in Borneo. *Biodiversity and Conservation, 7*(5), 651–671. https://doi.org/10.1023/a:1008808519096

Pro Regenwald. (2010). *Appendix 1- Sarawak NCR land dispute cases involving logging and other issues* (pp. 1–8). Pro Regenwald.

RSPO. (2019). Revision of RSPO New Planting Procedure (NPP) 2015 in alignment with the RSPO Principles and Criteria (P&C) 2018. In *Roundtable for sustainable palm oil.* https://rspo.org/revision-of-rspo-new-planting-procedure-npp-2015-in-alignment-with-the-rspo-principles-and-criteria-pandc-2018/

Sarawak Report. (2014, January 20). *Sarawak oil palm owners show true colours—and let Taib's cat out of the bag!* SR. https://www.sarawakreport.org/2014/01/sarawak-oil-palm-owners-show-true-colours-and-let-taibs-cat-out-of-the-bag/

Land Code—Chapter 81 (1958 Edition), (2022). Sarawak State Government, State Attorney-General's Chambers.

Singapore Institute of International Affairs. (2017). *Peatland management & rehabilitation in Southeast Asia: Moving from conflict to collaboration* (pp. 1–20). SIIA.

Stone, R. (2007). Can palm oil plantations come clean? *Science, 317*(5844), 1491.

Tan, K. T., Lee, K. T., Mohamed, A. R., & Bhatia, S. (2009). Palm oil: Addressing issues and towards sustainable development. *Renewable and Sustainable Energy Reviews, 13*(2), 420–427. https://doi.org/10.1016/j.rser.2007.10.001

Taylor, D., Hortin, D. W., Parnwell, M. J. G., & Marsden, T. (1994). The degradation of rainforests in Sarawak, East Malaysia, and its implications for future management policies. *Geoforum, 25*(3), 351–369. https://doi.org/10.1016/0016-7185(94)90036-1

TRAFFIC International. (2004). *Forest law enforcement and governance in Malaysia in the context of sustainable forest management* (pp. 1–13). International Tropical Timber Council.

US Department of Agriculture. (2023, October). *Oil Palm explorer.* Foreign Agricultural Service; International Production Assessment Division. https://ipad.fas.usda.gov/cropexplorer/cropview/commodityView.aspx?cropid=4243000

van Noordwijk, M., Purnomo, H., Peskett, L., & Setiono, B. (2008). *Reducing emissions from deforestation and forest degradation (REDD) in Indonesia: Options and challenges for fair and efficient payment distribution mechanisms.* World Agroforestry Centre—ICRAF Southeast Asia Regional Office.

Varkkey, H. (2016). *The haze problem in Southeast Asia: Palm oil and patronage.* Routledge.

Varkkey, H., Tyson, A., & Choiruzzad, S. A. B. (2018). Palm oil intensification and expansion in Indonesia and Malaysia: Environmental and socio-political factors influencing policy. *Forest Policy and Economics, 92*, 148–159. https://doi.org/10.1016/j.forpol.2018.05.002

Wijedasa, L. S., Jauhiainen, J., Könönen, M., Lampela, M., Vasander, H., Leblanc, M.-C., Evers, S., Smith, T. E. L., Yule, C. M., Varkkey, H., Lupascu, M., Parish, F., Singleton, I., Clements, G. R., Aziz, S. A., Harrison, M. E., Cheyne, S., Anshari, G. Z., Meijaard, E., … Andersen, R. (2017). Denial of long-term issues with agriculture on tropical peatlands will have devastating consequences. *Global Change Biology, 23*(3), 977–982. https://doi.org/10.1111/gcb.13516

Yong, C. (2006). *Forest governance in Malaysia: An NGO perspective* (pp. 1–40). FERN.

4

ASSEMBLAGE OF OIL PALM GOVERNANCE AND LAND-USE CHANGES IN AN ISLAND ENVIRONMENT

The case study of the Pulot watershed in Palawan Province, Philippines

Michael D. Pido, May C. Lacao, John Francisco A. Pontillas, Francisca R. Dimaano and Rodolfo O. Abalus Jr

Introduction

Oil palm in tropical Asia

Over the last half-century, oil palm (*Elaeis guineensis*) cultivation has continuously increased in Asia's tropical belt. Although West African in origin, the crop's bio-physical characteristics are well suited for cultivation in Southeast Asia and countries in this region pioneered the growth of the oil palm industry. Economic profit is the primary reason for its expansion. Cramb and Curry (2012) describe palm oil as the most productive of all current oil crops and Sheil et al. (2009) estimate that palm oil produces three to eight times greater yield of oil per area cultivated than any competing oil crop. Hence, it has become the world's most widely used vegetable oil, with nearly 10 percent of the world's permanent croplands now given over to oil palm (Koh & Wilcove, 2008).

In tropical Asia, the Philippines is still a minor contributor to palm oil production compared to neighbouring Indonesia and Malaysia which dominate the ranks of global producers. Indonesia ranks first, accounting for 46 percent of the world market (US Department of Agriculture, 2023), contributing around 3.5 percent of its Gross Domestic Product (GDP) (Tandra & Suroso, 2023). Malaysia is the world's second-largest producer of palm oil, contributing about 24 percent of

DOI: 10.4324/9781003459606-5

global production (US Department of Agriculture, 2024). The crop's expansion in tropical regions has contributed to the emergence of a number of major multinational companies which dominate global production, including companies based in Indonesia, Malaysia, and Singapore (O'Reilly & Varkkey, 2020).

Enrolling smallholders in the palm oil boom

Over the last 30 years, the mobilisation of smallholder plots in large-scale agribusiness ventures have been a dominant strategy in the Philippines' national development planning. In this sense, strategies in the Philippines reflect the global movement in rural development in many parts of the Global South, involving the development of cash cropping, often on supposedly vacant, idle, and marginalised lands. Such strategies are premised on development discourses which stress the need to generate employment, alleviate poverty, and increase agricultural productivity via measures which serve to incorporate an amorphous class of smallholders into global value chains for commodity crops. Often operating under contract farming arrangements, such strategies tend to support movements away from local economic circuits towards more extensive global value circuits via contract arrangements with larger intermediary companies serving global markets (see Barrett et al., 2012; Groenewald et al., 2012). In the Philippines, the national government has prioritised certain "high-value crops" for widespread production and investment. Oil palm was classified as one such high-value crop, alongside rubber, coffee, assorted fruits, vegetables, and some ornamental plants. Hence, oil palm became a priority agricultural commodity for both domestic and potential export markets.

Under President Fidel V. Ramos' (1992–1998) administration, the promotion of oil palm cultivation formed part of the Agriculture and Fisheries Modernisation Act of 1997 (Republic Act or RA 8435). This national law provided a legal framework to modernise the agricultural sector through private investment. During the time of President Joseph Ejercito Estrada (1998–2001), oil palm was again given emphasis in agricultural policies. President Estrada's unfinished six-year presidential term (which was supposed to last until 2004) was continued by President Gloria Macapagal Arroyo; oil palm remained a key crop in her administration's Philippine Agriculture and Modernisation Plan 2001–2004. This continued under the Arroyo administration's 2005–2010 Medium Term Philippine Development Plan (MTPDP). This aimed to develop at least 2 million hectares of new agri-business. During her term, oil palm continued to be regarded as a source of economic growth and the "overall the best environmentally-friendly option for eradicating rural poverty while reducing dependence on imported edible oils". Her administration facilitated the growth of the palm oil industry by pushing through the Biofuels Act of 2006 (RA 9367) and related legislation that gave corporations tax holidays and fiscal incentives to support the crop.

The administration of President Benigno C. Aquino III likewise promoted the oil palm industry, principally in Mindanao and Bohol in the Visayan region. His MTPDP 2011–2016, targeted the promotion of "long-term financing for long-gestating crops such as coconut, rubber, oil palm, coffee, cacao, and fruit trees similar to Indonesia, Malaysia, and Thailand" (Government of the Philippines, 2011). The cumulative impact of these policies was a steady, but by no means spectacular, increase in the area of land given over to palm oil cultivation in the Philippines. From just shy of 25,227 hectares in 2003, the area planted with oil palm in the Philippines was approximately 73,460 hectares by April 2013 (Pamplona, 2013).

The administration of President Rodrigo Duterte similarly supported the promotion of oil palm, although it is not explicitly stated in the MTPDP 2016– 2022. Under his watch, the oil palm industry was referred to as the "Sunshine Industry", with plans to cultivate another 1 million hectares of oil palm, mainly in Mindanao. Given palm oil's increasing demand for use as food, cosmetic ingredients, and biofuel, its cultivation was again billed as a means of achieving multiple developmental goals including the alleviation of poverty, reduction of armed conflict, and increased agricultural revenue. The current administration of Ferdinand Marcos Jr. appears to be ambivalent about oil palm. The Philippine Development Plan 2023–2028 has no specific mention of the crop while the three-year Agriculture Development Plan 2023–2025 which aims to intensify production and export, does not include oil palm in the Department of Agriculture's High Value Crops Development Programme.

Aim of the chapter

While the benefits of palm oil have been widely deployed to support its expansion in the Philippines, the crop is also widely criticised in international literature. It has been linked to issues of environmental degradation (Mizuno et al., 2016); modes of economic organisation which reproduce colonial forms and its failure to deliver balanced development, which in turn would provide widely spread benefits (Pye & Bhattacharya, 2013); a number of other social costs and dis-welfares are associated with the crop such as land grabbing and human rights failures (Cramb & McCarthy, 2016). Furthermore, within Southeast Asia, the crop has become the centre of tensions between different nations, due to the transboundary impacts of air pollution linked to large-scale forest fires; in addition to flows of labour, capital, and wealth which are often seen as exploitative (Varkkey, 2012). However, while these issues may be broadly understood and discussed in the literature on palm oil policy, few studies have explored any of these questions in relation to the development of oil palm in the Philippines. In the absence of such work, questions relating to the likely costs and benefits of the industry in the Philippines and of their distribution in spatial, environmental, and human terms remain unanswered.

This chapter seeks to begin the task of exploring these questions, at once engaging with the first, second, and fourth research questions of the book: how the different governance arrangements enable and legitimise the expansion of oil palm; how power differentials affect oil palm governance; as well as how conflict is governed and moderated in the oil palm sector. In order to do so, we present findings from a case study of oil palm development in Palawan since 2003. The Pulot watershed in Palawan's southern mainland provides the focus of the study, which, in particular focuses on land-use changes brought about by oil palm cultivation. Palawan was not originally identified as an area for oil palm, thus state policies by no means provide an adequate explanation of how palm oil has come to be adopted in this area. Palawan therefore provides us with an interesting case through which to explore how diverse human and non-human entities and multiple processes contribute to the ways in which locations and communities come to be incorporated into an oil palm assemblage. We adopt elements of assemblage theory to understand the complexity of the oil palm industry in Palawan.

The chapter begins by briefly outlining how we employ assemblage in our analysis. Having done so, we next describe the development of the Philippines' oil palm industry over the last three decades. We then examine the development of oil palm farming in Palawan Province. This is followed by taking a look at the institutions and/or organisations involved in the province's oil palm industry. Subsequently, the chapter discusses land-use change—in relation to oil palm cultivation—in the Pulot watershed within the context of sustainable development: economic profitability (indicated by business profitability at the level of the firm and household), social acceptability (defined by the individual firm and household positive or negative perception of the business), and environmental friendliness (measured in terms of the positive or negative impact of the business to the environment). The last section provides a reflection regarding the future of the oil palm industry in Palawan.

Conceptualising the Philippine palm oil industry: An assemblage perspective

The assemblage approach is used to provide a conceptual framing through which to explore the development and current shape of the oil palm industry in Palawan Province, the Philippines. Assemblage theory refers to a set of related approaches which are informed by the work of French thinkers, Deleuze and Guattari (1988), and subsequently developed by others. Rather than a rigid theoretical approach, assemblage theory may be best understood as an ontological framework which perceives social reality as being constituted of and as formations of human and non-human entities and/or objects which are brought together for a period of time, and in so doing, enact certain functions or effects. A key feature of assemblage theory is that it stresses the contingent and partial nature of these formations and the associations between objects that make them up, stressing instead the contingency and inherent fragility of any social-material formation and the work that goes into maintaining them. A critical feature of this approach is its emphasis on "relations of exteriority", the idea that the links that bind the elements of any assemblage–the meanings, understanding and knowledge that define these links–are inherently fragile and prone to fracture dissolution and reassembly in entirely different ways and assemblages. In this sense, any social formation or assemblage is constantly being (re)produced through the activities of its constituent entities, which are in turn relationally placed in respect of other elements in the assemblage. As Manuel Delanda (2016, p. 10) puts it, "We need to conceive of emergent wholes in which the parts retain their autonomy, so that they can be detached from one whole and plugged into another one, entering into new interactions".

Assemblages are both complex and constantly being produced, comprising both form and function. In this context, the way in which assemblages change and the effects that they produce are difficult to predict. Assemblage theory has come to be widely employed by environmental researchers such as Murray Li (2002, 2007), and developed by Müller (2015) as a means for describing and exploring complex interactions between human and non-human objects. The concept has also been widely adopted in policy studies, where it is particularly valued in studies which seek to understand the ways in which complex policies are worked out through multiple interactions and sites. The approach is suited to studying complex situations in which a range of human and non-human objects are collectively linked to significant transformations in the environments and landscape since it pays as much attention to how natural and man-made objects and technologies impact human entities as it does the action of human entities on the non-human.

In this context, the process of governance might be understood as involving practices in which some human entities seek to capture and arrange other elements of an assemblage in ways which advance their projects (see Marsden et al., 2018). A key feature being that these efforts are always partial, contingent, and at least notionally open to challenges and subversion by different human and non-human actors in the assemblage. In recent years, assemblage approaches have been employed to describe the emergence of salmon-producing regions in Chile (Blanco et al., 2015), freshwater fisheries and communities in Indonesia (Thornton et al., 2020), and Scottish small farming landscapes (Sutherland & Calo, 2020).

Assemblage theory offers an alternative approach to understanding socio-economic practices in ways that facilitate an exploration of issues of scale, livelihood, power, and ordering (O'Reilly & Varkkey, 2020). The assemblage, as opposed to systems approach/thinking, describes both the process and the result of processes through which actors and other heterogeneous entities are brought together to serve certain functions in a particular time period.

This chapter attempts to apply this approach in the interpretation of data generated via a desk review of existing literature and selected interviews. In comparison to Indonesia and Malaysia, the key oil palm literature about the Philippines in general and Palawan in particular is relatively sparse. However, there are a number of significant contributions, many of which have been incorporated into our analysis. They include the following: the Ancestral Land/Domain Watch (ALDAW) (2014); Barcia (2017); Batugal (2013); Larsen et al. (2014, 2018); Lerom (2015);

Montefrio et al. (2014); Montefrio (2015); Montefrio and Dressler (2016); Pamplona and Pamplona (2014); the World Bank (2016); and Villanueva (2011). We also analysed social media and information provided via the official websites of Philippine government offices. We also undertook unstructured interviews with selected government officials and developed maps generated from a spatial analysis of the Pulot watershed over three time periods in which the development of oil palm plantations discussed in this study occurred.

Oil palm development in Palawan province

Palawan Province, situated in Western Philippines, was not in the original list of provinces identified for developing oil palm plantations (Figure 4.1). Its territorial land area covers some 1.5 million hectares, comprising over a thousand islands. Often dubbed as the Philippines' "last ecological frontier", the province has unique bio-geographic features and has very rich biodiversity. Its flora and fauna are more closely associated with that of the island of Borneo than the rest of the Philippines. The entire province was declared as the United Nations Educational, Scientific, and Cultural Organisation's (UNESCO) Man and Biosphere Reserve in 1990. Moreover, in 1992, the Strategic Environmental Plan for Palawan Act (RA 7611) was enacted as a special law to guide the province's sustainable development. There are two World Heritage Sites in the province: (1) the Tubbataha Reefs Natural Park and (2) the Puerto Princesa Underground River National Park.

Palawan's economy is highly dependent on three sectors: (1) mining and oil and gas; (2) agriculture, including fisheries; as well as (3) tourism. The province's tourism is primarily nature-based, attracting a huge number of domestic and international tourists. During the COVID-19 pandemic, the tourism sector was severely impacted due to the limited arrival of tourists up until 2022. As mentioned, the focus is nature tourism with resorts and the so-called agri-tourism sector still at the incipient stage. Agriculture in Palawan focuses on the cultivation of crops and the raising of livestock. Development challenges in the province include limited-service industries and a high population growth rate due to in-migration. In this context, oil palm has been promoted for agricultural development with a poverty alleviation focus. Not unsurprisingly, given population changes and the promotion of agriculture and mining, land use change is prominent in the province, with decreasing forest cover throughout the years, correlated with increasing built-up areas and agricultural plantations. In this context, the environment faces developmental pressures. The 1.2-million-hectare main island of Palawan is particularly ecologically-vulnerable, having steep highlands and limited lowlands.

In 2000, oil palm started to be promoted nationally as a priority agricultural crop for cultivation. In 2003, the Provincial Government of Palawan engaged with the Agusan Plantation Group of Companies, to hold a number of multi-stakeholder fora, during which various government agencies voiced their interests and pledged support to implement the proposed oil palm project. By 2004, the Palawan Palm Oil Industry

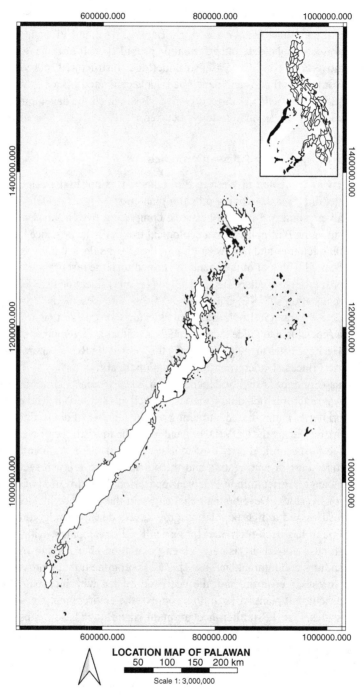

FIGURE 4.1 Location map of Palawan Province, Philippines.

Source: PCSD.

Development Council (PPOIDC) was created (Provincial Ordinance No. 739-04). In 2005, oil palm was included in the Comprehensive Development Plan of Palawan as a plantation crop. As an emerging production area, some 100,000 hectares were targeted for cultivation in the southern municipalities of mainland Palawan. In 2005–06, Palawan State University offered a short course on Oil Palm Production Technology (OPPT) in collaboration with the University of Southern Mindanao, which provided technical training on the crop. The Palawan Palm & Vegetable Oil Mills, Inc. (PPVOMI) and its sister company, Agumil Philippines, Inc. (AGPI), started their operations in 2006. The oil palm growers started planting in 2007 and, four years later, harvesting commenced in 2011 (Larsen et al., 2014). Since then, the harvesting of fresh fruit bunches (FFBs) in plantation areas have continued until the present time.

In 2013, the oil palm growers consisted of the following: 134 smallholder cooperatives, two additional cooperatives, and four self-financed farmers (Larsen et al., 2014). The 134 cooperatives were spread out in five municipalities (Table 4.1). The two additional cooperatives are the 25-member cooperative started up by Cavite Ideal International Construction and Development Corporation (CAVDEAL), which originally served as a building contractor for AGPI and San Andres, a 35-member cooperative chaired by the director of the Manila-based company Capital Oil Refinery. Many members of the cooperatives that were consulted for this study were migrant settlers, while the labour force recruited by cooperatives and independent farmers appears to be dominated by indigenous peoples (Larsen et al., 2014). These migrants typically come from other parts of the country, with the intention of looking for better economic opportunities in Palawan.

Figure 4.2 depicts the spatial distribution of oil palm development areas in Palawan's southern municipalities. Oil palm licenses were issued for some 15,469 hectares to be used for cultivation and this hectarage represents about 2 percent of the total land area (Larsen et al., 2014).

TABLE 4.1 The 14 smallholder cooperatives engaged in oil palm cultivation

Municipality	Cooperatives
Aborlan	1 Aborlan Small Coconut Multi-Purpose Cooperative
Quezon	2 Couples for Christ FAMICO Multi-Purpose Cooperative
	3 IKBA Oil Pam Producers Cooperative
	4 Aramaywan Farmers Multi-Purpose Cooperative
Española	1 Labog Agri-Based Multi-Purpose Cooperative
	2 Tapisan Oil Palm Growers Multi-Purpose Cooperative
	3 Malalong Multi-Purpose Cooperative
	4 White Palm Producers Cooperative
	5 Golden Palm Agro Industry Multi-Purpose Cooperative
Brooke's Point	1 Calasaguen-Maasin Oil Palm Growers Multi-Purpose Cooperative
	2 Tagbikal Oil Palm Cooperative
Bataraza	1 Sandoval Oil Palm Growers Multi-Purpose Cooperative
	2 Sumbiling Oil Palm Producers Cooperative

Source: Provincial Cooperative Development Office, June 2013.

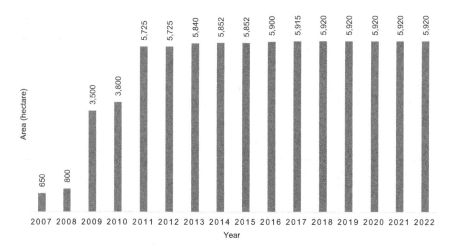

Municipality	Barangays	Barangay area (ha)	Palm oil concession area (ha)	Fraction of land (%)	Fraction of alienable and disposable land (%)
Aborlan	Mabini, Iraan, Sagpangan	12 000	3 500	30%	99%
Quezon	Isugod, Panitian, Tagusao, Aramaywan	39 000	1 700	4.4%	15%
Narra	Princess Urduja	2 600	650	25%	82%
Sofronio Española	Iraray, Pulot Interior, Labog, Punang	29 000	4 600	16%	54%
Brooke's point	Pangobilian, Maasin, Calasaguen, Samariñana	26 000	930	3.6%	12%
Rizal	Iraan	7 800	1 000	13%	44%
Bataraza	Sandoval, Tarusan, Igang-Igang	11 000	3 100	27%	91%
Total		130 000	15 000	10%	33%
New acquisitions (average of informal estimates)			15 000		

FIGURE 4.2 Spatial distribution of oil palm developments in Palawan's southern municipalities.

Source: DENR; Larsen et al., 2014, p. 13.

FIGURE 4.3 Extent of area devoted to oil palm plantation in Palawan Province, Philippines, from 2007 to 2022.

Source: PSA.

Note that there are discrepancies in the data/information from various government sources. The Philippine Statistics Authority (PSA) estimated the oil palm production area in Palawan in 2022 to be 5,920 hectares (Figure 4.3), while the Philippine Coconut Authority (PCA) estimate was approximately 9,040 hectares (PCA, 2023) (Table 4.2).

TABLE 4.2 Extent of area devoted to oil palm plantation in Palawan Province, Philippines, as of 2022

Municipality	Barangay	Classification			Area (ha)	Number of oil palm trees			No. of farmers	Production of FFB/ha (tonnes)
		Inland/ Upland	Inland flat	Coastal flat		Bearing	Non-bearing	Total trees		
PALAWAN					9,040.12	907,209	260,032	1,167,241	1,693	
Brooke's Point					2,229.15	241,612	44,101	285,713	194	10MT
	Calasaguen	✓	✓	✓	1,817.86	191,374	41,695	233,069		
	Maasin	✓	✓	✓	377.75	47,481	870	48,351		
	Pangobilian				15.54	1,989		1,989		
	Ipilan	✓	✓		6.00		768	768		
	Mambalot	✓	✓		6.00	768		768		
	Samarinana		✓		6.00		768	768		
Sofrinio					1,925.71	224,225	21,371	245,596	574	10MT
Española	Pulot	✓	✓		1,160.22	145,404	2,212	147,616		
	Interior									
	Iraray	✓			386.03	39,660	9,750	49,410		
	Punang	✓			266.59	24,714	9,409	34,123		
	Labog	✓			112.87	14,447		14,447		
Batarraza					3,499.86	264,638	193,968	458,606	94	7.5MT
	Sandoval	✓	✓		2,457.86	254,138	67,137	321,275		
	Ocayan	✓	✓		191.04	5,112	20,604	25,716		
	Sarong	✓	✓		57.71		9,009	9,009		
	Igang-igang	✓	✓		166.39		22,237	22,327		
	Iwahig	✓	✓		136.70	4,508	14,630	19,138		
	Culandanum	✓			455.16	880	55,814	56,694		
	Rio-tuba			✓	35.00		4,447	4,447		

(Continued)

TABLE 4.2 (Continued)

Municipality	Barangay	Classification			Area (ha)	Number of oil palm trees			No. of farmers	Production of FFB/ha (tonnes)
		Inland/ Upland	Inland flat	Coastal flat		Bearing	Non-bearing	Total trees		
Quezon					902.00	114,861	592	115,453	498	7.5MT
	Tagusai	✓			276.63	35,408		35,408		
	Alfonso XIII	✓	✓		226.84	29,035		29,035		
	Aramaywan	✓	✓		102.54	13,125		13,125		
	Isugod	✓	✓		105.65	13,523		13,523		
	Berong	✓			190.34	23,770	592	24,362		
Aborlan					447.31	57,254	0	57,254	331	7MT
	Magbabadil	✓	✓		200.03	25,603		25,603		
	Sagpangan	✓	✓		120.40	15,411		15,411		
	Iraan	✓	✓		126.88	16,240		16,240		
Rizal					36.09	4,619	0	4,619	2	7MT
	Iraan	✓			15.09	1,931		1,931		
	Panalingaan	✓			21.00	2,688		2,688		

Source: PCA, 2023.

Assemblage of oil palm governance in Palawan

In Palawan, we identified seven categories of key human actors involved in efforts to shape the governance of the palm oil assemblage (Box 1) in the Pulot watershed, the case study site. These players are contextualised here as an assemblage following the frameworks of researchers (e.g. Müller, 2015; Murray Li, 2007, 2002) and those who applied it in the context of oil palm governance (e.g. O'Reilly & Varkkey, 2020). Many of the descriptions of these organisations/institutions were taken from Larsen et al. (2018, 2014). The assemblage of organisations/institutions involved in oil palm governance in Palawan.

BOX 1   THE ASSEMBLAGE OF ORGANISATIONS/INSTITUTIONS INVOLVED IN OIL PALM GOVERNANCE IN PALAWAN.

1 International actors
2 National government agencies
3 Local government units
4 Civil society organisations
5 Private sector
6 Councils, associations, and working groups
7 Other actors/players

International actors

Several organisations operate at the international level. The few described here may influence the Philippine's oil palm industry and in effect are part of the Philippine palm oil assemblage.

The Roundtable on Sustainable Palm Oil (RSPO) is a global partnership to make palm oil sustainable. The RSPO claims to be "transforming the sector by bringing together stakeholders across the supply chain to develop and implement global standards for producing and sourcing certified sustainable palm oil" (RSPO, 2018). Established on 21 November 2015, the Council of Palm Oil Producing Countries (CPOPC) is an intergovernmental organisation for palm oil producing countries (CPOPC, 2022). Jointly founded by Indonesia and Malaysia, the Philippines is yet to become a member. While these organisations do not appear to have any direct linkage with oil palm producers in Palawan province, they have been hugely influential in promoting discourses associated with the developmental benefits of palm oil and of the possibility of developing the industry sustainably whilst simultaneously supporting a developmental trajectory for the industry in which large corporate entities occupy a lead role.

International non-governmental organisations (NGOs) are equally involved in advocating for a sustainable oil palm industry. Conservation International 'works across the oil palm value chain to create benefits for nature and people, incorporating sustainable palm oil production as part of an integrated landscape approach' (Conservation International, 2019). The World Wildlife Fund (WWF) "envisions a global marketplace based on socially acceptable and environment-friendly production and sourcing of palm oil" (WWF, 2019). Interestingly, while both CI and WWF have conservation works in Palawan, these NGOs do not have initiatives that are specific to oil palm.

National government agencies (NGAs)

Seven national government agencies (NGAs) are key entities in oil palm development and management in the Philippines. Foremost among these is the PCA, which "is the sole government agency that is tasked to develop the industry to its full potential in line with the new vision of a united, globally competitive and efficient coconut industry" (PCA, n.d.). Its mandate is to promote the rapid integrated development and growth of the coconut industry, and more recently, the palm oil industry. Structurally, the PCA has regional offices (including in the four other provinces of Marinduque, Occidental Mindoro, Oriental Mindoro, and Romblon) and a provincial office in Palawan.

At the national level, the PCA is guided by the updated Philippine Palm Oil Industry 2014–2023 Road Map which aims to develop the oil palm industry through gainful production, processing, and marketing of oil palm products and by-products that complement the coconut industry. Its target by 2023 was to plant at least 300,000 hectares nationwide and add milling capacity of about 500 tonnes per hour. This road map was an offshoot of the earlier Philippine Palm Oil Industry Development Plan 2004–2010 and the draft Policy Framework for the Development of the Palm Oil Industry. It is important to note that the road map stresses that oil palm expansion shall only utilise idle and unproductive lands, something that is worth keeping in mind when the development of the industry in Palawan is considered. More recently in 2023, the Philippine Palm Oil Industry Roadmap (2024-2033) was published by PCA in collaboration with the Department of Trade and Industry and PPDCI (PCA, Department of Trade and Industry, & PPDCI, 2023).

The Department of Agriculture (DA) takes charge of the overall planning and policymaking process in the agriculture and fisheries sector at the national level. The DA identifies high-value commercial crops for different regions and provinces of the Philippines. Hence, its mandate covers various terrestrial crops, including oil palm. As a regulatory body, the DA has regional and provincial offices to carry out its mandate. The PCA is an NGA attached to the DA.

The Department of Environment and Natural Resources (DENR) oversees and administers environmental management, conservation, and development at the national and local levels. Within the DENR are several natural resource management bureaus, such as the Environmental Management Bureau and the Forest Management Bureau (FMB). The policies formulated by the DENR and its bureaus are implemented by DENR Regional Offices, the Provincial Environment and Natural Resources Offices (PENROs) within each province, and the Community Environment and Natural Resources Offices (CENROs) within the municipalities. The Pulot watershed is within the municipality of Sofronio Española, which is under the CENRO in Brooke's Point. The DENR issues permits to the indigenous peoples to enable them to gather non-timber forest products (NTFPs). It was the DENR which passed a directive in 2004 that qualified oil palm as a reforestation species; however, its effectiveness at mitigating soil erosion has been contentious.

The Palawan Council for Sustainable Development (PCSD) is mandated to promote sustainable development in Palawan province. In the Philippines, only Palawan has this arrangement. Republic Act No. 7611 or the Strategic Environment Plan (SEP) for Palawan Act, legislated on 19 June 1992, provided for the adoption of a comprehensive framework for the sustainable development of Palawan, compatible with protecting and enhancing the natural resources and endangered environment of the province. The SEP guides the public and private sectors in conservation, development planning, policy formulation, and in regulating the entry of development projects. The SEP Law created the PCSD, a multi-sectoral planning, policymaking, and coordinating body responsible for the governance, implementation, and policy direction of the SEP. In 2018, the Man and the Biosphere Council at the UNESCO Paris headquarters and the UNESCO National Commission of the Philippines recognised PCSD as the Onsite Management Authority of the Palawan Biosphere Reserve, under the MAB Programme of UNESCO. They acknowledged PCSD for its effort to ensure that the Palawan BR continuously serves as a model for biosphere reserves in the country as well as the World Network of Biosphere Reserves. The PCSD is the prime mover of biodiversity conservation and sustainable development, and the builder of a knowledge-based society in Palawan. By virtue of RA 7611, it is also vested with the mandates and functions from other environment-related laws as implemented in the province of Palawan. The PCSD is an NGA attached to the DENR.

The DENR and PCSD have an influece on the development of oil palm in Palawan in terms of regulatory processes. The DENR and PCSD issue SEP Clearance and Environmental Compliance Certificate permits, respectively. The PCSD issued SEP Clearance (No. POP-032510-020) to PPVOMI on 25 March 2010 for its Integrated Palm Oil Plantations Development, Production, and Processing Project, while the coverage of the PCSD clearance includes the establishment of a nursery and oil palm mill in a 13-hectare area, which is located in Barangay Maasin, Brooke's Point (Larsen et al., 2014). The PPVOMI's Environmental Compliance Certificates were issued for a palm oil mill in Barangay Maasin on 2 July 2010 and for the sites in the relevant municipalities between September 2008 and

February 2009. The above-cited certificates were issued by the regional office of the Environmental Management Bureau, which is one of the operating bureaus under the DENR.

The National Commission on Indigenous Peoples (NCIP) is the lead NGA that is responsible for protecting the rights of indigenous peoples or indigenous cultural communities. Attached to the Department of Social Welfare and Development, it maintains a provincial office in the capital city of Puerto Princesa. The NCIP is a significant player in the oil palm assemblage due to the alleged encroachment of oil palm cultivation on tribal ancestral lands and the use of indigenous people as farm labourers. The so-called tribal ancestral lands are either covered by a Certificate of Ancestral Domain Title or a Certificate of Ancestral Domain Claim.

The Indigenous People's Rights Act (Republic Act No. 8371 of 1997) requires the developers to secure Free Prior and Informed Consent (FPIC) for certain development projects to be undertaken within ancestral domains. Oil palm cultivation/ plantation is among those development projects where an FPIC is required (Larsen et al., 2018). Hence, the NCIP is mandated to evaluate project proposals and applications for land title processes by other agencies. Certificates of Ancestral Domain Claims are issued under the rules and procedures of the DENR Department Administrative Order 02 of 1993 (DAO No. 02-93). Some complaints were submitted to the NCIP about alleged negligence regarding the issuance of FPICs associated with the establishment of palm oil projects. The NCIP in Puerto Princesa for example, did not review requirements for FPIC since the palm oil project was not considered to infringe on lands covered by the Certificate of Ancestral Domain Claim or the Certificate of Ancestral Domain Title (Larsen et al., 2014).

The Department of Agrarian Reform (DAR) is partly involved because some of the farmers involved in oil palm planting are agrarian reform beneficiaries who are DAR clientele. Beneficiaries have been granted individual titles called the Certificate of Land Ownership Award for a maximum of three hectares. The Comprehensive Agrarian Reform Law of 1988 (RA No. 6657), as amended, paved the way for land redistribution of previous state farms and private estates (*haciendas*) through the Certificate of Land Ownership Award. Companies such as AGPI thus obtain land through leases, contract farming, or direct purchase from beneficiaries. From the DAR's perspective, the contract arrangements between agrarian reform beneficiary farmers and AGPI are considered beneficial as they help compensate for the lack of financial credits and market access that otherwise hamper the opportunity to make use of newly awarded land (Larsen et al., 2014). Of the 14 smallholder cooperatives (listed in Table 4.1) engaged in palm oil outgrower schemes, three are assisted by DAR.

The Department of Labour and Employment (DOLE) is involved due to labour-related issues in oil palm cultivation. Part of the DOLE's mission is "to protect workers and promote their welfare". The labour force recruited by oil palm cooperatives and independent farmers appears to be dominated by indigenous peoples, more specifically the *Tagbanua* indigenous members who come from the southern mainland.

Workers are primarily engaged in either land maintenance or fruit harvesting roles. As a regulatory agency, minimum wage rates are defined by the DOLE according to the industry sector and the economic area. The rate of pay for labourers in oil palm cultivation in rural Palawan is set at PHP210 per day (Wage Order No. IV-B-06). Given the current currency exchange rate, this roughly translates to about USD4.50. Based on a series of NGO reports (e.g. ALDAW, 2013a, 2013b; Neame & Villarante, 2013; Dalabajan, 2011; Villanueva, 2011; Barraquias-Flores, 2010), some labour concerns brought to the DOLE's provincial office in Palawan include labour rights infringements (under the Philippine Labour Code as contained in the DOLE Department Order 18-A) and a lack of social security benefits.

The Cooperative Development Authority (CDA) is primarily responsible for promoting the sustained growth and full development of cooperatives in the Philippines. All of the smallholder oil palm cooperatives in Palawan are registered with the CDA as mandated by the Philippine Cooperative Code (Republic Act No. 9520) of 2008. These cooperatives explored the option of uniting in a federation (legal entity under the Cooperative Code). However, this notion did not materialise due to the extensive travel coordination required among the prospective members and the inability to raise the required capital of PHP1.5 million to register a federation with the CDA.

Local government units (LGUs)

In the Philippine context, the local government units or so-called LGUs have three-tier hierarchical layers. From the top, these are: the province, municipality (city), and village (or *barangay*).[1] Hence, the LGUs involved in this context are the provincial government of Palawan and the municipality of Sofronio Española, where the bulk of the Pulot watershed is located. The palm oil project was duly endorsed by *barangays*, municipalities, and the provincial government. As legislative bodies, members of the *Sangguniang Panlalawigan* (Provincial Board), *Sangguniang Pambayan* (Municipal Board), and *Sangguniang Pambarangay* (Village Council) provided council resolutions to this effect.

The LGUs' interests in properly developing and managing their land resources are driven by the fact that these resources contribute directly and significantly to food production, livelihood opportunities, and the general well-being of residents. Regulatory mechanisms are operationalised through legal ordinances, delivery of essential services, and undertaking specific programmes and projects. Control options available to local governments include, among others: authority/permit to use the land for agricultural development, such as oil palm; land/water use planning and zoning; habitat conservation and restoration (such as reforestation of denuded watersheds); development of infrastructure and facilities; and provision of agriculture extension services and livelihood training programmes.

The appointment of an agriculturist is a mandatory position for provincial governments and an optional position for municipal governments. The agriculturist is responsible for agricultural services, provision of technical assistance and development of agricultural plans/strategies. The Provincial Agriculturist Office covers province-wide concerns, while the Municipal Agriculture Office covers the agriculture concerns within the municipality of Sofronio Española. The municipal LGU has management responsibilities for its land resources and municipal waters (marine waters within 15 kilometres from the shoreline) within its territorial boundaries.

Aside from the Provincial Agriculturist Office, other provincial-level offices are involved in oil palm issues. These include the Provincial Legal Office and the Provincial Cooperative Development Office which support the cooperatives in clarifying the contractual relationship with AGPI. The Provincial Government sets out the priority crops in the Comprehensive Provincial Development Plan. Under the then governorship of Governor Joel Reyes, oil palm was not originally included as a priority crop in Palawan. Nevertheless, his administration actively facilitated the launch of the palm oil project in the province through active linkages with the national government and international investors. Moreover, the provincial government launched a financial subsidy and technical support programme to plant oil palm on vacant lands. The current provincial administration, however, no longer promotes oil palm as a priority crop. Its agriculture priorities now centre on other high-value crops such as coffee, cacao, and Napier grass.

Civil society organisations (CSOs)

Civil society organisations (CSOs) are taken here to include both the people's organisations and NGOs. The Philippine Constitution defines people's organisations as "bona fide associations of citizens with demonstrated capacity to promote the public interest and with identifiable leadership, membership, and structure" (Role and Rights of People's Organisation, 1987). These include, but are not limited to, the following: ALDAW, the Environmental Legal Assistance Centre (ELAC), Inc., and the Palawan NGO Network, Inc. (PNNI). ALDAW is a "Philippines-based advocacy campaign network of indigenous peoples defending their ancestral land and resources from mining corporations, oil palm companies, top-down conservation schemes, and all forms of imposed development" (ALDAW, 2011). ALDAW submitted a letter on 7 June 2013 to the then Philippine President Aquino. This correspondence called for an in-depth investigation into the alleged human rights violations perpetrated by oil palm companies and called for a nationwide moratorium on oil palm expansion. As an environmental NGO, ELAC is "committed to helping communities uphold their constitutional right to a healthful and balanced ecology" (ELAC, 2023). Established in 1991, PNNI is a coalition of 39 people's organisations and NGOs in Palawan with the "motivation of presenting a broader NGO consensus that would carry more weight in the policymaking process" (PNNI, 2023).

These people's organisations and NGOs have raised opposition concerning both the introduction and expansion of oil palm cultivation in southern mainland

Palawan. Among others, concerns relate to the negative environmental impacts, labour practices, and corporate responsibilities. Work-related matters include discouragement in establishing labour unions, breaches of the legally-mandated minimum wage, use of child labour, and non-payment of social security benefits. These environmental and socio-economic concerns are reflected in a series of NGO reports (e.g. ALDAW 2013a, 2013b; Neame & Villarante, 2013; Dalabajan, 2011; Villanueva, 2011; Barraquias-Flores, 2010).

The private sector

Private interest groups engaged in Palawan's oil palm industry include both Filipino and foreign entities. Foremost is the PPVOMI (which is 60 percent Singaporean-owned and 40 percent Filipino) and its sister company, AGPI (which is 75 percent Filipino-owned and 25 percent Malaysian; Larsen et al., 2018, 2014). The parent company is Agusan Plantations, Inc., which is domiciled in Malaysia. PPVOMI and AGPI are registered with the Security Exchange Commission in accordance with the Philippine Corporations Code.

The PPVOMI operates the sole oil palm mill in Palawan, which is situated in the *barangay* of Maasin, in the directly adjacent (southward) municipality of Brooke's Point. PPVOMI Construction equipment is rented from CAVDEAL. Its seedlings are supplied by the Papua New Guinean company, New Britain Palm Oil Limited. Meanwhile, AGPI is the contractor in the cultivation of oil palm. AGPI receives corporate finance assistance from First Consolidated Bank and the Land Bank Countryside Development Foundation, Inc., which is a non-profit subsidiary of the Land Bank of the Philippines (Larsen et al., 2018). The First Consolidated Bank is a private, independent development savings bank with headquarters in Tagbilaran City, located in the province of Bohol. AGPI has principally accessed land for cultivation through either lease agreements or contract arrangements with farmers in so-called outgrower schemes (Larsen et al., 2018, 2014). This outgrower model seems to be built on the Agusan Group's experiences from Mindanao and even from Malaysia (Nozawa, 2011). Farmers have been engaged in contracts with AGPI either through cooperatives or as self-financed independent landowners (Larsen et al., 2014). AGPI lists four self-financed farmers who deliver FFB to the mill in Maasin.

There are 14 smallholder palm oil growing cooperatives as listed earlier in Table 4.1. At least two additional cooperatives are engaged in oil palm cultivation. These include the 25-member Palawan Evergreen Oil Palm Cooperative initiated by the staff of CAVDEAL, the infrastructural contractor of AGPI, and San Andres, a 35-member cooperative chaired by Mr. Koh, a businessman based in Manila who is also director of the company Capital Oil Refinery that produces coconut oil. CAVDEAL and San Andres initially purchased seedlings from AGPI, but without marketing agreements, and are thus freed of obligations to deliver FFB to the PPVOMI mill in Maasin.

All smallholder cooperatives engaged in oil palm—with varying organisational backgrounds and membership compositions—are registered with the CDA. Some cooperatives were already operating, while others were explicitly established for oil palm cultivation. As described earlier, the PCA had no direct hand when farmers formed cooperatives to engage in AGPI's outgrower scheme. Many smallholder cooperatives act as multi-purpose cooperatives, with many members not directly engaged in oil palm farming. As such, they provide loans to members for farming-related expenses.

Councils, associations, and working groups

At the national level, the Philippine Palm Oil Development Council, Inc. (PPDCI), was established by the PCA to coordinate national-level efforts on oil palm. The PPDCI is governed by a board with representation from the government, as well as companies and growers (PPDCI, 2013). As such, the PPDCI provides guidance and oversight with the oil palm councils at the provincial lower levels.

The creation of the PPOIDC was spearheaded by the Provincial Government on 13 January 2004. The PPOIDC's mandate is to promote, monitor and ensure proper regulation of Palawan's oil palm industry. Its membership includes provincial government staff, representatives from research institutions, and the private sector. Therefore, the PPDCI organisationally provides guidance to the PPOIDC.

The 14 cooperatives form the Association of Oil Palm Growers in Southern Palawan. Despite the existence of PPOIDC, the members perceive that such an association serves as a more flexible organisational platform to pursue their interests. The association has acted as a liaison with the provincial government offices. Moreover, it filed several resolutions and requests for support, including a request for tax exemption to alleviate their financial distress and a request for assistance with legal counsel for contract negotiations with AGPI (Larsen et al., 2014).

The Protected Area Management Board (PAMB) of the Mt. Mantalingahan Protected Landscape (MMPL) somehow played a role in "limiting" oil palm expansion. The MMPL was established as a protected area by virtue of Presidential Proclamation 1815 on 23 June 2009. Covering some 120,457 hectares, it straddles the territorial jurisdiction of the five southernmost mainland Palawan's municipalities of Bataraza, Brooke's Point, Quezon, Rizal, and Sofronio Española. As described earlier, oil palm cultivation in Palawan is concentrated in these five municipalities where indigenous peoples also abound. The land claim in 2012 of one oil palm cooperative was within MMPL's territorial jurisdiction, in which the cooperative tried to secure the required Protected Area Community-Based Resource Management Agreement from the PAMB of MMPL. The PAMB rejected converting the Community-Based Forest Management Agreement license to a Protected Area Community-Based Resource Management Agreement to allow oil palm cultivation by the cooperative members. Had it been allowed, the areas planted with oil palm would have been considerably more extensive.

The Working Group on Oil Palm Concerns was established through the initiative of a number of provincial NGOs. The Working Group advocated for a moratorium to further expand oil palm cultivation and backtracking on illegal cultivation in forest land or indigenous peoples' ancestral land (Larsen et al., 2018). Communal irrigation associations for water use in rice fields also exist in the Pulot watershed.

Other actors/players

Government-owned and controlled corporations or lending institutions also play a crucial role in the oil palm industry, notably the Land Bank of the Philippines (LBP). The description of LBP is largely drawn from Larsen et al. (2014) and Larsen et al. (2018). As a universal bank, its special focus is on serving the needs of farmers and fishermen. The LBP earmarked some PHP1.5 billion to support the financial needs of more than 6,000 oil palm growers (Larsen et al., 2014). The LBP, through its non-profit subsidiary called the Land Bank Countryside Development Foundation, Inc., provides corporate finance to AGPI. The funding is channelled through the Food Supply Chain Programme, in support of the National Government's efforts to attain food self-sufficiency and increase agricultural productivity (Gonzales, 2010).

The LBP provides financial support to the cooperatives under the Development Advocacy Programme. The contract between the LBP and the 14 oil palm cooperatives in Palawan stipulates that the Loan/Line Agreement is for a period of ten years, with a four-year grace period on the loan and interest. As a form of collateral, the LBP holds land titles of a selected number of cooperative members/key officers. The granting of loans is also premised on the condition that the cultivation solely takes place in previously idle and unproductive agricultural lands.

Three state universities and colleges have been involved in terms of research and training concerning oil palm. These are publicly-funded academic institutions. In June 2004, Palawan University organised a training course on oil palm cultivation/management in collaboration with the University of Southern Mindanao. Academics from the University of Southern Mindanao served as trainers covering topics such as nursery establishment, pest management and fertilisation. Some 30 trainees successfully completed the training course, of which 10 were subsequently engaged in the oil palm industry. The Western Philippine University has likewise been engaged in some oil palm-related research, having prepared a report about the palm oil industry in Palawan (Lerom, 2015). The Palawan State University, the University of Southern Mindanao and the Western Philippine University are three institutions under the Commission on Higher Education, an NGA under the President's Office, which supervises tertiary-level education.

Oil palm cultivation and land use changes in the Pulot watershed

Location of the Pulot watershed

The Pulot watershed is situated in the southern part of the province of Palawan and is located across three municipalities (Figure 4.4). This watershed is adjacent to the province's lone oil mill plant, which is located in the municipality of Brooke's Point. The largest part of the watershed's is located within Sofronio Española (and in the *barangays* of Labog, Pulot Interior, Pulot Shore, Pulot Center, and Panitian), followed by Brooke's Point (*barangay* Calasaguen) on the southwestern side, the smallest portion is within Quezon (the *barangays* of Calumpang, Quinlogan, and Sowangan) on the northwestern side. The watershed has a total area of approximately 18,084 hectares.

Land use change

Land cover change analysis of the Pulot Watershed showed the extent of forest degradation. Anthropogenic activities are the primary drivers of land cover change, which in most cases, reduces biodiversity and ecosystem services. Using a NAMRIA-generated land cover change map covering the periods of 2005, 2010, and 2015 for a World Bank Pilot Ecosystem Account for Southern Palawan, a comparative analysis of the Pulot watershed forest cover between 2005 and 2010 shows a significant decrease of closed forest from 6,206 hectares in 2005 to an estimated 91 hectares in 2010 and has been totally lost in 2015 (Figure 4.5). Open forest has increased more than threefold from 2,139 hectares in 2005 to approximately 7,872 hectares in 2010 and reduced to 7,670 hectares in 2015. Brushland areas have gradually decreased from 5,492 hectares to 4,138 hectares and 4,107 hectares in the same period.

The rest of the closed forest has been converted to open forest. Later, portions of this open forest may have been converted into oil palm production areas (perennial crops). Areas devoted to perennial crops have increased more than threefold from 1,730 hectares in 2005 to 5,169 hectares in 2015. The established oil palm plantations within the Pulot watershed totalled about 901 hectares in 2010 and increased to 1,012 hectares in 2015. Coconut plantation areas established have been estimated at 1,315 hectares in 2010, and slightly increased to 1,455 hectares in 2015. Species diversity, population numbers, as well as the type and quantity of ecosystem services may potentially be affected by changes in forest/wooded land formation and composition.

The pressure to increase food production, alongside economic gain, has propelled agricultural expansion in the Pulot watershed. Perennial crop areas have increased by almost 300 percent between 2005 and 2010. Oil palm has emerged

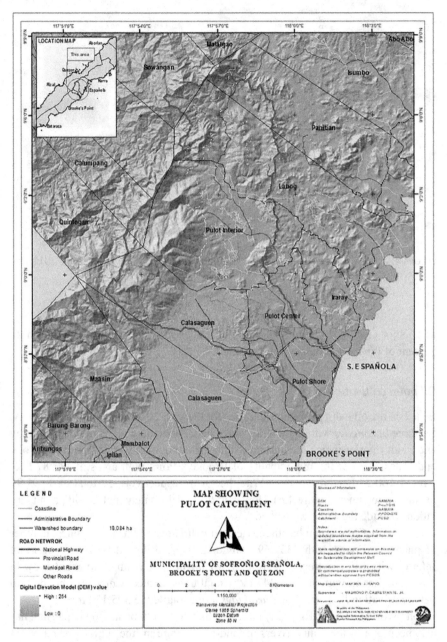

FIGURE 4.4 Location map of the Pulot watershed in southern mainland Palawan province, the Philippines.

Source: PCSD.

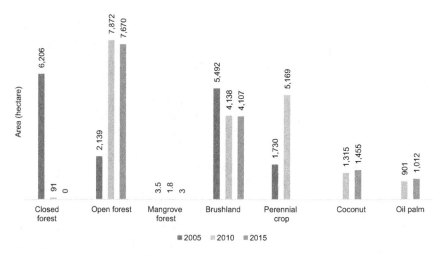

FIGURE 4.5 Land cover change in the Pulot watershed for periods 2005, 2010 and 2015.

as a major crop in the Pulot watershed in the last decade, and covers close to 20 percent of the total perennial crop area.

Oil palm cultivation

Based on the oil palm plantation statistics for the year 2022 obtained from the PCA, the three *barangays* within the Pulot watershed that planted oil palm include Labog and Pulot Interior which are located within the municipality of Sofronio Española; and Calasaguen of the municipality of Brooke's Point with areas of 112.87 hectares, 1,160.22 hectares, and 1,817.86 hectares, respectively—a total of 3,091.95 hectares altogether (Figure 4.6). However, these values may include oil palm plantations outside the Pulot watershed.

Barangay Calasaguen has the largest oil palm area and has the highest number of planted oil palm with 233,069 individual trees, followed by *Barangay* Pulot Interior with 147,616 trees (Figure 4.7). Meanwhile, oil palm cultivation is low in *Barangay* Labog with only 14,447 individual trees planted. Currently, the total oil palm cultivation in these three *barangays* alone totalled to 395,132 trees.

The World Bank (2016, p. 21) indicated that "oil palm plantation expansion has also been one of the main drivers of land-use change in [the] Pulot watershed (in Southern Palawan)". By law, natural forests and those on steep slopes cannot be converted into oil palm plantations. Much of the forestlands belong to the ancestral or tribal lands of the indigenous peoples. Areas on steep slopes and forested areas are considered as core zones under the Environmentally Critical Areas Network, the main strategy or zoning scheme in Palawan. As such, the network provides a graded system of protection and development control in the province.

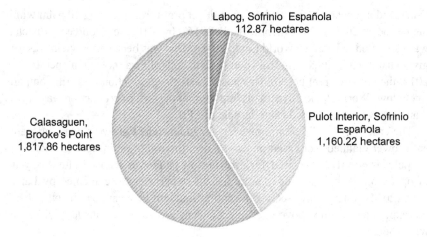

FIGURE 4.6 Areas (in hectares) planted with oil palm within the Pulot watershed.

Source: PCA, 2023.

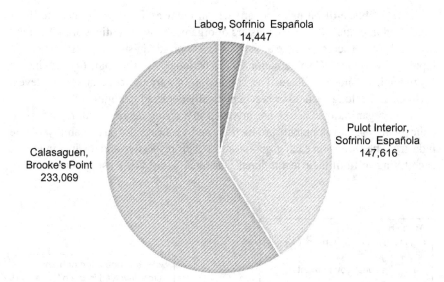

FIGURE 4.7 Number of oil palm trees planted within the Pulot watershed.

Source: PCA, 2023.

Economics of palm oil in Palawan

Palm oil producers in the Pulot watershed are not homogeneous. A clear distinction exists between those who are contractually tied to AGPI and depend on that company's support to develop their plantings and more independent growers. In 2013, smallholders in the former category interviewed indicated that they felt oil palm planting was not financially profitable (Larsen et al., 2014). In the case of the Pulot watershed in Española, the growers did not get profit from planting oil palm within the period of 2010 to 2014 (World Bank, 2016). In 2010, the resource rent/hectare was estimated to be zero (World Bank, 2016). This was because the yield was low, given that the oil palms were still young and the oil mill was not yet in operation. In 2014, the resource rent/hectare was negative as the production was still comparatively low (World Bank, 2016). This implies that the cost of production was greater than the revenue generated from the sale of FFB.

The case may be different, however, for independent growers. One independent grower claimed that based on experience, the return on investment (ROI) for oil palm is positive. The ROI for growing oil palm was realised after six years of operation. It is important to note that the experiences documented by Larsen et al. (2014) were generalised for the entire Southern Palawan. On the other hand, the experience of the above-cited independent grower was outside of the Pulot watershed.

Discussion

In the assemblage of oil palm governance in Palawan in general—and the Pulot watershed in particular—seven clusters of organisations/institutions are intricately involved. These actors operate at various levels or administrative scales, from the topmost as follows: (1) international; (2) national; (3) regional; (4) provincial; (5) municipal; and (6) village. The interaction of various actors at various levels between and among each other is schematically depicted in Figure 4.8.

These organisational actors are involved in varying degrees with regard to oil palm planning, development, production, and ensuring the sustainability of the industry. The entities in each group promote their respective and distinct interests, each trying to fulfil their institutional mandates. However, these groups are very

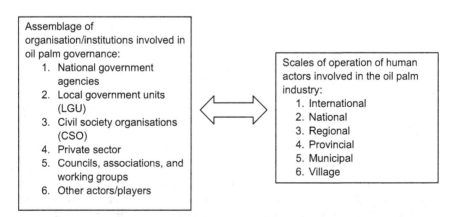

FIGURE 4.8 Cluster of organisations and levels of involvement in the oil palm industry.

diverse, with varying degrees of power and influence, differing levels of invest-ment in the oil palm assemblage, and different, sometimes contradictory aims and interests. Some government agencies, for example, have a focus on the expansion and growth of oil palm to support economic objectives, while some have a stake in promoting more conservation-facing priorities and an interest in restricting land use change to oil palm. Yet others have a much more specific interest in protecting the rights and interests of various groups, such as indigenous communities and palm oil plantation company employees, as well as groups whose interests may directly run counter to those of the palm oil companies. Furthermore, these inter-ests are not fixed, they have evolved over time, responding to shifting ideas about the oil palm itself and its place in global and national policy contexts.

In assemblage terms, these human actors, along with non-human entities such as the land and soil in the catchment, competing crops and others are engaged in assemblage activities. Through this, the role of specific parts of the catchment, and its land in particular, are constantly being redefined and negotiated over time as they fall in and out of different assemblages and are brought into new sets of relationships with other entities as a result.

On one side are some entities involved in the promotion of palm oil, such as the nationally-mandated agency, the PCA, which is responsible for administering the development of oil palm in Palawan. As such, through the PCA, Palawan must con-tribute to the realisation of the original national target of 300,000 hectares planted with oil palm. In doing so, the PCA tends to be involved in measures to incorporate greater areas of Palawan into the oil palm assemblage, as a factory floor for the large-scale production of palm oil in monoculture. As of 2022, the organisation has enjoyed partial success. The spatial survey indicates that 20 percent or so of the Pulot catchment's cultivable lands are now under palm oil. This task has been largely achieved through the successful de-linking of large areas of land from pre-vious "natural", traditional or mixed farming lands and their re-assembling as part of an assemblage arrayed around the intensive production of palm oil. The con-tribution of Palawan as a service area supporting national goals was about 9,040 hectares in 2022 and involved 1,693 farmers. Some 1,167,241 oil palm trees have been planted. Out of this total, 907,209 trees are fruit-bearing, while 260,032 are non-bearing. The PCA's provincial office provides technical assistance to the farm-ers (particularly the farmers involved in small cooperatives), as well as the pro-vision of limited sacks of free fertilisers to these farmers. Note that the PCA's provincial office in Palawan has played an active role in oil palm development only in relatively recent times. It was not actively involved in the early 2000s until the controversy regarding land issues with cooperatives arose. The numerical indica-tors of success are not un-typical of similar discourses connected to the "success" of palm oil elsewhere, in which prominence is given to the sheer scale of the area covered, the numbers involved, and the scale of production, equating success with production scale as opposed to measures of the actual impact on people and com-munities who tend to feature solely as "producers".

Yet it is also clearly the case that the incorporation of these areas into the oil palm assemblage has only been achieved on a partial basis. In part, this can be associated with changes in policy at the national and provincial levels. The DA, which has a provincial office that is involved in Palawan's agricultural development, no longer "actively" promotes oil palm as a priority, high-value terrestrial crop. This reflects changing agricultural policies and perhaps a rebalancing between competing organisations with the DENR, charged with safeguarding Palawan's natural environment, working to ensure that oil palm cultivation does not encroach on forest lands. In the case of this organisation, it can point to the negative environmental impacts of the industry, such as the pollution due to the use of inorganic inputs (pesticides and fertilisers) and terrestrial biodiversity loss as a result of oil palm expansion, to press for restrictions on palm oil expansion, effectively establishing an alternative "conservation assemblage" in which it seeks to incorporate areas of the Pulot catchment in different relations with other human and non-human components in this assemblage. However, its PENRO and CENROs do not have sufficient numbers of trained personnel to maintain the assemblage organised around these environmental priorities. This leaves the door open for encroachment and an undermining of its activities by entities in the palm oil assemblage. Meanwhile, seeming to fall somewhere between these positions, the PCSD is mandated to promote the triad of sustainable development in oil palm plantations: to provide the maximum economic benefits, make it an acceptable crop to the farmers, and encourage environment-friendly cropping practices. The PCSD has a multi-sectoral Scientific Advisory Panel which provides scientific and technical advice for sustainable development concerns. In other words, the PCSD seeks to redefine relationships between entities within the palm oil industry in ways which result in reduced environmental damage and a better distribution of benefits. Were it successful, then this might go some way towards neutralising the notion of an alternative "conservation assemblage" by making its aims coalesce with those of a reformed palm oil assemblage.

However, this of course assumes that sustainable palm oil is both possible and corresponds with the needs and wants of many other human and non-human entities and that they will become enrolled in a reformed palm oil assemblage. Such a situation might never apply to the NCIP which is most concerned that oil palm cultivation does not infringe on tribal ancestral lands without the FPIC of the indigenous peoples. It almost certainly will not apply to indigenous peoples themselves.

Private interest groups in Palawan's oil palm industry, including Filipino and foreign nationals aiming to maximise their profit and income, have pushed for palm oil, linking it strongly to a range of socio-economic benefits. Particularly in the early phase of its development, they had the benefit of public agencies and discourses that supported the crop, as a greener and more efficient source of fuel and food than others, and as an unparalleled high-value form of agriculture. In so doing, they drew on precedents from the Malaysian and Indonesian industry. Efforts to transfer the "success" of this model can be found in the contract grower agreements which are very similar to those under which "supported" smallholders operate in those countries. The description of the business relationship between these oil palm cooperatives and AGPI is taken from Larsen et al. (2014). The smallholder cooperatives have entered into Production, Technical, and Marketing Agreements with AGPI that specify the terms and conditions for the business relationship. Under this scheme, the cooperatives provide the land and organise labour while AGPI provides seedlings and technical know-how (Larsen et al., 2014).

The contracts bind the parties for 25 years, during which the cooperative must deliver FFB to PPVOMI, in accordance with specified quality standards. Note that the 25-year contract roughly coincides with the productive lifetime of the oil palm tree. The sales price is not fixed, but instead regulated through an FFB pricing formula based on these three parameters: (1) AGPI's oil extraction ratio, (2) the average selling price, and (3) average processing costs. A Management Services Agreement sets out the terms and conditions for AGPI's management of the project on the land provided by the cooperatives. Both contracts contain provisions, which place the control with AGPI and the majority of financial and managerial risks with the cooperatives. For instance, AGPI is entitled to take over management (on the basis of sole opinion), if the project is not managed to the satisfaction of AGPI. It is scarcely surprising given the terms of these contracts that this was a route the companies favoured and equally unsurprising that smallholders in these schemes state that they have struggled to generate income. LGUs promoted oil palm as a high-value crop to significantly contribute to livelihood opportunities and increase the income of rural populations. They have multiple other concerns including safeguarding the rights (and labour) of indigenous peoples, the protection of their ancestral lands and environmental protection in general, with regard to the negative impacts of oil palm development.

Various oil palm-related councils and livelihood-based associations are involved at various levels of governance. The PPDCI operates at the national level; while the PPOIDC promotes, monitors and helps regulate Palawan's oil palm industry. The PAMB of the MMPL disallowed the use of ancestral lands within its territorial jurisdiction for cultivation by an oil palm cooperative. The Working Group on Oil Palm Concerns advocates for a moratorium on further expansion of oil palm plantations and the illegal cultivation of forest lands belonging to indigenous peoples. The Pulot Irrigators Association is concerned that oil palm cultivation is competing with water use, particularly in the growing of rice, which is the main staple crop in the province.

The government-owned lending institution, LBP, provides corporate financial assistance to AGPI and the 14 oil palm cooperatives. This is in support of the Philippine government's efforts to attain food self-sufficiency and increase agricultural productivity. Three institutions of higher education—Palawan State University, the University of Southern Mindanao, and the Western Philippine University—have been involved in research and training concerning oil palm.

The above paragraphs describe some of the human entities that have been incorporated into the nascent palm oil assemblages in the Palawan province with a focus on the

Pulot watershed. These organisational entities include national, provincial, and local governance bodies, smallholders, NGOs, and private companies. The introduction of farmer cooperatives, and in particular the development of an outgrower scheme, coincides with the development of industry infrastructure in the form of milling facilities and nurseries in providing mechanisms through which a range of producers, traders, and administrators are brought into and ascribed roles in the oil palm assemblage.

We see, in turn, how this assemblage has impacted and influenced varying and significant changes in the local landscape (physical, social and political). More specifically, local lands are decoupled from previous agricultural land uses of traditional crops or more biodiverse closed forest assemblages and incorporated into the palm oil assemblage as sites of palm oil production. The transformation of these areas and reconstitution within the palm oil assemblage, in turn, has wider implications for the range of objects and entities which had previously been linked with this land, with implications for the environment and use of natural resources.

We observe, however, that the oil palm assemblage is continuously in a state of flux, with the relationships between the different components of the assemblage being formed, reshaped, and reformed constantly. The relationships between these elements can be seen as being "viscose" as opposed to being fixed (Umans & Arce, 2014). Thus, in this context, we can think of governance as involving efforts by different, often competing, actors to "fix" or impose orderings on elements of the assemblage in ways that support the attainment of different objectives. This incomplete and ongoing practice is evidenced in the various texts and practices of actors in the assemblage, for example, in the nested and overlapping mandates within the government bureaucracy itself.

There are eight areas of concern about oil palm cultivation in the Pulot watershed, which may be considered as a microcosm of what happens in southern mainland Palawan. The first is obviously competition around the use of scarce freshwater resources. Note that freshwater from the Pulot watershed is primarily used for irrigation of agricultural food crops, particularly rice. Since oil palm cultivation uses substantial water, it competes for the water requirements of crops such as rice and corn. The available rainfall in mainland Palawan has only a mean of 1,780 mm/year (World Bank, 2016), which is simply not enough to meet the oil palm's water requirements. To address this deficit, the oil palm growers' have to source for additional supply, from either riverine sources or groundwater. For riverine sources, some farmers extract water using pumps and trucks to irrigate their crops. Alternatively, a few of them directly access irrigation water by constructing diversion canals to their farmlands, with a range of implications for other water users, the land itself, hydrology areas, and even soils. In this sense, we can see how the emergence of the oil palm assemblage in this area prompts the reorganisation of existing entities as well as the introduction of new ones, which impacts the environment in ways that have substantial consequences.

The second concern about the oil palm assemblage is biodiversity reduction. A palm oil plantation is essentially a monoculture. There is less plant diversity as compared to the previous assemblage of mixed cropping and intact forest in the watershed. This corresponds to lesser animal diversity. Wildlife is typically associated with mixed croplands; hence, animals such as monkeys, parrots, and doves are not ordinarily present in palm oil plantations. Note that the pollinator of oil palm is also an introduced species. Hence, it might become a potential pest later in Palawan. The *Elaeidobius kamerunicus* weevil plays a vital role in oil palm yield by transferring pollen between male and female flowers.

The third aspect is related to the impacts of increased fertiliser use. As a rule of thumb, fertilisers and their administration make up about 70 percent of the total cost of oil palm operations. The main fertilisers used generally consist of nitrogen, phosphorus and potassium. The most common brand utilised is "Atlas", which is typically packed in a 50 kg bag. In any fertiliser application, however, there is no 100 percent uptake by the plant. A certain proportion of the applied fertiliser is leached out into the environment. Since the oil palm does not take up all of the nitrogen, the nitrogen is eventually leached out to the following endpoints: groundwater, riverine systems within the Pulot watershed, and coastal marine areas of Sofronio Española.

Pesticide residue is the fourth aspect of oil palm cultivation. Pesticides are applied to address two cropping concerns: (1) to eliminate weeds and (2) to eradicate pests and plant diseases. The pathways in which pesticide spread into the environment is similar to that of fertilisers. "Roundup" is a popular brand of weed killer with glyphosate as an active ingredient. Pesticides are applied for creepers and vine weeding, circle weeding, path weeding, and selective weeding. There is no quantitative data about the amount of pesticides applied in oil palm cultivation in the Pulot watershed. Hence, we only provide two generic statements about two potential negative impacts of pesticide use: (1) the impact on non-target species, such as bees and butterflies; and (2) the development of resistance of the target pests. Hence, it implies that greater amounts will be applied to eradicate the pests. There may also be health hazards as some farmers reportedly do not wear the proper protective gear when spraying pesticides.

The fifth concern is about crop-associated pests. One of these is the rhinoceros beetle (*Oryctes rhinoceros*) which breeds in rotting wood on the plantation floor. The expansion of oil palm plantations in Palawan is reported to have accelerated infestations of the coconut leaf beetle (*Brontispa longissima*) (Larsen et al., 2014). This species of beetle is considered a major pest of both coconut and oil palm plantations (Wan Khairul & Idris, 2013). Other pests include the various species of bagworms (order *Lepidoptera*: family *Psychidae*) which are important leaf-eating pests of oil palm in Indonesia and Malaysia. Rats (*Rattus* spp.) are traditionally pests in coconut plantations, but these rodents also infest oil palm plantations by eating the leaf bases of young palms. Rats cause damage to mature palms yields by eating the ripening fruits on the bunches. These partially-eaten bunches will be sold at the mill at reduced prices, given that some oil is lost.

The sixth concern relates to oil palm by-products and waste utilisation. During the palm oil extraction process at the palm oil mill, several by-products and waste are generated. These are generally classified into five: (1) empty fruit bunches; (2) palm oil mill effluents; (3) steriliser condensate; (4) palm fibre; and (5) palm kernel shells. There are no quantitative estimates of these by-products and waste in the case of the Pulot watershed. A considerable management challenge is to determine how these so-called by-products and waste can have useful economic uses. The seventh concerns its impact on livelihoods, affecting those who do and do not choose to be involved in the industry. This is impacted by multiple factors; for example, oil palm's profitability is contingent on the seedlings. Some seedlings planted are not true hybrids. Thus, the quality of this second generation is lower compared to true hybrids. This means that planting palm oil not coming from the original parent will result in poor production.

The eighth area relates to food security. The substitution of oil palm for traditional food crops, such as rice and corn, may contribute to food insecurity among local consumers in the province.

Rethinking the oil palm industry in Palawan

Globally, the oil palm industry has remained lucrative, given the high demand for oil palm products in many countries. Reflecting on the development of the industry in Palawan over the last two decades (2003–2023), much has changed, especially in terms of Philippine government support.

In 2003, the provincial government of Palawan initiated the introduction of oil palm. At the time, the palm oil industry was a national priority during the administration of former President Gloria Macapagal-Arroyo (2001–2010). By 2004, the PPOIDC was created and, in the subsequent year, oil palm was included in the Comprehensive Development Plan of Palawan as a plantation crop. In 2006, the PPVOMI and its affiliate company, AGPI began their operations in the province and have continued up to the present. As of 2022, Palawan has 9,040 hectares planted with oil palm, representing about 3 percent of hectarage contribution to the national Philippine target.

The main difficulty in assessment is the limited research and publicly-available data about oil palm in Palawan. We do not have publicly-available and reliable data regarding the cost-benefit analysis and ROI on oil palm plantations. It may be argued that the claims made for palm oil as a vehicle for economic development and livelihood improvement have not fully materialised. The assemblage of oil palm governance in Palawan does not function in ways that deliver on these aspirations as discourse around the crop suggests. The actions of the assemblage players are not fully coordinated or integrated or indeed understood. Rather, different groups display different capacities to capture aspects of the palm oil assemblage in ways which correspond to their own projects and interests or indeed to appropriate elements of the palm oil assemblage into alternative assemblages. The case provided in this chapter shows that the ROI has remained positive given correct agronomic practices, as claimed by an independent grower. However, it appears that the triad of sustainable development has not been sufficiently attained in oil palm cultivation for other players as it is currently constituted in the area.

Reluctance to actively promote oil palm is in part due to socio-economic issues (such as profitability of oil palm cultivation among small-scale cooperatives as well as oil palm's competition for land and water with other agricultural crops) and environmental hazards such as water pollution and increase in the number of agricultural pests. Moreover, oil palm is no longer considered a priority crop by both the local and national governments.

Two major considerations may be worth pondering. The first consideration relates to the role of science in the development of the oil palm industry. There is still very limited scientific information to guide the management bodies responsible, and in some cases, disparate data as in the case of PSA and PCA as mentioned earlier in the chapter. Moreover, there is no central repository for this information. The Scientific Advisory Panel of the PCSD earlier declared a moratorium on further expansion of oil palm plantations pending further research related to socio-economic and environmental impacts.

The second concern relates to the institutional analysis of the oil palm assemblage. Much still needs to be done—on the part of the national government agencies, local government units, CSOs, the private sector, councils/associations, as well as the other actors/players—to make oil palm a truly viable sustainable industry. It may be high time to reconvene the PPOIDC. Although no longer promoted as a high-value agricultural crop, oil palm has characteristics that make it a "superior" agricultural perennial oil crop. Aside from a productive lifespan of up to 30 years, oil palm is relatively better than other oil crops as it yields up to five tonnes per hectare. It has a variety of uses in several industries: food, non-food, and biofuel. These are economic features and opportunities that may still be worth considering for the oil palm industry in Palawan. However, it is important to recognise the limitations of the crop, especially in terms of high-water demand which negatively impacts adjacent crops such as rice.

Note

1 Province is the equivalent of "state" in Malaysia; the municipality is equal to "district", while a rural *barangay* roughly translates to "*desa*" in Indonesia.

References

ALDAW. (2011). *Save Palawan: The Philippines' last frontier*. Forest Peoples Programme. https://www.forestpeoples.org/en/topics/extractive-industries/news/2011/03/news-alert-ancestral-landdomain-watch-aldaw-save-palawan-p

ALDAW. (2013). *The Palawan oil palm geotagged report—Part I. The environmental and social impacts of palm oil expansion on Palawan UNESCO man and biosphere reserve (The Philippines)* (pp. 1–94). ALDAW, Rainforest Rescue, and World Rainforest Movement.

ALDAW. (2014). *"Washing out diversity": The impact of oil palm plantations on non-timber forest products (NTFPs), indigenous people's livelihood and community conserved areas (CCAs) in Palawan (the Philippines)*. https://www.regenwald.org/files/en/ALDAW%20 NTFP%20OIL%20PALM%202014%20REPORT.pdf

Barcia, R. B. (2017, February 19). *Palm plantations threaten biodiversity in Palawan.* The Manila Times. https://www.manilatimes.net/2017/02/19/todays-headline-photos/ regions/palm-plantations-threaten-biodiversity-palawan/313093

Barraquias-Flores, T. (2010). *Oil palm cultivation in Palawan: Status of investments and impacts to communities and the environment.* Environmental Legal Assistance Centre.

Barrett, C. B., Bachke, M. E., Bellemare, M. F., Michelson, H. C., Narayanan, S., & Walker, T. F., (2012). Smallholder participation in contract farming: Comparative evidence from five countries. *World Development, 40*(4), 715–730. https://doi.org/10.1016/j. worlddev.2011.09.006

Batugal, P. (2013). *Philippine palm oil industry road map, 2014–2023.* PCA. https://pdf4 pro.com/view/philippine-palm-oil-industry-road-map-1fcf4b.html

Blanco, G., Arce, A., & Fisher, E. (2015). Becoming a region, becoming global, becoming imperceptible: Territorialising salmon in Chilean Patagonia. *Journal of Rural Studies, 42,* 179–190. https://doi.org/10.1016/j.jrurstud.2015.10.007

Conservation International. (2019). *Sustainable palm oil.* https://www.conservation.org/ projects/sustainable-palm-oil

CPOPC. (2022). *Our mission.* Council of Palm Oil Producing Countries. https://cpopc.org/ our-mission

Cramb, R. A., & Curry, G. N. (2012). Oil palm and rural livelihoods in the Asia-Pacific region: An overview. *Asia Pacific Viewpoint, 53*(3), 223–239. https://doi.org/10.1111/ j.1467-8373.2012.01495.x

Cramb, R. A., & McCarthy, J. F. (2016). *The oil palm complex: Smallholders, agribusiness and the state in Indonesia and Malaysia.* NUS Press.

Dalabajan, D. (2011). *Halting the bandwagon: The impacts of biofuel feedstock plantations in the province of Palawan.* Non-Timber Forest Protection Programme (NTFPP) & Environmental Legal Assistance Centre.

Delanda, M. (2016). *Assemblage theory.* Edinburgh University Press.

Deleuze, G., & Guattari, F. (1988). *A thousand plateaus: Capitalism and schizophrenia.* Athlone Press.

ELAC. (2023). *How we work: Helping communities defend the earth.* Environmental Legal Assistance Centre. https://www.elacphilippines.org/

Gonzales, I. (2010, October 5). *Landbank allots P50 billion for government food supply chain programme.* Philstar. https://www.philstar.com/business/2010/10/05/617638/ landbank-allots-p50-billion-government-food-supply-chain-program

Government of the Philippines. (2011). *Philippine development plan 2011-2016: Results matrices* (pp. 1–98). National Economic and Development Authority.

Groenewald, J. A., Klopper, J., & van Schalkwyk, H. D. (2012). Unlocking markets to smallholder farmers: The potential role of contracting. In H. D. van Schalkwyk, J. A. Groenewald, G. C. G. Fraser, A. Obi, & A. Tilburg (Eds.), *Unlocking markets to smallholders: Lessons from South Africa* (pp. 133–148). Wageningen Academic Publishers.

Harbinson, R. (2015, October 23). *Broken promises: Communities on Philippine Island take on palm oil companies.* Mongabay. https://news.mongabay.com/2015/10/broken-promises-communities-on-philippine-island-take-on-palm-oil-company/

Koh, L. P., & Wilcove, D. S. (2008). Is oil palm agriculture really destroying tropical biodiversity? *Conservation Letters, 1*(2), 60–64. https://doi.org/10.1111/j.1755-263x. 2008.00011.x

Larsen, R. K., Dimaano, F. R., & Pido, M. D. (2018). Can the wrongs be righted? Prospects for remedy in the Philippine oil palm agro-industry. *Development and Change, 50*(5), 1373–1397. https://doi.org/10.1111/dech.12416

Larsen, R. K., Dimaano, F., & Pido, M. (2014). *Emerging oil palm agro-industry in Palawan, The Philippines* (pp. 1–46). Stockholm Environment Institute.

Lerom, R. (2015). *Palm oil industry in Palawan (draft)*. Western Philippines University & Palawan Council for Sustainable Development.

Marsden, T., Hebinck, P., & Mathijs, E. (2018). Re-building food systems: Embedding assemblages, infrastructures and reflexive governance for food systems transformations in Europe. *Food Security, 10*(6), 1301–1309. https://doi.org/10.1007/s12571-018-0870-8

Mizuno, K., Fujita, M., & Kawai, S. (2016). *Catastrophe and regeneration in Indonesia's peatlands: Ecology, economy, and society*. NUS Press & Kyoto University Press.

Montefrio, M. J. F. (2015). Green economy, oil palm development and the exclusion of indigenous swidden cultivators in the Philippines. *Land Grabbing, Conflict and Agrarian-Environmental Transformations: Perspectives from East and Southeast Asia, 22.* https://www.iss.nl/sites/corporate/files/CMCP_22-_Montefrio.pdf

Montefrio, M. J. F., & Dressler, W. H. (2016). The green economy and constructions of the "idle" and "unproductive" uplands in the Philippines. *World Development, 79*, 114–126. https://doi.org/10.1016/j.worlddev.2015.11.009

Montefrio, M. J. F., Ortiga, Y. Y., & Josol, Ma. R. C. B. (2014). Inducing development: Social remittances and the expansion of oil palm. *International Migration Review, 48*(1), 216–242. https://doi.org/10.1111/imre.12075

Müller, M. (2015). Assemblages and actor-networks: Rethinking socio-material power, politics and space. *Geography Compass, 9*(1), 27–41. https://doi.org/10.1111/gec3.12192

Murray Li, T. (2002). Engaging simplifications: Community-based resource management, market processes and state agendas in upland Southeast Asia. *World Development, 30*(2), 265–283. https://doi.org/10.1016/s0305-750x(01)00103-6

Murray Li, T. (2007). Practices of assemblage and community forest management. *Economy and Society, 36*(2), 263–293. https://doi.org/10.1080/03085140701254308

Neame, A., & Villarante, P. (2013). Overview of the palm oil sector and FPIC in Palawan, Philippines. In M. Colchester & S. Chao (Eds.), *Conflict or consent? The oil palm sector at a crossroads* (pp. 201–231). Forest Peoples Programme.

Nozawa, K. (2011). *Oil palm production and cooperatives in the Philippines* (pp. 1–38). University of the Philippines School of Economics. https://econ.upd.edu.ph/dp/index. php/dp/article/view/681/147

O'Reilly, P., & Varkkey, H. (2020). Palm oil governance in different locations: Using the assemblage approach to understand a "complex" sector. *International Review of Modern Sociology, 46*(1–2), 1–17.

Pamplona, P. (2013). Philippine agriculture and the high incidence of poverty and unemployment. *Agriculture, 17*(9), 60–64.

Pamplona, P., & Pamplona, A. G. (2014). The Philippine palm oil industry: Moving toward a brighter future for the small landholders. *PPDCI and Province of Sarangani Business Forum.*

PCA. (n.d.). About us. Philippine Coconut Authority. https://pca.gov.ph/index.php/about-us/ overview

PCA. (2023). *Extent of area devoted to oil palm plantation in Palawan Province, Philippines, in 2022.* PCA Provincial Office. Unpublished report.

PCA, Department of Trade and Industry, & PPDCI. (2023). *Philippine palm oil industry roadmap (2024-2033)* (pp. 1–41). PCA.

PNNI. (2023). *Our story.* Palawan NGO Network Inc. https://www.pnni.org/about

Pye, O., & Bhattacharya, J. (2013). *The palm oil controversy in Southeast Asia: A transnational perspective.* Institute of Southeast Asian Studies.

Role and rights of People's Organisation, 15 (1987).

RSPO. (2018). *How RSPO helps.* Roundtable on Sustainable Palm Oil. https://rspo.org/

Sheil, D., Casson, A., Meijaard, E., van Noordwijk, M., Gaskell, J., Sunderland-Groves, J., Wertz, K., & Kanninen, M. (2009). *The impacts and opportunities of oil palm in Southeast Asia: What do we know and what do we need to know?* Center for International Forestry Research. https://doi.org/10.17528/cifor/002792

Sutherland, L.-A., & Calo, A. (2020). Assemblage and the "good farmer": New entrants to crofting in Scotland. *Journal of Rural Studies, 80*, 532–542. https://doi.org/10.1016/j.jrurstud.2020.10.038

Tandra, H., & Suroso, A. I. (2023). The determinant, efficiency, and potential of Indonesian palm oil downstream export to the global market. *Cogent Economics & Finance, 11*(1). https://doi.org/10.1080/23322039.2023.2189671

Thornton, S. A., Setiana, E., Yoyo, K., Dudin, Yulintine, Harrison, M. E., Page, S. E., & Upton, C. (2020). Towards biocultural approaches to peatland conservation: The case for fish and livelihoods in Indonesia. *Environmental Science & Policy, 114*, 341–351. https://doi.org/10.1016/j.envsci.2020.08.018

Umans, L., & Arce, A. (2014). Fixing rural development cooperation? Not in situations involving blurring and fluidity. *Journal of Rural Studies, 34*, 337–344. https://doi.org/10.1016/j.jrurstud.2014.03.004

US Department of Agriculture. (2023). *Indonesia palm oil: Historical revisions using satellite-derived methodology* (pp. 1–11).

US Department of Agriculture. (2024). *Palm oil 2023 world production.* USDA Foreign Agricultural Service. https://ipad.fas.usda.gov/cropexplorer/cropview/commodityView.aspx?cropid=4243000

Varkkey, H. (2012). Patronage politics and natural resources: A historical case study of Southeast Asia and Indonesia. *Asian Profile, 40*(5), 438–448.

Villanueva, J. (2011). Oil palm expansion in the Philippines: Analysis of land rights, environment and food security issues. In M. Colchester & S. Chao (Eds.), *Oil palm expansion in South East Asia: Trends and implications for local communities and indigenous peoples* (pp. 110–216). Forest Peoples Programme & Perkumpulan Sawit Watch.

Wan Khairul, A. W. A., & Idris, A. B. (2013). Field incidence on Brontispa longissima (Gestro), an invasive pest of coconut. *AIP Conference Proceedings, 1571*(1), 355–358. https://doi.org/10.1063/1.4858682

World Bank. (2016). *Pilot ecosystem account for Southern Palawan* (pp. 1–118). Wealth Accounting and the Valuation of Ecosystem Services. https://www.wavespartnership.org/sites/waves/files/kc/WB_Southern%20Palawan%20Tech%20Report_FINAL_Nov%202016.pdf

WWF. (2019). *What is palm oil? Facts about the palm oil industry.* World Wildlife Fund. https://www.worldwildlife.org/industries/palm-oil

5

THAILAND'S PALM OIL

Evolving from domestic smallholder centrism to sustainable exports

Khor Yu Leng and Nithiyah Tamilwanan

Introduction

Thailand is currently the world's third-largest palm oil producer, with a 4 percent share of global output. Production in Thailand is spatially concentrated in a few areas. 60 percent of production is in the mid-south provinces of Surat Thani, Krabi, and Chumphon (OAE, 2022). Thailand is situated close to the core palm oil-producing countries, Malaysia and Indonesia, and the industry includes many of the same value chain actors and processes found, not only in Indonesia and Malaysia but also in other palm oil-producing countries. Yet, as is the case in studies conducted elsewhere and in this volume, a detailed examination of the relationships between similar players in the palm oil value chain (farmers, traders, mills, politicians) engaged in similar activities in different national, regional, and local jurisdictions reveal that the configuration of national palm oil assemblages and the outcomes (in terms of land management, environmental impact, palm oil governance, as well as the distribution of costs and benefits along the value chain) may exhibit significant variation.

Of immediate notice in the case of the Thai palm oil industry is its supposed "farmer centrism" and more specifically its "smallholder centrism". In contrast with the abiding global image of palm oil, the oligarchic, corporate-dominated industry most commonly associated with Indonesia and Malaysia (see for example, Cramb & McCarthy, 2016), the Thai oil palm industry is dominated by smallholders. Currently, 70 percent of Thai production comes from 407,225 independent smallholders (DIT, 2023). The experience of the Thai industry thus supports the findings and observations of other contributions to this volume, contrary to the perceptions of many, both inside and outside the industry—including those who regard it as a development opportunity and those who regard it as an environmental

DOI: 10.4324/9781003459606-6

curse—there are many different ways to do oil palm. For those interested in constructively engaging with the industry, this observation is highly significant, implying that the shape and impact of national palm oil industries cannot be simply assumed through reference to the way in which the industry has evolved elsewhere. This, in turn, may offer a wider scope for exploring how policies and industry practices can be adapted to simultaneously meet the needs and wants of producing countries and the global thirst for palm oil while meaningfully addressing the environmental and social concerns and problems associated with the industry.

In this context, the case of Thailand, a country that shares land and sea borders with the industry's centres of production and technology, allows for a particularly interesting exploration of these questions. Does the fact that the Thai industry differs so significantly in its development and current operation demonstrate the scope that exists for different nations to shape national palm oil value chains in ways that vary from what are often perceived to be indelible and inherent features of the industry? Or is it simply a reflection of the historical point at which Thai palm oil finds itself, a reflection of the relatively small size of the industry in that country, which time and industry growth will inevitably erase? Simply put, what can the Thai example tell us about the extent to which local agency and policymaking shape national oil palm industries in ways that reflect local priorities, needs, and wants as much as global trends?

It is such questions that the assemblage perspective proposed in the introduction to the present volume may help us to address. Assemblage was absorbed into the social sciences having first been developed by French philosophers Deleuze and Guattari (1988), with later elaboration by others (in particular De Landa, 2006). The approach was driven by a dissatisfaction with "stable" ontologies that conceived social reality in terms of relatively fixed, discrete objects. While such ontologies are fundamental to many approaches to understanding and ordering social life, assemblage theorists are particularly concerned with the failure of such conceptualisations of the social world to fully capture and accurately describe its complexity (see Müller, 2015). Instead, theorists interested in the concept propose assemblage as a new metaphor in which social order is not conceived as an inherent property of the things and structures that make up the social, but rather a function of dynamic and fluid processes which bring together "heterogeneous entities so that they work together for a certain time" (Müller, 2015). Assemblage conceptualises complex social phenomena as the effect of dynamic processes whereby human and non-human objects are brought together to serve certain purposes for an indeterminate period of time, in contingent ways and with unpredictable consequences. This perspective and approach eschew simplistic ideas about cause and effect within formal organisational contexts. Instead, it stresses the need to be attuned to the contingent and unpredictable ways in which a range of actors and entities produce new territorialised assemblages, including both formal and informal arrangements and relationships.

In our case, the Thai oil palm industry can be regarded as an assemblage that is (re)produced through the interactions of a number of human individuals and bodies, non-human plant and animal species, and non-living physical objects. Understanding this assemblage requires a consideration of how the interactions between entities shape the whole, avoiding preconceived ideas about what the key variables involved are. In this framing, the Thai palm oil industry and its properties (including its value) are understood as the contingent outcome of processes through which relations between human and non-human actors (including farmers, nitrogen fertiliser, traders, oil palms and local politicians) are forged, broken, reforged, and renegotiated over time. The outcome of this process is not necessarily stable or fixed, but is continuously (re)produced. Such frameworks may be discomfiting to analysts and industry commentators whose analytical focus presumes that industries consist of stable, formally-defined, and proscribed entities such as palm oil companies, farmers' groups and political parties—particularly if the connections between these are conceptualised as concrete, stable, exclusive systems; supply chains, governance hierarchies, and markets, articulated via rational socio-legal, socio-political, or economic media (money, contracts, and electoral politics for example). However, from the point of view of policy researchers and others interested in understanding how local palm oil industries in different countries are constituted and maintained, the advantages of such an approach are obvious. In the case of the Thai oil palm industry, there is significant *prima facie* evidence that notable variation exists between the way the industry has evolved in that country when compared to its development in neighbouring states. Assemblage provides us with a heuristic, which is particularly useful for probing and expanding upon this evidence and theorising the processes through which actual connections between different human and non-human entities in Thailand (re)produce a uniquely Thai oil palm industry.

The purpose of this chapter is thus two-fold, to chart the recent development of the Thai industry and to use this information to explore the practices of assemblage that have brought this situation about. In doing so, we touch upon some of the key issues explored in this book: how do different governance arrangements enable and legitimise expansion; and how do power differentials affect oil palm governance? We draw on existing industry data to provide an accurate description of how the industry has evolved, as well as to gain some insight into its current trajectory. Having done so, we employ secondary sources and data derived from in-depth interviews with people connected to that industry in order to flesh out the picture painted by the statistics to shed light on some of the "practices of assemblage" within the industry. In particular, we are interested in exploring how Thailand has been able to make a space for itself in the global palm oil industry whilst retaining its own distinct industry arrangements.

The remainder of the chapter is thus structured as follows. In the next section, we provide a concise overview of key themes in the current literature concerning the Thai oil palm industry. After this, drawing on secondary data, a detailed description of recent trends and the current trajectory of the Thai oil palm industry is presented, including a discussion of production facts and statistics, trade and pricing policies.

The next section considers the dynamics that underlie these statistics; we identify various stakeholders and other entities that have influenced the palm oil assemblage in Thailand. The concluding section will discuss the future prospects of the Thai oil palm assemblage and its relationship with the global palm oil industry.

Literature

The chapter draws on literature concerning the palm oil industry in Thailand as well as data derived from a range of qualitative and quantitative sources. Our literature review explored current knowledge and debates concerning the Thai oil palm industry and included an examination of industry reports, reviews by financial institutions, national policy papers, news articles, and academic papers. In the case of the latter, the review is based on English-language material. This decision was taken due to constraints on resources and difficulties in accessing material in Thai and other languages. We fully recognise that a further and more exhaustive exploration of the literatures in these languages would have added to our analysis. Unsurprisingly, our review confirmed the relatively small amount of attention that the industry has attracted in comparison to the Malaysian and Indonesian oil palm sectors (Figure 5.1). While this is entirely understandable given the relative size of the industry in each country, it does illustrate how the "dominant" model for the industry is "reproduced" in the literature.

In the current academic literature, key papers discuss characterisations of the industry's early expansion. Thongrak et al. (2011) described the socio-economic characteristics of participating farming households in Krabi, Surat Thani, and

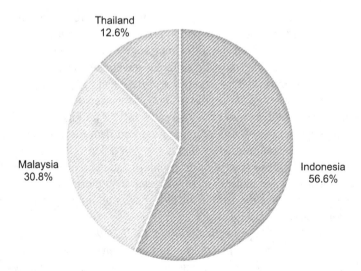

FIGURE 5.1 Literature on palm oil governance. 182 papers found on Google Scholar on "palm oil" and "palm oil governance" for the three selected countries.

Chumphon. Of note, they found that palm oil constituted part of a mixed income strategy of farmers who had previously lived and worked on the same land (up to 60 percent of total average earnings of USD13,000 each year). However, as is discussed below, land has also been reallocated to smallholders as well as in concessions to larger plantation companies in Thailand. Dallinger (2011) highlighted further differences between the industry in Thailand and elsewhere in Southeast Asia. Noting the absence of the "supported" smallholder schemes seen in Indonesia and Malaysia, Dallinger suggested that the absence of larger producers and the limited presence of corporate estates, alongside an undersupply of fresh fruit bunches (FFBs), has enabled Thai farmers and intermediaries to negotiate higher prices. Nupueng et al. (2022) confirmed these findings, underscoring the bargaining power held by intermediaries in supply chains that were not certified by the Roundtable on Sustainable Palm Oil (RSPO), as opposed to in RSPO-certified supply chains, which were shorter and more transparent with intermediaries excluded. Another theme that was featured in the literature concerned the industry's outlook. Sowcharoensuk (2023), for example, forecasts lower yields in 2024–2025 due to El Niño climate conditions, with improvements likely in 2026, in response to rising demand from the food and oleochemical industries. The same paper suggests that despite changes in palm oil-based biodiesel policies, biodiesel demand will also increase. Nupueng et al. (2018) discussed the need for re-planting initiatives to enhance production and market liberalisation to support a move away from unstable biodiesel policies, which are dependent upon FFB, cooking oil prices, and CPO inventories.

Another topic which features in the academic literature concerns land use changes and rights. Saswattecha et al. (2016) found that the oil palm area planted between 2000 and 2012 in Krabi, Nakhon Si Thammarat, and Surat Thani had replaced 65 percent of rubber and only 6 percent of forest, mangrove, peat, and wetland area. Srisunthon and Chawchai (2020) detected that 10 percent of the wetland area in Narathiwat was converted to oil palm in 2009–2016. Pobsuk (2019) examined issues of land rights, providing case studies of how oil palm farming communities within the Southern Peasants' Federation of Thailand (SPFT) had fought against palm oil companies to obtain titles to land. Issues relating to land use change and rights also feature in newspapers; Hubbell (2021) reported on the illegal presence of private companies on state-owned land, highlighting instances of intimidation and murder of SPFT villagers in Klong Sai Pattana, Santi Pattana, and Khao Mai Pattana. Stokes (2017) detailed how private investors and politicians were using oil palm farmers as proxies to encroach onto protected peat forests in Chumphon.

Trends in the distribution of production

Following our literature review, we undertook a secondary analysis of quantitative data from national and international sources, focusing in particular on the past decade. Information concerning a number of relevant industry variables was examined. This included data on harvested areas, demographic and economic information concerning oil palm farmers, FFB production, CPO prices, and trade flows. This data was employed to gain insight into previous and current industry trends.

Since 2013, the harvested area of oil palm plantations in Thailand has increased from just over 600,000 hectares to almost one million hectares in 2022, a compound annualised growth rate of 5.0 percent per year. Key points relating to geographic distribution, patterns of industry development, and yield growth are illustrated in Figure 5.2. There are 132 oil palm mills in the country (DIT, 2021). However, because the popularly used Global Forest Watch (2022) Universal Mill List only counts about half of this number, the representation in Figure 5.2 is indicative but incomplete.

More than 85 percent of the country's oil palm production is located in the mid-south, along with a proportionate amount of mill capacity, as the oil palm FFB has to be speedily processed. The provinces of Surat Thani, Krabi, and Chumphon supply 60 percent of FFB. On the value chain (Figure 5.3), there is limited information available on palm kernel crushing mills, but these should be near mills or refineries (a total count of 22 nationwide). Several refineries, including the top three, are mostly located in or around Bangkok and Chonburi, with some in the southern province of Chumphon. The export of Thai CPO to India and elsewhere in 2023 was largely via Phuket, which has a deep-sea port. There are 15 biodiesel plants, with the top five mostly located near Bangkok (and some in the mid-south, for example, New Biodiesel). In 2023, biodiesel plants were running with a low 30 percent utilisation rate and by all accounts, this segment is believed to have a negative outlook in the context of Thailand's alternative energy policy outlook (Participant 7, Thai palm oil policy committee member, October 23, 2023; Sowcharoensuk, 2023).

The Thai palm oil upstream sector is dominated by smallholders and pro-smallholder policies, in contrast to its neighbours, Indonesia and Malaysia. Thai palm oil mills and refineries are also smaller sized on average. But despite the big palm duo's corporate-concessionaire (efficiency-driven) dominant model, the key CPO per hectare yield outcomes of all three countries have apparently converged and are startlingly similar now.

Outside of this core mid-south, upstream and milling infrastructure remains limited. There are a few mills in the central region, with some oil palms opportunistically planted in citrus zones and benefiting from existing irrigation, yielding over 30 FFB tonnes/hectare (Participant 7, Thai palm oil policy committee member, October 23, 2023) and at least one in the northeast (see Figure 5.2). There are fewer mills in the country's "deep south" and by 2021, only 1.2 percent of Thai FFB came from Yala, Pattani, and Narathiwat.

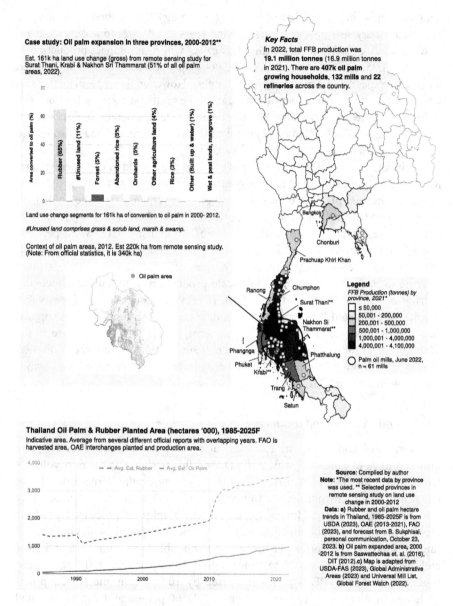

FIGURE 5.2   Map of oil palm areas, mills, expansion, and land-use change in Thailand.

Yield, volumes, and productivity

FFB yield across Thailand had grown to about 19 million tonnes in 2022, from approximately 12.4 million tonnes in 2013, while CPO production reached 3.6 million tonnes in 2022 (OAE, 2022; DIT, 2023). As is the case in other producing countries, yield in Thailand is sensitive to changing climate conditions, notably the extended dry seasons during the El Niño years of 2015–2016 and 2018–2019. Sowcharoensuk (2023) forecasts that this effect will again impact production in 2024 and 2025, estimating that FFB and CPO supply will decline by 2 percent and 4 percent, respectively due to extended droughts. However, they speculate that this may be offset if oil palm expansion policies in the northeast come to fruition. The impact of the trend observed by Sowcharoensuk (2023), whereby high prices are incentivising smallholders to cultivate oil palm instead of rubber, may also reduce the impact of El Niño-linked productivity falls.

A critical measure of the efficiency of oil palm is the yield of CPO per hectare. Recent data shows that Thailand's aggregate palm oil yield has risen to 3.5 CPO tonnes per hectare plateau. Figure 5.3 suggests convergence with the equivalent figures for Indonesia, Malaysia, and Colombia. This observation is particularly telling in the context of global discussions of the industry. CPO yield per hectare is often regarded as a proxy for industry efficiency. Smallholder palm oil producers are frequently presumed to be less efficient, a fact which is often referred to in the literature on small-scale production in Indonesia, especially (Sari et al., 2021; Soliman et al., 2016). However, the data here suggests that despite an estimated 90 percent of all production in Thailand via this supposedly sub-optimal agronomic model, the industry's productivity is close to the Malaysian and Indonesian benchmark, challenging assumptions about the performance of Indonesian and Malaysian corporate-centric as opposed to smallholder production models.

Trade trends for palm oil products

Between 2018 and 2022, the total and unit value volumes of Thai oil palm products have expanded significantly. During this period, CPO export volumes have risen from 260,000 to 870,000 tonnes, and crude palm kernel oil (CPKO) has nearly doubled from 60,000 to just over 110,000 tonnes. Unit prices have also risen dramatically. However, the export volumes of processed products are not growing at equivalent rates, nor are the unit prices achieving similar rates of increase. The price per tonne of CPO doubled between 2018 and 2021 and increased further in 2022 to almost USD1,300 per tonne. In the same year, processed palm oil (PPO) prices increased by 107 percent and CPKO prices increased by 63 percent, reaching about USD1,500 per tonne. Palm kernel meal (PKM) was fetching between USD200 and USD212 per tonne (a 108 percent increase). Processed palm kernel oil (PPKO) and other value-added ingredients achieved much smaller increases; PPKO increased by 39 percent, and animal feed ingredients such as palmitic and stearic acids, salts and esters were about USD2,000 per tonne, up by 10 percent.

In the same period, volumes of key palm oil imports have remained limited. In 2022, imports of palm kernel shells (PKS) approached half a million tonnes, up

Upstream

Harvested area, 2021, 2023E: 970k, 1.0 million ha
Area, 5-year and 3-year CAGR: 3.6% and 2.4% per year
Dominated by mid-south provinces, with 77% of national area in six:
Chumphon, Krabi, Nakhon Si Thammarat, Phanga, Surat Thani, Trang. In
Krabi, Nakhon Si Thammarat, Surat Thani (51% of area), land use change:
6% from forest, mangrove, peat & wetlands, 65% from rubber, 2000–2012.

FFB of oil palm, 2021, 2023E: 16.9, 19.3 million tonnes
2011: est. 70% smallholders, 30% plantations. 2022, Univanich: est. 90%.
10% of CPO produced from smallholders and own estates crops.

Traders & transport of FFB. Mostly via collection centers/ramps
(separating & cleaning), run mostly by intermediaries and some by mills.

Smallholders, 2022: 407k households on est. 814k ha
from OAE est. ave. 2 ha per household; growth via self-funding, bank
funding and Rubber Authority of Thailand incentives.
Plantation corporations: area, not reported. Author est. 150k (data above).

CPO yield, 2022: Est. 3.5 tonnes/ha, a convergence to the same national
aggregate yield outcome as Indonesia, Malaysia & Colombia, which boast
bigger (corporate) efficient segments and better agronomic potential.

Downstream processing

Domestic CPO consumption, 2022: 2.17 million tonnes
(Domestic demand is considered to be maximised)

To produce biodiesel: 0.92 million tonnes (42%)

To produce RPO: 1.25 million tonnes (58%).

22 refinery plants, 2022: With annual capacity of 2.5 million tonnes of
refined palm oil (average 115k tonnes each) and 1.0 million tonnes of RPO
production. Smaller capacities may allow Thai supply to more easily meet
EUDR traceability needs than large refineries in Indonesia and Malaysia.

15 biodiesel plants, 2022: With annual capacity of 10.26 million litres/day
and production of 3.81 million liters/day. But this sector is falling out of
favour due to its high subsidy cost plus policy for EV and rail transport.

Mid-stream processing

CPO processing, 2022: 132 palm oil mills
With 5.6 million tonnes CPO capacity. Typical mill size, est. ave. 45 tonnes/hour (FFB
processing) for annual capacity of 42k CPO tonnes from 212k FFB tonnes.

CPO production, 2021, 2023E: 3.04, 3.45 million tonnes
With 13 depository warehouses (for CPO & other products) with real-time monitoring of
stocks and 1+ month buffer stock criteria. Mainly for cooking oil and mandated biodiesel
(with no/limited international supply chain and less sensitive to certification).

**Est. 60% Thai palm oil mills are equipped with methane capture. Biogas engines
may be up to 10MW,** boosted with the use of sugarcane biomass.

CPKO production, 2021E, 2023E: author est. 0.45, 0.52 million tonnes
With PK from mills to PK crushing plants to extract CPKO.
For manufacturing, applications for home & personal care; specialty fats for food (with
international company and/or supply chains. More sensitive to certification need).

**Other products include: Shell & fiber for biomass, kernel meal for animal feed,
waste oil from sludge & waste water.**

Shipping & trade

Imports, 2019, 2021, 2022: 52k, 25k, 25k tonnes for refined oils & its products
Limited volume & declining trend; mostly from Indonesia and some from Malaysia. Plus,
palm kernel meal and palm kernel shells, 2022: 200k and 490k tonnes.

Exports, 2019, 2022, 2023E: 0.62, 1.25, 0.88 million tonnes
Recent large expansion of CPO export almost entirely to India. Adds to small rising RPO
exports to Myanmar and other Asian importers. Notable policy trend away from domestic
price and biodiesel subsidies toward export promotion depending on stocks situation and
with longer term value-added goal. With burgeoning interest in compliance with EUDR
and international standards; including enquiries to RSPO from farmers and millers.

Source: Compiled by author. Note: The Thai palm oil upstream sector is dominated by smallholders and pro-smallholder policies, in contrast to its neighbours, the nearby core palm producers Indonesia and Malaysia.
Thai palm oil mills and refineries are also smaller sized on average. But despite the big palm duo's corporate-concessionaire (efficiency driven) dominant model, the key CPO per hectare yield outcomes have
apparently converged and are startlingly similar now. Data: Author's estimates; interviews with Participants 5, 6, 7, 8, and 9 (October 19, 23, November 1 and December 7, 2023); Customs Department of Thailand
(2023), Dallinger (2011), DIT (2023), OAE (2020-2023), OAE (2020), Sowcharoensuk (2022, 2023), USDA-FAS (2023).

FIGURE 5.3   Thailand's palm oil value chain.

from a quarter million; and PKM imports doubled to nearly 40,000 tonnes. Imports of other animal feed ingredients grew but at a lower rate than PKM imports. Imports of other oil palm products declined. The unit prices of Thai imports for 2022 show rates of increase on a par with those achieved for exports. PPO import prices rose to USD1,350 per tonne (up from USD860 in 2021); PKM was USD200 per tonne. Animal feed ingredients rose to USD2,240 per tonne (up 12 percent from 2018). The almost equivalent prices achieved for imports and exports suggest that the patterns in these may reflect individual buyer behaviours and local factors rather than aggregate responses to price signals.

Trends in the total value of imports and exports in and out of Thailand are depicted in Figure 5.3. Sowcharoensuk (2022) cites industry players reporting that between 2020 and 2022, the value of exports of key palm products has risen from 11 percent to almost 30 percent of the value of total production, with the remaining volumes used for domestic consumption and biodiesel production. This increase in exports resulted in an 18 percent decline in CPO buffer stocks between 2020 and 2021 (Sowcharoensuk, 2022). As of the end of 2022, these inventories had increased again to 101.6 percent. This resulted in stocks of some 0.35 million tonnes rising above the buffer inventory target of 0.25–0.3 million tonnes. Sowcharoensuk (2023) attributes this to a softening of domestic consumption and biodiesel demand. Up to around 2012, 70 percent of domestic CPO supply was channelled to the food, oleochemical, and refined palm oil industries (the key Thai manufacturing segments include pharmaceuticals, ice cream, sweetened condensed milk, instant noodles, soap, and skincare); this consumption steadily declined over 10 years, to just under 50 percent by 2020, before rebounding to 60 percent in 2023. Inversely, supply to the biodiesel industry increased from about 30 percent in 2010 to almost 60 percent in 2020, subsequently declining to 40 percent in 2022 (Sowcharoensuk, 2023). Domestic consumption is considered "maxed out" now (Participant 7, Thai palm oil policy committee member, October 23, 2023), giving rise to an increased desire to develop export markets.

Since 2017, the Thai surplus in the palm oil trade has continued to grow substantially in value terms. With the exception of a dramatic reduction in 2015–2016 (likely due to El Niño climate conditions and domestic inventory priorities), CPO is by far the largest component of Thai palm oil export earnings (see dark solid line in Figure 5.4). Currently, this appears to be increasingly the case. In 2022, Thai palm exports were worth USD1.5 billion, whereas imports were only worth USD120 million. Exports of CPO in that year surged as traders opted to export at rates that were favourable to those possible in well supplied and policy-constrained domestic markets. In contrast, the value of exports of CPKO and PPO experienced only small increases in 2022, while exports of processed PKO dropped. In 2022, PKS was the top import at USD60 million.

The expansion in Thai palm oil exports, evident since 2017, has been dominated by CPO (see Figure 5.4). The surge in CPO exports in 2022, in particular, went primarily to the Indian market. This occurred when Indian buyers had issues sourcing via Indonesian supply chains when shortages of domestic supplies of

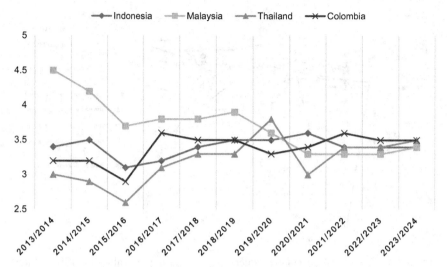

FIGURE 5.4   Yield trends of the top four palm oil producing countries. CPO per hectare yield in Thailand has caught up with others over ten years despite its reliance on farmers. Compiled by authors using USDA-FAS (2023) data.

cooking oil in that country were caused and exacerbated by Indonesian export and domestic supply policies. A three-week Indonesian export ban (Luxton & Ng, 2022) resulted from political pressures.[1] This, in turn, led to an increased need for importers in India to source supplies elsewhere. This opened the door for Thai CPO. In 2023, India and Thailand formally signed a memorandum of understanding (MOU) covering palm oil. Sudhakar Desai, President of the Indian Vegetable Oil Producers' Association (IVPA), commented on how Thailand had helped meet Indian demand during the Indonesian ban, stating that Thailand has emerged as a natural supplier of oil palm to India (Vinayak, 2023). Highlighting the logistical advantage of exporting from Thailand, BV Mehta, Executive Director of SEA of India, said a voyage from Phuket to any southern port in India was hardly 4–5 days, while the equivalent trip from Malaysia or Indonesia would take around 10–12 days. Regardless of the actual significance of this journey duration as a trade barrier, fostering trade links with the Thai palm oil sector is certainly consistent with current Indian policies aimed at diversifying supply to enhance the country's edible oil security (see for example Nema & Ansari, 2023). While growth in Thai exports benefitted from the opening up of readily accessible markets in India. Policy changes in Thai biodiesel and inventory management also helped, effectively leading to more CPOs being made available for export (Sowcharoensuk, 2023). At the same time, the country's already small volume of palm oil imports has fallen, being tightly regulated via levies tied to domestic inventories. Effectively, therefore, the National Oil Palm Policy Committee has implemented a protectionist policy regarding imports via the Public Warehouse Organisation (the country's sole palm oil importer).

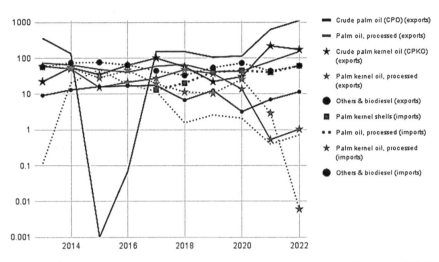

FIGURE 5.5 Selected Thai palm products, export and import value (in USD million) between 2013 and 2022. Note: Solid lines for export products and dashed lines for import products. Others comprise palm kernel meal & animal feed ingredients. Palm kernel shells emerged as an import item in 2017, 180,000 tonnes and are now about half a million tonnes, worth about USD60 million (see grey short-dash line with square icons). Compiled by authors using data from the Thailand Customs Department (2022) and UN COMTRADE (2022), accessed July 7, 2023.

Domestic prices and policy

Local wholesale prices in Thailand diverge significantly from those in Malaysia and Indonesia. Historically, Thai prices have tended to be higher than those found in the core producing countries (Figure 5.5). This reflects the significant involvement of the state, effectively providing a degree of protection to Thai oil palm farmers (causing mill owners to complain bitterly on occasion, and even shut down mills on negative margins). In recent years, palm oil pricing policies have changed in ways that have reduced the level of support Thai domestic CPO prices enjoy. For example, since February 2019, the Department of Internal Trade (DIT) has allowed the price of bottled refined palm oil to float freely and for the market to set its retail price. In the past, a price ceiling of 42 Baht (USD1.26) per bottle was set (Sowcharoensuk, 2022). Following the changes instituted in 2019, retail prices troughed in 2019 (at just over 30 baht (USD0.96) per one-litre bottle, while wholesale prices were several baht lower) before increasing sharply. In mid-2022, these shot up to about 70 baht (USD1.96) before settling at 45 baht (USD1.26) by September 2023 (with minimal margins between the two prices). The cumulative impact of these measures is that the prices achieved by Thai CPO producers are converging with the international (Malaysian) benchmark. Based on our desk survey, we found that Thai CPO prices were 28 percent higher than the benchmark in

2020, 9–10 percent higher between 2021 and 2022, and only 4 percent higher for most of 2023. In comparison, Indonesia's domestic CPO price estimates were 91 percent versus Malaysia's benchmark in 2020, 79 percent in 2021, 75 percent in 2022, and 87 percent in most of 2023. Pro-farmer prices in Thailand were a major contrast from policy in the core countries. With the rising comparative efficiency of Thai farmers (and heightened international commodity prices, for instance CPO at USD800 versus USD500 per tonne from late 2022 versus before mid-2020), it appears that Thai policymakers are now shifting to a new regime in which the level of local price support is reduced and additional financial, administrative, and political support for exports is offered. It certainly appears to be the case that Thai farmers are efficient enough to compete in export markets. For example, Thai palm oil production cost was 17.6 baht per kg (USD528 per tonne CPO) in 2022 (estimates from FFB cost in Sowcharoensuk, 2022). That same year, the Malaysian palm oil production cost was estimated at RM2,000–RM2,300 (USD440–USD506) per tonne CPO (Participant 10, corporate analyst, February 21, 2024). Thai CPO prices remained at or above the levels achieved by Malaysian producers, suggesting that Thai production is very competitive. While the cost basis may differ between the estimates, cost convergence towards Malaysia may give players and policymakers

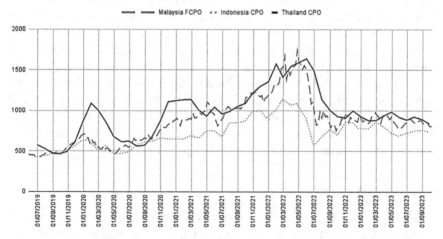

FIGURE 5.6   Domestic CPO prices Thailand, Malaysia and Indonesia 2019–2023 (USD/ tonne). Note: Solid line for Thai domestic CPO monthly price (Department of Internal Trade), dashed line for proxy benchmark for Malaysia CPO, the FCPO weekly price (Bursa Malaysia derivatives, based on its first position futures which is a key international benchmark) and dotted line for Indonesia weekly domestic CPO price (weekly tender prices from state-owned enterprise involved in commodity trading, PT Kharisma Pemasaran Bersama Nusantara or KPBN). Compiled by authors using data from KPBN (2022), Thailand Department of Internal Trade (2023), and Bursa Malaysia (2022), accessed July 10, 2023.

in the Thai oil palm industry cause for concern over time, having the potential to place further downward pressure on Thai CPO prices.

Actors in the Thai oil palm assemblage: Upstream, downstream, and policy players

While Thailand remains a relatively small player in the global palm oil industry, the analysis presented above shows that this role is growing. Since 2017, Thailand has seen a rise in the volume and value of oil palm exports and a decline in the value of both imports and the proportion of total production consumed domestically (of PPO and PKO in particular). A combination of push factors in the form of available stocks and domestic policies, and pull factors in the form of expanding and improving overseas markets and prices are serving to foster the country's emergence and development as an oil palm exporter. Furthermore, the analysis presented above provides no strong evidence that the different structure of the palm oil sector in Thailand adversely impacts the industry. Currently, yield rates and prices in Thailand are broadly consistent with those achieved by producers in Malaysia and Indonesia, if not better. Furthermore, it is clearly the case that smallholders in Thailand have demonstrated an ability to adopt the crop, integrating their production into the downstream elements of the value chain despite lacking the kinds of "supported smallholder schemes" deemed vital to the development of the palm oil sector in Malaysia and Indonesia.

Having discussed the Thai palm oil industry's performance, in this section we focus more detail on some of the key players within the industry. We identify and describe actors involved in the upstream and downstream parts of the palm oil industry, from primary and secondary production, through processing, sale and distribution, as well as those linked to the industry via the policy practices tied to it. Having identified the actors involved, we then consider how these actors have contributed to the current shape of the Thai oil palm assemblage.

Farmers/Primary producers

Smallholders dominate Thai oil palm production, and as of 2021, 85 percent of the smallholders were located in the mid-south, of which 60 percent were from the provinces of Surat Thani, Krabi, and Chumphon (OAE, 2022). While the imprecision of the term "smallholder" makes direct comparison challenging, Thai smallholders broadly appear to own and manage larger areas of land than that are comparable to those of their Indonesian and Malaysian counterparts (averaging 11 hectares each), but with average smallholder oil palm plantings that are less than those reported for either Malaysia or Indonesia. In 2019, the Office of Agricultural Economics (OAE) reported an average of two hectares of oil palm per household (Bureau of Agricultural Economic Research, 2020). This is smaller than the average of 2.4 hectares claimed for Indonesian smallholders (Badan Pusat Statistik,

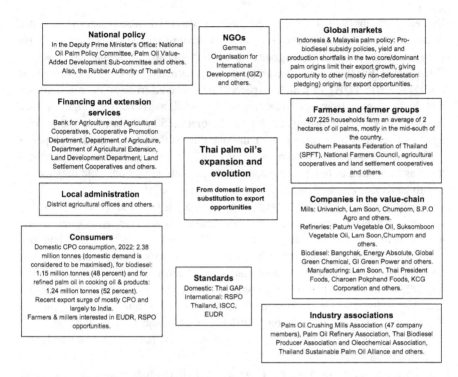

National policy
In the Deputy Prime Minister's Office: National Oil Palm Policy Committee, Palm Oil Value-Added Development Sub-committee and others. Also, the Rubber Authority of Thailand.

NGOs
German Organisation for International Development (GIZ) and others.

Global markets
Indonesia & Malaysia palm policy: Pro-biodiesel subsidy policies, yield and production shortfalls in the two core/dominant palm origins limit their export growth, giving opportunity to other (mostly non-deforestation pledging) origins for export opportunities.

Financing and extension services
Bank for Agriculture and Agricultural Cooperatives, Cooperative Promotion Department, Department of Agriculture, Department of Agricultural Extension, Land Development Department, Land Settlement Cooperatives and others.

Thai palm oil's expansion and evolution

From domestic import substitution to export opportunities

Farmers and farmer groups
407,225 households farm an average of 2 hectares of oil palms, mostly in the mid-south of the country.
Southern Peasants Federation of Thailand (SPFT), National Farmers Council, agricultural cooperatives and land settlement cooperatives and others.

Local administration
District agricultural offices and others.

Companies in the value-chain
Mills: Univanich, Lam Soon, Chumporn, S.P.O Agro and others.
Refineries: Patum Vegetable Oil, Suksomboon Vegetable Oil, Lam Soon, Chumporn and others.
Biodiesel: Bangchak, Energy Absolute, Global Green Chemical, GI Green Power and others.
Manufacturing: Lam Soon, Thai President Foods, Charoen Pokphand Foods, KCG Corporation and others.

Consumers
Domestic CPO consumption, 2022: 2.38 million tonnes (domestic demand is considered to be maximised), for biodiesel: 1.15 million tonnes (48 percent) and for refined palm oil in cooking oil & products: 1.24 million tonnes (52 percent). Recent export surge of mostly CPO and largely to India.
Farmers & millers interested in EUDR, RSPO opportunities.

Standards
Domestic: Thai GAP
International: RSPO Thailand, ISCC, EUDR

Industry associations
Palm Oil Crushing Mills Association (47 company members), Palm Oil Refinery Association, Thai Biodiesel Producer Association and Oleochemical Association, Thailand Sustainable Palm Oil Alliance and others.

FIGURE 5.7   The assemblage of organisations and entities connected to the Thai palm oil sector development in different ways. Compiled by the authors.

2023) and Malaysian certified smallholders (Wan, 2022), and significantly smaller than the Federal Land Development Authority (FELDA) allocation which averaged four hectares (Khor et al., 2015). The extent and significance of these differences are however difficult to establish, given the loose and in some cases politicised use of the term smallholder to apply to a wide range of different types and scales of palm oil farming enterprises in different national territories (see for example Jelsma et al., 2017; Lee et al., 2013). The picture is further complicated by the challenges of obtaining comparable statistics and local context. For example, Thai families have an incentive to divide the ownership of lands between different family members as the government provides support services to landowners who have less than three hectares (Participant 6, representative of a Thai palm oil company, October 19, 2023).

Some useful insights into the Thai smallholder sector are to be found in a 2010 baseline study of 500 oil palm smallholders in the south and east of Thailand (Thongrak et al., 2011). This focused on a selection of larger farmers in the provinces of Krabi, Surat Thani, and Chumphon. On average they had four family members, 15 years of planting experience, and an average of almost 11 hectares per family, with seven hectares of oil palms. This study found that smallholders rely on household labour (93 percent), but the majority also used hired labour (81 percent). A large portion of these labourers were likely landless tribal people from the north of Thailand, as well as poor migrants from Cambodia and Myanmar (Colchester & Chao, 2011). For these larger farms, households earned about USD13,000 per year (USD1 = 35.75 THB), of which 60 percent was derived from oil palm production and the remainder from other farming sources, such as rubber planting, livestock raising, and vegetable/fruit tree growing. Among the key factors that had encouraged smallholders to commence oil palm planting on their land include high prices of FFB, which pointed to high returns; the relative ease in cultivation and management as compared to other crops; appropriate climate conditions; and the propensity to generate early and ongoing returns (Thongrak et al., 2011).

The behaviour of smallholders adopting palm oil in Thailand appears to embody an entrepreneurial approach to crop selection and income generation within multi-income livelihood strategies. Thai smallholders plant other crops, and also work off-farm in trade and other businesses, pursuing a multiple-income livelihood strategy (Participant 6, representative of a Thai palm oil company, October 19, 2023). Something that has been widely observed in smallholder communities more generally, their choice of crops is based on their potential to generate income as opposed to for their direct consumption as part of subsistence strategies. This is consistent with smallholder practices elsewhere, where oil palm planting is part of a mixed enterprise (O'Reilly et al., 2020). It suggests, among other things, that decisions regarding oil palm adoption involve their incorporation into existing mixed livelihoods based on their income-generating potential when compared to alternatives. This thinking is reflected in the farmers' decisions to plant oil palm.

Government actors and supply side measures

Oil palm smallholders in Thailand are mostly self-funded (Participant 6, representative of a Thai palm oil company, October 19, 2023). However, the government has also incentivised planting through policies that have involved expansionary initiatives and the provision of technical support to grow supply, with smallholders making widespread use of these supports to augment their own resources. For example, in 2007, the Ministry of Agriculture and Cooperatives (MOAC) and Ministry of Energy introduced the Thai Oil Palm and Palm Oil Industries Development Plan 2008–2011, which aimed to increase new plantings by 80,000 hectares annually, raise oil extraction rates to 18.5 percent, and grow FFB yield to 21 tonnes per hectare (Dallinger, 2011). Support for oil palm cultivators include measures to make farming inputs more accessible through special credit cards introduced by the government and the Bank for Agriculture and Agricultural Cooperatives (BAAC) in 2015 (Nupueng et al., 2022). The BAAC also provides loans for oil palm farming management and the purchase of relevant assets such as vehicles and land. Thongrak et al. (2011) found that almost 70 percent of farmers relied on loans from the BAAC. Land Settlement Cooperatives (LSC), a type of agricultural cooperative established after farmers with insufficient or no land were allocated land, also appear to be important. LSCs provide credit services, offer secure access to markets, a stable supply of affordable inputs, and tailor-made fertilisers, in efforts to encourage smallholders to adopt RSPO practices (Rodthong et al., 2020).[2] The Rubber Authority of Thailand has also played a key role, providing incentives to farmers of approximately USD2,756 per hectare over a three year period for rubber to oil palm conversions. This would be sufficient to cover the coswt of immature plantings (Participant 5, representative of a Thai palm oil company, October 19, 2023).

Besides finance, the Thai state has acted to provide technical support to smallholders. Once smallholder plantations or cooperatives are established, smallholders are occasionally visited and offered assistance by a number of government bodies. These include the Cooperative Promotion Department (CPD), the Land Development Department (LDD), the Department of Agriculture (DOA), and the Department of Agricultural Extension (DOAE) (Rodthong et al., 2020). These government bodies fall under the purview of Thailand's MOAC. The CPD is responsible for promoting and supporting cooperatives, as well as improving learning and training to enhance operational efficiency. The DOAE provides services and disseminates knowledge on modern agricultural practices to raise farm income and standards of living. The LDD is responsible for land use matters, including land use change as a result of oil palm expansion. District agricultural officers and CPO mills also play a significant role in smallholder expansion through knowledge sharing and training. From available studies, it is unclear how often these departments visit farmers or assess the impact extension services have on farmers' output, income, and livelihoods. However, it is noteworthy that in the Thai case, a range of state bodies are directly involved in the delivery of technical support. Such advice includes issues linked to commercial and land management as well as the provision of agricultural advice. This is in marked contrast to the situation in the core countries, where the delivery of such support is often linked to participation in "supported" schemes and where much of the "commercial" management expertise in an area remains vested in plantation companies; in the case of Indonesia, and/ or FELDA in the case of Malaysia—essentially via private corporate mill owners or government-linked corporate mills arranged through a monopolistic licensing system aimed at reducing excessive competition in the mill sector. Notably, Thai oil palm mills are on average smaller than those in Malaysia and Indonesia and lack their industry heft (Participant 7, Thai palm oil policy committee member, October 23, 2023).

Many of the support strategies made available by the Thai state and governmental entities have targeted smallholders directly. At least on paper, the justification for this approach is that palm oil was regarded as a high-value crop to support smallholder income growth. This emphasis differs in subtle ways from Malaysian and Indonesian policies and appears to have emphasised the development of the industry as a pro-smallholder ideological programme. Evidence from our interviews and from comparisons with supported schemes elsewhere suggest that in terms of the scope of freedom of action they enjoy, the situation of Thai small-holders may compare favourably to that of smallholders in Malaysia and Indone-sia, particularly to participants in supported schemes in those states there that are tied to specific sources for inputs and sales via a single large oil palm company. Many observers have noted that Thai smallholders have many choices available to them (Nupueng et al., 2022; Participant 5, representative of a Thai palm oil com-pany, October 19, 2023; Participant 7, Thai palm oil policy committee member, October 23, 2023). Specifically, they are not tied to a single buyer or mill, so the model adopted here seems, in principle at least, to be more appropriate to the entre-preneurial and commercially opportunistic smallholder practices mentioned above. The Thai palm oil assemblage appears to incorporate a dense network of enabling bodies and actors, in the form of industry players, state, and parastatal bodies, as well as non-governmental organisations (NGOs) (Figure 5.7). The absence of tied arrangements and possible availability of multiple buyers means that so long as prices remain reasonable, Thai palm oil growers have the option to pick and choose where they buy inputs and sell their products. It might also be noted that as part of a mixed farm economy, they may not solely rely on FFB sales for income, giving them a degree of security and negotiating room that is not available to those who are entirely dependent on oil palm.

National monitoring, oversight, and demand side measures

The state provides monitoring, oversight, and industry regulation. At the district level, statistics on agricultural output are collected and collated into a national data-base on economic crops. However, in 2010, only 61 percent of oil palm farmers had registered in this system, pointing to a possible reason for Thailand's relatively poor database on oil palm cultivation at the regional and national levels (Thongrak et al., 2011). Some improvement has been reported in recent years, driven by efforts to enhance farm data through insurance schemes, incentives and disincen-tives (Participant 2, representative of an NGO promoting sustainable palm oil in Thailand, September 18, 2023).

The Thailand National Oil Palm Policy Committee is responsible for oversight of the industry, enhancing competitiveness and maintaining stability through the development of policies, allocations of harvest between household and industry needs, import controls and price interventions (Sowcharoensuk, 2022). The Deputy Prime Minister usually acts as chairperson, while other officials are from the DIT. Its members are made up of representatives from the Crushing Mill and Refinery Associations (Participant 7, Thai palm oil policy committee member, October 23, 2023). The Deputy Prime Minister's Office also includes the Palm Oil Value-Added Development Sub-committee and other relevant bodies. The industry is also regu-lated by other government entities, such as the Ministry of Industry, which sup-ports the food and oleochemical sectors, as well as the Ministry of Energy which oversees biodiesel production. These bodies, along with the MOC and MOAC, have largely facilitated the expansion of the palm oil industry through supportive policy implementation, pricing of products, altering biodiesel blending rates, monitoring tank storage, and reducing illegal imports from Malaysia.

A critical element of the state's regulatory efforts is their use of demand-side measures to maintain prices and stimulate additional demand. The Central Committee on the Prices of Goods and Services (operating under the DIT, MOC) sets the purchase price of FFBs (with reference to oil content) and the price of bottled refined palm oil, allowing prices to move with the cost of inputs. The price for CPO is set with reference to the cost of inputs and CPO prices on world markets (Sowcharoensuk, 2022). The financial drain of supporting prices (in the past) was considered significant (Participant 7, Thai palm oil policy committee member, October 23, 2023), but the totality of costs and who paid (fiscal, consumer, other) is hard to find. Among the costs were these: (1) in 2022/2023, the MOC is estimated to have incurred costs of USD177 million to support the yearly price guarantee scheme for farmers, with the minimum price of fresh palm fruits set at USD110 per tonne (Arunmas, 2022a); (2) the MOC supports CPO exporters at USD55 per tonne, premised upon stocks exceeding 300,000 tonnes and domestic prices exceeding that of the international market (Arunmas, 2022b); and (3) biodiesel deadweight losses.

Another aspect of state involvement in demand management has been via biodiesel production. In 2008, the Ministry of Energy introduced the Renewable Energy Development Plan (REDP) 2008–2022 through a mandated B2 blending rate up to 2010 and B5 by 2022 (Nupueng et al., 2018). This plan aimed at achieving 20 percent renewable energy consumption, with biodiesel demand reaching 4.5 million litres per day nationwide by 2022. The REDP explicitly targeted an increase in oil palm plantation area to 400,000 hectares by 2012. Subsequent revisions of this plan established the Alternative Energy Development Plan (AEDP) 2012–2021, which set biodiesel demand at 7.2 million litres per day and targeted 880,000 hectares of oil palm area nationwide by 2021. Through this plan, the government had hoped to balance biodiesel production with CPO supply (Nupueng et al., 2018). The AEDP was then further revised, creating the AEDP 2015–2036, increasing the total renewable energy consumption to 30 percent and biodiesel demand to 14 million litres per day. As a result, the targeted oil palm area increased again to 1.6 million hectares by 2036, with close to 6 million tonnes of CPO production annually and a balance supply of 4 million tonnes (before deducting export volumes). As a complement to this policy, the government upped its support to smallholders, distributing palm seeds, providing initial funding, discounting fertiliser costs, and establishing a central market to purchase FFBs (Srisunthon & Chawcha, 2020).

However, three years later, the Thai Cabinet and National Energy Policy Office (NEPO) revised the AEDP for 2018–2037, reducing biodiesel consumption targets by almost half to 8 million litres per day. These adjustments were made in light of the Power Development Plan 2018–2037, which reduced the proportion of biodiesel and ethanol consumption as the government shifted its focus to the promotion of electric vehicles (EVs) and rail transport (Sowcharoensuk, 2022). The changes were also reflected in targets for CPO volumes, which contracted to an average of 5 million tonnes annually, with remaining stocks at 3.5 million tonnes (before deducting export volumes and consumption by the oleochemical industry).

In general, the Thai biodiesel policy has proven expensive, its benefits unclear, made possible via costly cross-subsidies from petroleum products via the State Oil Fund (International Institute for Sustainable Development, 2013). A new National Energy Plan (NEP) is being floated, which will include a new AEDP that would further reduce annual biodiesel consumption targets (Prasertsri, 2022). Chenphuengpawn (2012) reported that allocations to prop up biodiesel had generated a deadweight loss of USD318 million between 2007 and 2011 and the Thai government enacted the State Oil Fund Act 2019 to control its liabilities and limit subsidies (to biodiesel and oil palm). Removing these subsidies has proven challenging; in September 2022, the Energy Minister announced that subsidies would remain in place until September 2024 in order to support farmers and encourage the use of greener fuels (Bangprapa, 2022). The hope is that the Thai palm oil sector can gain in value-add and become a self-supporting export driver without the need for expensive support from the Thai state, which would rather allocate fiscal support to rice and sugar farmers, a larger voter base than oil palm farmers (Participant 7, Thai palm oil policy committee member, October 23, 2023).

Corporations, refineries, and downstream

Plantation areas cultivated by Thai palm oil companies are modest in comparison to the estates of large companies in Indonesia and Malaysia, where corporate plantations are estimated at 9 million hectares and 4 million hectares respectively (Badan Pusat Statistik, 2023; Malaysian Palm Oil Board, 2022). In 2007, average land holdings by oil palm corporations in Thailand were estimated at only 796 hectares (Dallinger, 2011). Plantation estates are largely located in the mid-south provinces of Surat Thani, Krabi, Chumphon, and Nakhon Si Thammarat. The public-listed Univanich Palm Oil Company Public Limited pioneered Thailand's first oil palm plantations in 1969. The company owns and manages six plantation estates, covering an area of 5,971 hectares, almost 4,850 hectares of which are RSPO-certified. Univanich also has five mills, producing 230,000 tonnes of CPO annually (Univanich, 2022).

In contrast to Malaysia and Indonesia, where private companies have received very large concession areas, including tens to hundreds of thousand hectares (and sometimes in forested areas which have not previously been used for agriculture), such areas have not been available in Thailand. This may be due to the relatively greater capacity of Thai smallholders to exert political influence, and the availability of readily convertible rubber small holdings, meaning that less land is available to allocate to these companies. Furthermore, and as a consequence of this, there is less scope for them to assume a role in supporting smallholders via the nucleus, estate, and other supported smallholder schemes that have been widely developed in other countries. The absence of "tied" smallholders with contractual obligations to supply their FFB to specific mills undoubtedly results in a situation whereby the bargaining position of smallholders in Thailand is strengthened.

Further down the value chain are refineries. Few are vertically integrated, and as such, Lam Soon-UPOIC and Chumporn Palm Oil Industry Public Company Limited are notable. Lam Soon Plc has a 70 percent shareholding in United Palm Oil Industry Company Limited (UPOIC), with 3,800 hectares of planting area, all of which is RSPO-certified (Lam Soon Plc, 2022). In 2022, Lam Soon's Trang mill produced almost 200,000 tonnes of CPO and its Bangpoo refinery produced 192,000 tonnes of refined palm oil. Chumporn has over 3,300 hectares of oil palm plantations, almost 3,200 hectares of which are RSPO-certified, and a refining production that requires 180,000 tonnes of CPO per year (Chumporn, 2022)

Other refiners include Patum Vegetable Oil, Suksomboon, Pitak Palm Oil, and Morakot (owned by Malaysia's Sime Darby Plantations). As Thai vegetable oil refiners have historically produced for the domestic market, observers note that they are smaller in scale than the giant export-driven refineries of Malaysia and Indonesia (Participant 7, Thai palm oil policy committee member, October 23, 2023). Downstream players manufacture for food and feed applications (including Lam Soon, Thai President, Charoen Pokphand Foods, KCG Corporation, GFPT Public Company Limited, and Chaveevan International Foods; plus Associated British Foods, Ajinomoto, Cargill, Ezaki Glico, Mars, Namchow and other foreign players, as well as ingredient suppliers like Fuji Oil), home and personal care (Beiersdorf, Namchow, Kao, and suppliers such as Badische Anilin und Soda Fabrik), and also biodiesel producers (including Global Green Chemical, Patum, New Biodiesel, BBGI Biodiesel—formerly Bangchak, AI Energy, Energy Absolute, and GI Green Power) which may supply the energy and chemical producers (state-owned big three PTT Global Chemical, ThaiOil, and IRPC Public Company Limited, Chevron-owned Star Petroleum Refining, ESSO, and state-founded Bangchak Petroleum and RPCG Public Company Limited).

Other organisations

NGOs and parastattsals involved in the palm oil industry in Thailand include overseas entities that reflect the broader global tendency for such organisations to focus on sustainability and "global" environmental goals rather than productivity and "national development goals", a particular focus being the drive for "sustainable" palm oil. The German Organisation for International Development (GIZ) has worked closely with the Thai government for over a decade to enhance sustainable palm oil production among smallholders. Along with private companies, the GIZ supported the first RSPO pilot project for smallholders, leading to the certification of 412 farmers in 2012 (Nupueng et al., 2022). These growers were the world's first RSPO independent smallholder group. At present, the GIZ continues to work with Thai authorities such as the DOA and DOAE to increase RSPO certification among farmers, collaborating with the private sector to increase demand for sustainable palm oil, formulating policy towards a "jurisdictional approach" for sustainable production and developing, monitoring, and reporting mechanisms for greenhouse gas reduction potential among smallholders who apply sustainable practices (GIZ, 2018).

In 2022, the RSPO and the Thailand Environment Institute launched the Thailand Sustainable Palm Oil Alliance (TSPOA), a platform dedicated to enhancing cooperation among stakeholders along the supply chain and promoting sustainable palm oil, from the demand and supply sides (RSPO, 2022). In its listed objectives, the Alliance aims to leverage policies and set standards for sustainable production and consumption towards international requirements. It is unclear how effective their efforts have been over the past years; however, at the same time, the close involvement of a number of large Western firms in the sustainable palm oil drive in Thailand has led to concerns about greenwashing. While their objectives, if effectively carried out, could help farmers and corporations in meeting the European Union Deforestation Regulation (EUDR) standards, in line with experience elsewhere, there is significant uncertainty over the cost and benefits that accrue to smallholders from certification (see for example De Vos et al., 2021). Efforts to promote RSPO certification are unlikely to prosper if uncertified markets are available to producers and the relative payoff from certification is neither clear nor sufficient to justify the additional effort (particularly where palm oil is only a part of a broader income strategy).

Discussion

The preceding sections highlight significant differences and similarities between the Thai oil palm sector and those in Malaysia and Indonesia. While the big two producers have much larger areas of suitable land, better climate conditions (while Thailand is at a higher latitude with a longer dry season), and more capacity to meet the growing requirements for the oil palm on a sufficient scale to make the crop a viable option, Thailand does contain suitable areas to cultivate oil palm. In addition, all three countries share similar historical and developmental challenges, with large rural populations for whom agricultural development offers an important mechanism to improve livelihoods and wellbeing. They also share specific areas in which current patterns of land use are perceived as being economically sub-optimal when compared to the returns attainable via palm oil. Given these facts, it is scarcely surprising that Thailand has sought to adopt the oil palm, and has done so on such a significant scale that it was able to step in to fill in the gap left by Indonesian suppliers when that country paused exports to one of the key markets, India, in 2022.

The oil palm assemblage in Thailand bears many similarities to that of Malaysia and Indonesia. The landscape impacts of the crop are similar—in the form of large areas of palm oil and its supporting infrastructure of roads and palm oil processing facilities. In the previous section, we briefly examined some—by no means all—of the various human actors in the Thai oil palm industry, including smallholders, government bodies, policymakers, palm oil companies, millers, and international bodies. Again, many of these actors are to be found in all three countries. The similarities extend to some key economic data. Yield per hectare is unexpectedly similar, with Thailand achieving better than expected yields (despite its agronomic and production scale disadvantages), and its farmers enjoying higher prices per tonne than those in the core countries.

However, as we look at the Thai oil palm assemblage, it is clear that there are features that appear to diverge significantly from those found in the Malaysian and Indonesian oil palm industries. This not only concerns relationships between human agents within these assemblages, where perhaps such variation is most reliably to be expected, but also in relation to the multiple non-human actors that are necessarily transformed and re-defined via their incorporation into the palm oil assemblage. The most obvious, in this context, might be the impact on the land itself. Oil palm requires large swathes of land, lending to distinctive monoculture plantations. Actors in the Malaysian and Indonesian oil palm assemblage have come together to make this possible on a very large scale, and in the process, implying as a necessity the emergence of very large organisations from both the private and the public sectors. By contrast, patterns of land use change in Thailand differ from what has been observed in Indonesia and Malaysia. As opposed to the large tracts deemed necessary for a highly centralised commercial palm oil industry in those countries, Thai palm oil plots are smaller and appear to have involved changes in agricultural land use by current landholders to a very large extent with comparatively limited (although by no means absent) evidence of the large scale conversion of wetlands and forests.

Evidence from Thailand suggests that there has been relatively limited direct deforestation, compared to Indonesia and Malaysia. A 2016 study examined land use changes in the Krabi, Surat Thani, and Nakhon Si Thammarat (Tapi River basin) provinces, an area that has almost 496,000 hectares of oil palm plantations and produced 51 percent of Thailand's FFB output in 2021 (OAE, 2022). In this key oil palm zone, there were direct land use (gross)[3] changes of 1 percent of wet/peat/mangrove land, 5 percent of forest, and 11 percent of unused (marsh, grass, swamp) land between 2000 and 2012; so, by 2012, oil palm had displaced some areas of abandoned rice fields (5 percent) and orchards (5 percent), but the majority of land use change or 65 percent of new oil palm had displaced rubber plantations (Saswattecha et al., 2016). This contrasts with figures for the same land use changes in Indonesia, where forest conversion accounted for some 18 percent of a much higher area of land between 2001 and 2010, with agricultural conversion accounting for just over half that of the Thai rate (33 percent).

These figures appear to suggest that the key landscape impact of the oil palm assemblage in Thailand has been its incorporation of areas previously utilised for rubber-centric farming[4] into one that includes oil palm alongside other crops and livelihood activities. The literature suggests that this process has been led by farmers, relatively small processing companies, and low-level intermediaries supported by the state. Whether accurate or not, the perception that Thai palm oil is smallholder-led and chiefly involves rubber-to-oil palm conversions has simultaneously spared Thailand from the deforestation narratives associated with Malaysian and Indonesian palm oil and the pervasive accusations of land grabbing that have dogged the industry in both those countries. As such, the Thai industry is well-placed to claim to be a provider of sustainably-sourced palm oil.

For similar reasons, to date, Thailand can also claim a good record in relation to another contentious aspect of its Southeast Asian neighbours' oil palm assemblages: the use of peat and wetlands in oil palm production (Stokes, 2017; Dallinger, 2011). Nevertheless, we have found evidence of some oil palm encroachment onto Thailand's limited areas of peat swamps and wetlands. In the south, for example, the United Nations Development Programme (UNDP) found that 1,000 hectares of the Kuan Kreng peat swamp, in Nakhon Si Thammarat, had been converted into oil palm plantations by 2014, lowering the groundwater table to 20–70 cm below the soil, and causing fires across the province which affected a further 3,200 hectares of peatlands (UNDP, 2014). Also, in the deep southern province of Narathiwat, bordering Malaysia, oil palm planting converted 10 percent (2,446 hectares) of the Princess Sirindhorn Wildlife Sanctuary wetland area between 2009 and 2016 (Srisunthon & Chawchai, 2020). During the same period, indirect land use change linked to oil palm expansion had caused displaced rubber to be planted on approximately 5 percent of the wetland (1,128 hectares) (Isranews, 2016). However, the rates are much less alarming than those of Indonesia and Malaysia, and there is evidence that steps have been taken to reverse these encroachments (Isranews, 2016).

The Thai palm oil assemblage has also enjoyed a somewhat different relationship to questions of land rights compared to Malaysia and Indonesia. In these two countries, questions of land grabbing have been frequently raised in a context where corporate players were the recipients of very large concessions, and transmigration programmes have been seen as fundamentally undermining traditional forms of land ownership and use (Widyatmoko & Dewi, 2019). As is the case in Indonesia and Malaysia, land titling problems across farming communities in Thailand are an important and fraught issue. However, the issue and its working out in Thailand has a different history and trajectory. Protests and uprisings in the early 1970s pressured the government to pass the Agricultural Land Reform Act 1975 which established the Agricultural Land Reform Office (ALRO), tasked with redistributing private and state-owned unused land to families with insufficient or no land for agricultural purposes (Pobsuk, 2019). From 1975 to 2003, ALRO had successfully allocated 3.7 million hectares of state-owned land, particularly degraded forests, to 1.5 million farmers who received freehold titles or legal use-rights (Dallinger, 2011).

Prior to this legislation, the Thai government had enacted the National Reserved Forests Act 1964; Section 16 of the Act made it possible for the government to hand out parcels of degraded forest to the private sector in the form of 30-year plantation concessions. This, in theory, would allow the commercial exploitation of land by corporate players on a larger scale. A coordinator at Land Watch Thailand claims that officials had used these provisions to sell lands to powerful individuals for as little as USD0.30 per *rai* (equivalent to 0.16 hectares), who then began cultivating oil palm (Hubbell, 2021). These plantations have remained—and even expanded— beyond the 30-year limit, leading to lengthy and dangerous land disputes between plantation companies and farmers. Hall (2011, p. 837) argues that the government facilitated the expansion of these oil palm concessions on state-owned land to enhance "export-oriented crops under boom conditions". However, interviews undertaken with processors and industry figures in key regions during this research suggest that this was an issue for a minority of areas and farmers.

This is not to suggest that land disputes are not present in the Thai oil palm story. Pobsuk (2019) and Hubbell (2021) cite long-running disputes in the south. In 2003, the Southern Poor People Network (SPPN) was established and they discovered expired palm oil concessions on more than 11,200 hectares (Pobsuk, 2019). As a result, the network occupied these lands, demanding that the government repossess them from companies and distribute titles to landless farmers. Later that year, armed officials cracked down aggressively on SPPN groups. Learning from the shortcomings of their predecessors, the SPFT was formed in 2008 with the aim of establishing a land rights movement in the Surat Thani province (Pobsuk, 2019). Landless farmers and wage labourers joined the Federation, galvanising collective action in efforts to obtain community rights to land and natural resources. As a consequence, a flurry of land rights campaigns has occurred with three out of SPFT's five farming communities embroiled in land disputes with palm oil companies (Hubbell, 2021).

These campaigns have led to serious conflict and violence. On 20 October 2020, land rights defender and SPFT member, Dam Onmuang, survived an assassination attempt carried out while he was on security duty at the community's checkpoint (Business & Human Rights Resource Centre, 2021a). In August 2021, the Wiang Sra Provincial Court of Surat Thani found Somphon Chimrueng guilty of the attempted assassination and sentenced him to 14 years and 4 months of imprisonment. Chimrueng was claimed to be "formerly employed by a private oil palm company" (Business & Human Rights Resource Centre, 2021b). Onmuang's case was not an isolated incident. Over the years, SPFT farming communities across key zones have faced similar threats and attacks, alleged by rights groups to have been orchestrated by oil palm corporations (Hubbell, 2021). Some communities, such as those living in the Klong Sai Pattana village, continue to suffer from lethal violence, despite obtaining permission from ALRO and the courts to settle in these plantation areas that were illegally occupied by corporations (Hubbell, 2021). However, both the violence and relatively small scale of these disputes—when compared to the volume of land involved in oil palm—suggest that these disputes are a relatively limited (albeit a clearly unwelcome feature of the Thai oil palm assemblage) and that smallholders have demonstrated considerable capacity to organise effectively and curtail the encroachments of large companies, and with some success.

Conclusion—The oil palm assemblage in Thailand

The key elements of the Thai oil palm industry are often depicted in terms of their position in a linear sequence, linking actors and entities involved in the primary production (those people and objects involved in the development and cultivation of commercial palm oil trees and FFBs) through to those involved in the harvesting and transportation of the fruits and their primary processing into "milled products" (oils, biomass, and waste), and subsequent secondary processing, distribution, and sales. Materials and goods in the form of different products at different points in their transformation, from living biomass to saleable commodities, are generally transported "down" this "value chain" (see Figure 5.6), while value in the form of money is transported up the same chain. In assemblage terms, what we see is a range of both human and non-human actors brought together in order to "produce" this chain with relationships between the different entities linked via mediations of products, knowledge, money, and power.

In the case of Malaysia and Indonesia, for example, constraints linked to the time between harvest and processing, between technical know-how and cultivation, and between the scale of production and income, have been credited with the way the industry has evolved in both countries. Large-scale production is seen as the means of securing the technology, economies of scale, and processing infrastructure necessary for the industry to prosper. Indeed, this dynamic is sometimes portrayed as the industry's defining characteristic, shaping relationships between the physical (land, climate, chemicals), biological (seeds, trees, pests, and fruits), and technical artefacts (mills and processing machinery), as well as human entities; farmers, plantation owners, and processors.

For Malaysia and Indonesia, these ideas have been put forward as the underpinning factor which has resulted in the shape that the industry has taken. In the case of the main producers, it has been the crop and its demands—for land, prepared and shaped in certain ways, alongside the technical and physical inputs that have been perceived as shaping connections between significant human actors. In particular, these have shaped an industry in which power and influence are vested in large corporate entities and an array of parastatals and state organisations geared to supporting and promoting a particular corporate-centric approach to oil palm. However, this tends to obscure the complexity of the practices of assemblage and numerous forms in which links between entities may be expressed. Unlike Malaysia and Indonesia, where smallholders are often attached to corporations and their mills, farmers in

Thailand commonly act independently from mills and in fact, farmer cooperatives have even established their own mills with government support (Colchester & Chao, 2011). Due to the lack of large plantation companies, crushing mills in Thailand are heavily reliant on FFB supply from smallholders. As mill licensing is less monopolistic in Thailand than in Malaysia and Indonesia (and capacity tends to be greater than supply in all three countries, especially during low-yielding seasons), Thai farmers and trade intermediaries (who collect and deliver FFB from farmers to mills) appear to have more power over pricing negotiations with mill owners.

A lack of formal arrangements and contracts with the mills further enhanced their ability to negotiate FFB prices. This is especially true for intermediaries between farmers and mills, who control harvesting, collection, and sales in non-RSPO-certified supply chains. This lack of supply, overcapacity and use of intermediaries have resulted in an emphasis on quantity over quality, with mills only conducting visual inspections at a later stage in the supply chain, when stating base prices (Nupueng et al., 2022).

It is noteworthy that, despite these differences and the absence of large corporate players, rates of productivity and prices in Thailand equal—and in some instances exceed—those of the Indonesian and Malaysian industries. Simply put, the interventionist Thai approach—which has seen domestic demands and inventory management as top priorities, and permits the government to step in to ensure price stability and stock levels, in addition to providing direct support to smallholders—seems to have "worked" for Thailand, at least to the extent that the industry has facilitated the transition of many farmers away from the less profitable rubber production, catered to domestic needs and maintained prices. Farmer CPO yields per hectare outperform the benchmark aggregate for corporations, their tied smallholders, and independent farmers in Malaysia and Indonesia.

In the same vein, it can be argued that Thailand has had fewer issues related to environmental degradation and human rights violations compared to other major producer countries. This situation reflects the different histories of the respective countries. In Malaysia and Indonesia, the palm oil industry is regularly associated with the large-scale conversion of previously unused lands and is also tethered to post-colonial development narratives which have led to extraordinarily large export-centred palm oil assemblages. Indeed, in Malaysia in particular, oil palm has assumed an ideological significance, where the conversion of "unproductive" land into profitable and productive plantations is celebrated. This contributed to the development of ambitious and extensive support schemes for supported smallholders and the aggressive push for large-scale production on previously non-agricultural lands. By contrast, in Thailand, the industry has never assumed the transformative role ascribed to it as in these countries. Schemes such as Indonesia's Nucleus Estate Scheme (NES) and Malaysia's FELDA settlements were not established in Thailand (Dallinger, 2011). Instead, in Thailand, a more modest industry has emerged in which oil palm was integrated into what were, in many cases, settled agricultural landscapes, where autonomous smallholders were able to act independently from mills and corporations, relying on informal arrangements along the supply chain to sell their produce (Nupueng et al., 2022). This reflects an assemblage and assemblage practices in which relationships between some of the key actors are fundamentally different from those found in Malaysia and Indonesia.

This is evident in the rural policies of successive Thai regimes which have tended to favour measures to support smaller producers. During Thaksin Shinawatra's premiership, the rural population benefited from expansionary and populist economic policies encompassing subsidies, low-interest loans, and debt moratoriums—often referred to as "Thaksinomics". Ricks and Laiprakobsub (2021) observed that these policies had largely benefitted rice farmers in Shinawatra's north and northeast strongholds while excluding palm oil and rubber cultivators in the south from receiving support. However, in 2004, the Prime Minister's efforts to enhance renewable and alternative energy sources, with an emphasis on palm-based biodiesel, played a significant role in encouraging families to cultivate oil palm as a secondary source of income.

It is important to understand that oil palm smallholders in Thailand are motivated by income diversification, besides the ease of oil palm cultivation and subsidies, as opposed to subsistence. Critically, the success of oil palm in Thailand appears to be tied to its capacity to deliver on these different goals for different actors, with smallholders occupying a much more influential place than they have done in Malaysia and Indonesia. The shape of the Thai oil palm assemblage is reflected in how it entwines into and synergistically supports existing smallholder livelihoods and broader Thai policy priorities.

It would of course be grossly simplistic to suggest that comparisons of national yield statistics of CPO tonnage per hectare can be taken as a straight comparison of the relative yield productivity in large-scale versus small-scale production. Indonesian commentators have provided plenty of evidence that its significant smallholder and small plantation sectors do worse on this variable than its large-scale producer, and that this perhaps accounts for the narrowing of national performance between that country and Thailand. However, the critical point here is that this in itself illustrates the variation that occurs within national palm oil assemblages. It raises the point that different national decisions about the distribution of power, management of land, markets, and agricultural know-how, linked to different national patterns of production and cost-benefit distributions, could result in overall outcomes that only slightly vary at the national level. Ultimately, it suggests that the way in which national palm oil assemblages are organised and to whose benefit, are as much about political choices and value judgements as they are about technical efficiency and market forces.

Notes

1 Very broadly, Indonesian suppliers found it more profitable to export than supply locally at the policy-constrained price. The export base was restricted as suppliers found it hard to show compliance with domestic market obligations. There was fallout for several individuals accused (three senior corporate executives, a top trade ministry official and

a policy advisor), with arrests and convictions on evading regulations, and subsequently there were asset seizures at the related corporations.

2 Smallholders who had access to these extension services were 96 percent more likely to adopt RSPO practices.

3 Gross change is the area converted to oil palm, while net change is the area converted to oil palm (gross) minus the area of oil palm converted to other land use types. The same applies when measuring rubber replanting. Unused land consists of grass and scrub land, and marsh and swamps.

4 We employ the term rubber-centric to suggest a mixed system in which rubber played a role as the main cash commodity produced in the area, as opposed to a monoculture "plantation" dominated system.

References

Arunmas, P. (2022a, October 20). *Palm oil subsidy scheme to seek cabinet approval.* Bangkok Post. https://www.bangkokpost.com/business/2418398/palm-oil-subsidy-scheme-to-seek-cabinet-approval.

Arunmas, P. (2022b, September 13). *Ministry seeks B11.4bn to aid farmers.* Bangkok Post. https://www.bangkokpost.com/business/general/2390106/ministry-seeks-b11-4bn-to-aid-farmers

Badan Pusat Statistik. (2023). Indonesia oil palm statistics 2022 (Volume 16). https://www.bps.go.id/id/publication/2023/11/30/160f211bfc4f91e1b77974e1/statistik-kelapa-sawit-indonesia-2022.html

Bangprapa, M. (2022, September 9). *Biofuel subsidies extended for two more years.* Bangkok Post. https://www.bangkokpost.com/business/2388330/biofuel-subsidies-extended-for-two-more-years.

Bureau of Agricultural Economic Research. (2020). Agricultural production and marketing potentials of oil palm in the collaborative farming scheme. Ministry of Agriculture & Cooperatives. https://www.oae.go.th/.

Bursa Malaysia Derivatives. (2022). *Crude palm oil futures.* Investing.com. https://www.investing.com/commodities/palm-oil-historical-data.

Business & Human Rights Resource Centre. (2021a, February 22). *Thailand: Members of the Southern Peasants' Federation of Thailand (SPFT) face violence and evictions amid land disputes with palm oil companies.* https://www.business-humanrights.org/en/latest-news/thailand-members-of-the-southern-peasants-federation-face-violence-and-evictions-amid-land-disputes-with-palm-oil-companies/

Business & Human Rights Resource Centre. (2021b, September 16). *Thailand: Criminal Court convicts former employee of oil palm company for the attempted murder of land rights defender.* https://www.business-humanrights.org/en/latest-news/thailand-criminal-court-convicts-former-employee-of-oil-palm-company-for-the-attempted-murder-of-land-rights-defender/

Chenphuengpawn, J. (2012). Deadweight loss of alternative energy pricing policy. *Applied Economics Journal, 19(1)*, 1–23.

Chumporn. (2022). *Annual report.* https://www.cpi-th.com/public/upload/ir/files/file-29032023-14334003.pdf

Colchester, M., & Chao, S. (2011). Oil palm expansion in South East Asia: An overview. In M. Colchester & S. Chao (Eds.), *Oil palm expansion in South East Asia: Trends and implications for local communities and indigenous peoples* (pp. 1–23). Forest Peoples Programme and SawitWatch.

Cramb, R., & McCarthy, J. (2016). *The oil palm complex: Smallholders, agribusiness, and the state in Indonesia and Malaysia*. NUS Press.

Dallinger, J. (2011). Oil palm development in Thailand: Economic, social, and environmental considerations. In M. Colchester & S. Chao (Eds.), *Oil palm expansion in South East Asia: Trends and implications for local communities and indigenous peoples* (pp. 24–51). Forest Peoples Programme and SawitWatch.

De Vos, R., Suwarno, A., Slingerland, M., Van Der Meer, P. J., & Lucey, J. (2021). Independent oil palm smallholder management practices and yields: Can RSPO certification make a difference? *Environmental Research Letters, 16(6)*. https://doi.org/10.1088/1748-9326/ac018d

DeLanda, M. (2006). Deleuzian social ontology and assemblage theory. In M. Fuglsang (Ed.), *Deleuze and the social* (pp. 250–266). Edinburgh University Press.

Deleuze, G., & Guattari, F. (1988). *A thousand plateaus: Capitalism and schizophrenia*. Athlone Press.

DIT. (2021). *Agricultural production trends*. Ministry of Commerce. https://agri.dit.go.th/.

DIT. (2023). *CPO grade a price*. Ministry of Commerce. https://pricelist.dit.go.th/main_price.php?seltime=month.

GIZ. (2018). *Sustainable palm oil*. https://www.giz.de/en/worldwide/76234.html

Global Forest Watch. (2022). *Universal mill list*. https://data.globalforestwatch.org/documents/gfw::universal-mill-list/about.

Growers' incomes based on Department of Internal Trade information. (2023). [Interview with B. Sukphisal].

Hall, D. (2011). Land grabs, land control, and Southeast Asian crop booms. *Journal of Peasant Studies, 38(4)*, 837–857. https://doi.org/10.1080/03066150.2011.607706.

Hubbell, D. (2021, August 3). *There has been blood*. Eater. https://www.eater.com/22589445/palm-oil-thailand-plantation-spft-jiew-kang-jue-pattana.

International Institute for Sustainable Development. (2013). *A citizens' guide to energy subsidies in Thailand*. IISD. https://www.iisd.org/gsi/sites/default/files/ffs_thailand_czguide.pdf.

Interview with Participant 1, Thai political economist. (2023, September 5).

Interview with Participant 2, representative of an NGO promoting sustainable palm oil in Thailand. (2023, September 18).

Interview with Participant 5, representative of a Thai palm oil company. (2023, October 19).

Interview with Participant 6, representative of a Thai palm oil company. (2023, October 19).

Interview with Participant 7, Thai palm oil policy committee member. (2023, October 23).

Interview with Participant 8, representative of an NGO promoting sustainable palm oil in Thailand. (2023, November 1).

Interview with Participant 9, mill specialist. (2023, December 7).

Interview with Participant 10, corporate analyst. (2024, February 21).

Isranews. (2016, September 26). *Uncovering the issue of encroaching on 1,400 rai of the Kaching peat swamp forest in Pathio District*. Isranews Agency. https://www.isranews.org/isranews-scoop/item/50315-report_patew_26959.html.

Jelsma, I., Schoneveld, G. C., Zoomers, A., & van Westen, A. C. M. (2017). Unpacking Indonesia's independent oil palm smallholders: An actor-disaggregated approach to identifying environmental and social performance challenges. *Land Use Policy, 69*, 281–297. https://doi.org/10.1016/j.landusepol.2017.08.012

Khor, Y. L., Saravanamuttu, J., & Augustin, D. (2015). Consulting study 12: The Felda case study. *High Carbon Stock*. https://doi.org/10.13140/RG.2.2.25267.66084.

KPBN. (2022, July 8). *KPBN CPO tender price, 7 July 2022 determined at Rp 6.755/ kg.* https://www.kpbn.co.id/id/news/715e8ba46f8bc29c24f3fbfdb8d96d71/harga-tender-cpo-kpbn-7-juli-2022-disepakati-rp-6755kg-

Lam Soon Plc. (2022). *Annual report.* https://lamsoon.co.th/wp-content/uploads/2023/03/56-1-ONE-REPORT_LST-Thai_final-2565.pdf

Lee, J. S. H., Ghazoul, J., Obidzinski, K., & Koh, L. P. (2013). Oil palm smallholder yields and incomes constrained by harvesting practices and type of smallholder management in Indonesia. *Agronomy for Sustainable Development, 34(2)*, 501–513. https://doi.org/10.1007/s13593-013-0159-4

Luxton, A., & Ng, B. (2022, June 21). *Palm oil protectionism—Cooking up a storm.* S&P Global. https://www.spglobal.com/marketintelligence/en/mi/research-analysis/palm-oil-protectionism-cooking-up-a-storm.html.

Malaysian Palm Oil Board. (2022). *Oil palm planted area 2022.* https://bepi.mpob.gov.my/images/area/2022/Area_summary2022.pdf.

Nema, S., & Ansari, A. (2023, November 10). *India's edible oil security at a cross-roads.* Hindustan Times. https://www.hindustantimes.com/ht-insight/economy/indias-edible-oil-security-at-a-crossroads-101699609402586.html

Nupueng, S., Oosterveer, P., & Mol, A. (2018). Implementing a palm oil-based biodiesel policy: The case of Thailand. *Energy Science & Engineering, 6(6)*, 643–657. https://doi.org/10.1002/ese3.240.

Nupueng, S., Oosterveer, P., & Mol, A. (2022). Governing sustainability in the Thai palm oil-supply chain: The role of private actors. *Sustainability: Science, Practice and Policy, 18(1)*, 37–54. https://doi.org/10.1080/15487733.2021.2021688.

OAE. (2022). *Agricultural production information—Oil palm.* Office of Agricultural Economics. https://www.oae.go.th/.

O'Reilly, P., Anshari, G., Sancho, J., Jaya, A., Antang, E., Antang, C., Evers, S., Evans, C., Wilson, P., Crout, N., Sjorgesten, S., Upton, C., & Page, S. (2020). Oil palm governance at the grassroots: How assemblage links oil palm, livelihoods, and local administration in an Indonesian village. *International Review of Modern Sociology, 46(1–2)*, 103–120.

Pobsuk, S. (2019). *Alternative land management in Thailand: A case study of the Southern Peasants' Federation of Thailand.* Focus on the Global South. https://focusweb.org/wp-content/uploads/2019/07/SPFT_CaseStudy_final-nl.pdf.

Prasertsri, P. (2023). *Biofuels annual* (Report no. TH2023-00340). United States Department of Agriculture—Foreign Agricultural Services. https://apps.fas.usda.gov/newgainapi/api/Report/DownloadReportByFileName?fileName=Biofuels%20Annual_Bangkok_Thailand_TH2023-0034.pdf.

Ricks, J., & Laiprakobsub, T. (2021). Becoming citizens: Policy feedback and the transformation of the Thai rice farmer. *Journal of Rural Studies*, 81, 139–147. https://ink.library.smu.edu.sg/cgi/viewcontent.cgi?article=4744&context=soss_research

Rodthong, W., Kuwornu, J., Datta, A., Anal, A., & Tsusaka, T. (2020). Factors influencing the intensity of adoption of the Roundtable on Sustainable Palm Oil practices by smallholder farmers in Thailand. *Environmental Management*, 66, 377–394. https://doi.org/10.1007/s00267-020-01323-3.

RSPO. (2022). *Supply chain certificate holders.* https://rspo.org/search-members/supply-chain-certificate-holders/

Sari, D., Hidayat, F., & Abdul, I. (2021). Efficiency of land use in smallholder palm oil plantations in Indonesia: A stochastic frontier approach. *Forest and Society, 5(1)*, 75–89. http://dx.doi.org/10.24259/fs.v5i1.10912

Saswattecha, K., Hein, L., Kroeze, C., & Jawjit, W. (2016). Effects of oil palm expansion through direct and indirect land use change in Tapi River basin, Thailand. *International Journal of Biodiversity Science, Ecosystem Services & Management, 12(4)*, 291–313. https://doi.org/10.1080/21513732.2016.1193560

Soliman, T., Lim, F. K. S., Lee, J. S. H., & Carrasco, L. R. (2016). Closing oil palm yield gaps among Indonesian smallholders through industry schemes, pruning, weeding and improved seeds. *Royal Society Open Science, 3(8)*. https://doi.org/10.1098/rsos.160292.

Sowcharoensuk, C. (2022). *Industry outlook 2022–2024: Palm oil industry*. Krungsri Research. https://www.krungsri.com/en/research/industry/industry-outlook/agriculture/palm-oil/io/oil-palm-industry-2022-2024.

Sowcharoensuk, C. (2023). *Industry outlook 2024–2026: Palm oil industry*. Krungsri Research. https://www.krungsri.com/en/research/industry/industry-outlook/agriculture/palm-oil/io/plam-oil-industry-2024-2026.

Srisunthon, P., & Chawchai, S. (2020). Land-use changes and the effects of oil palm expansion on a peatland in Southern Thailand. *Frontiers Earth Science*, 8. https://doi.org/10.3389/feart.2020.559868.

Stokes, D. (2017, March 24). *As Thailand ramps up its palm oil sector, peat forests feel the pressure*. Mongabay. https://news.mongabay.com/2017/03/as-thailand-ramps-up-its-palm-oil-sector-peat-forests-feel-the-pressure/.

Thailand Customs Department. (2022). Trade trends of selected palm products [Data set]. https://www.customs.go.th/

Thongrak, S., Kiatpathomchai, S., & Kaewrak, S. (2011). *Baseline study of the oil palm smallholders in the project areas*. GIZ Thailand. https://rspo.org/resources?category=proceedings-of-rspo-events-related-to-smallholders&id=4778

UN Comtrade. (2022). *Trade trends of selected palm products*. http://comtrade.un.org/

UNDP. (2014). *Request for proposal*. https://procurement-notices.undp.org/view_file.cfm?doc_id=24554

Univanich. (2022). *Annual report 2022*. https://univanich.com/wp-content/uploads/2023/04/ONEREPORTUVANE-1.pdf

USDA-FAS. (2023). *Palm oil area, yield and production by country*. Foreign Agricultural Services. https://ipad.fas.usda.gov/countrysummary/.

Vinayak, A. J. (2023, June 14). *Thailand comes to India's rescue with crude palm oil exports*. The Hindu BusinessLine. https://www.thehindubusinessline.com/markets/commodities/thailand-comes-to-indias-rescue-with-crude-palm-oil-exports/article66967308.ece

Wan, A. J. (2022). *Next generations of smallholders in Malaysia*. Malaysian Palm Oil Certification Council. https://www.mpocc.org.my/mspo-blogs/next-generations-of-small holders-in-malaysia#:~:text=Smallholders%20are%20the%20backbone%20of,followed%20by%20Sarawak%20and%20Sabah.

Widyatmoko, B., & Dewi, R. (2019). Dynamics of transmigration policy as supporting policy of palm oil plantation development in Indonesia. *Journal of Indonesian Social Sciences and Humanities, 9(1)*, 35–55. https://doi.org/10.14203/jissh.v9i1.139

SECTION 2
Latin America

6

ASSEMBLAGE OF SUSTAINABILITY GOVERNANCE IN THE COLOMBIAN OIL PALM SECTOR

Paul R. Furumo

Introduction

The rapid expansion of tropical commodity crops such as oil palm is the leading driver of global land use change and deforestation (Curtis et al., 2018). These land use changes often bring socio-ecological trade-offs and pose significant challenges for governance, considering that the outcomes of oil palm expansion vary greatly across different settings. In Southeast Asia, oil palm expansion has caused widespread forest conversion, whereas in Latin America, the most recent expansion has occurred on previously cleared lands, particularly cattle pastures (Furumo & Aide, 2017). With looming land constraints in Southeast Asia, Latin America has become the fastest-growing region of oil palm expansion globally and is expected to become a dominant region of future production (Pirker et al., 2016). Colombia is the largest palm oil producer in Latin America and has a major opportunity to lead on sustainability. The country has one of the lowest rates of forest conversion to oil palm plantations in the region and ample previously cleared lands surrounding existing plantations that have little overlap with biodiversity-rich habitat (Ocampo-Peñuela et al., 2018; Furumo & Aide, 2017; Garcia-Ulloa et al., 2012). Colombia is also pioneering public–private sustainability efforts in the sector to avoid future deforestation. However, seizing these opportunities to create a more sustainable oil palm industry will ultimately depend on effective policies and coordinated governance across multiple scales.

Governing the oil palm industry is a patchwork of interacting *enabling* and *regulatory* institutions situated at multiple levels within, above, and below the nation-state (Hamilton-Hart, 2014). Enabling institutions include markets for international trade and state policies put forth by governments seeking to expand rural development opportunities through oil palm cultivation (e.g., financial incentives,

DOI: 10.4324/9781003459606-8

land concessions). Regulatory efforts have largely been advanced by civil society and private actors through transnational initiatives that employ market-based mechanisms for more sustainable supply chain governance (Lambin et al., 2018). The leading industry standard is the Roundtable on Sustainable Palm Oil (RSPO), a multi-stakeholder commodity roundtable developed in the early 2000s. The RSPO is supported by a number of other market-based eco-certification standards targeting different aspects of palm oil production, such as habitat conservation (e.g., Rainforest Alliance), chemical inputs (e.g., the International Federation of Organic Agriculture Movements or IFOAM—Organics International), and carbon emissions (e.g., the International Sustainability & Carbon Certification or ISSC).

More recently, producers and traders have adopted commitments to eliminate deforestation from palm oil supply chains, initiated by the Consumer Goods Forum pledge in 2010 (Garrett et al., 2019). Although these initiatives began as corporate pledges, they have evolved into public–private implementation strategies supported by a number of transnational actor groups such as the Tropical Forest Alliance (TFA)—a global public–private partnership—and the New York Declaration on Forests (NYDF). Current sustainability efforts in the oil palm sector are thus increasingly embodied in hybrid governance arrangements forged by issue linkages among commodity production, forest conservation, rural development, and climate change agendas (Furumo & Lambin, 2020). The result has been a blurring of the traditional roles between state enablers and non-state regulators.

The extent to which states adopt, block, or circumvent transnational initiatives that address sustainability in the oil palm sector has led to complex regulatory interactions between public and private institutions at multiple levels (Pacheco et al., 2018). The rise of transnational governance has often been associated with a "hollowing-out" of the nation-state, in which governments are bypassed by new regulatory institutions (Jessop, 2013). However, a more accurate metaphor might be the "re-articulation" of the nation-state, rather than its retreat (Astari & Lovett, 2019; Andonova, 2013). For instance, the Indonesian government has sought to bypass regulation by the RSPO through the creation of its own national standard—Indonesian Sustainable Palm Oil (ISPO). The ISPO is modelled after the RSPO but is less stringent on land clearing, allowing deforestation in areas permitted by Indonesian law (Hospes, 2014). In other commodity sectors like Chilean timber, weaker national standards have also been developed to circumvent industry and civil society standards (i.e., the Forest Stewardship Council) with underperforming results (Heilmayr & Lambin, 2016). In the Colombian oil palm sector, the government led a national zero deforestation agreement with palm oil producers towards the goal of making all palm oil produced in Colombia deforestation-free by 2020 (Furumo & Lambin, 2020). Producers can use RSPO certification to demonstrate compliance with the national agreement in the form of transnational policy absorption (Lambin et al., 2020).

To better understand and theorise these interactions across scales (global–local) and actor domains (public–private, state–non-state), the notion of policy transfer has been supplanted with a *policy mobilities* orientation—i.e., policies are not merely copied from one setting and pasted into another, but undergo forms of translation, mutation, and reassembly during their adoption (McCann & Ward, 2013). This has brought assemblage thinking into discussions on policymaking (Savage, 2019; Prince, 2016). Assemblage thinking seeks to overcome the temporal and spatial scale issues challenging network perspectives of social movements (e.g., Social Network Analysis, Actor Network Theory) by merging distal connections with place-based conditions (Santo & Moragues-Faus, 2019; McFarlane, 2009). The notion of assemblage considers how heterogeneous sets of human (e.g., actors, institutions, policies) and non-human (e.g., land, ecosystems, disease) entities come together to produce emergent, irreducible wholes that serve a variety of interests and goals. Policy assemblages are best thought of as a process rather than an outcome; they are in a constant flux of destabilisation and reassembly ("de/re-territorialisation") as new actors and policies enter the mix.

To borrow from ecology, assemblages are defined as "a taxonomically related group of species that occur together in space and time" (Stroud et al., 2015, p. 4758). The term originates from community ecology, a field that examines the effects of biotic and abiotic features on the structure of communities of species. A "metacommunities" focus considers the interactions between communities at different scales through species dispersal mechanisms (Leibold et al., 2004; Wilson, 1992). Following this natural science analogue, assemblages of species (i.e., human entities) are the result of both habitat-specific local conditions (i.e., non-human entities), and interactions with other communities (i.e., assemblages) at different scales that reshape their composition (i.e., de/re-territorialisation). Species interactions across communities are analogous to the *multiplicity* of assemblages described in governance research (Briassoulis, 2019). An assemblage perspective thus facilitates our understanding of complex mechanisms across multiple scales of spatio-temporal organisation, whether focusing on a community of species or a policy domain and its actors.

This chapter adopts a "trans-local" assemblage perspective (McFarlane, 2009) to explore how enabling and regulatory policies from state and transnational institutions become articulated locally in Colombia. Palm oil production in Colombia occurs in four primary geographic zones—North, Central, East, and Southwest. Each zone features unique ecosystems, land use histories, cultures, and socioeconomic realities that influence the adoption and outcome of regulatory and enabling policies for oil palm expansion. The Colombian oil palm sector has historically been decoupled from international markets, oriented instead towards domestic food, industry, and biodiesel consumption. The institutions governing the sector are thus largely domestic, making for an interesting case study of the policy assemblage around sustainability adoption. This chapter focuses on the regional outcomes of (1) state- and sector-led enabling policies to expand domestic oil palm cultivation; and (2) the adoption of transnational regulatory governance (i.e., RSPO certification) to improve the sustainability of the sector. A focus on trans-local assemblages intends to "blur or bypass the scalar distinction between local and global" (McFarlane, 2009, p. 562). This builds on the concept of "regional assemblages" put forth by Allen and Cochrane (2007), which rejects the notion of regions as territorially fixed and governed according to political boundaries. Instead, regional assemblages are a tangle of public and private institutions, with embedded segments of central, regional, and local governments, and fragments of transnational authority. Hence, this chapter argues that the socio-ecological outcomes of oil palm expansion are the result of how enabling and regulatory policies are mediated by these trans-local assemblages.

This chapter engages the first and second research questions posed by the book: (1) how different governance arrangements enable and legitimise the expansion of the palm oil industry, and (2) how power differentials impact oil governance in Colombia. It seeks to examine the ways that different governance structures facilitate the expansion of the palm oil industry in the country, alongside the discourses at play that affect the sector's governance and its development. The next section of the chapter begins this examination by providing some background on the Colombian oil palm industry, including a description of the salient features of the Colombian oil palm complex—the assemblage of actors, institutions, and natural systems that support oil palm production (Cramb & McCarthy, 2016). Next, the major enabling policies advanced by the sector and state through different stages of oil palm development will be reviewed. How these institutions assume a more regulatory role through sustainability governance will be illustrated. Lastly, the chapter will focus on the most critical period of oil palm expansion since the early 2000s and on the political economy narratives advanced by enabling and regulatory policies, namely: (1) biodiesel demand, (2) peacebuilding through smallholder alliances, and (3) sustainability commitments to increase exports. The differential adoption and outcome of these policies across the trans-local assemblages of regional production zones will be considered.

Colombian oil palm complex

Colombia is the fourth largest palm oil producer in the world and the largest in Latin America, with 540,687 hectares planted as of 2018 (Fedepalma, 2019). Cultivation of oil palm in the Americas first began in the 1920s with interest from the United Fruit Company in developing new tropical cash cultivars to diversify banana holdings after the Fusarium wilt and Sigatoka disease outbreaks (Richardson, 1995). Experimental plots were first pioneered in Central America—Guatemala, Honduras, and Costa Rica. The global shortage of primary materials during World War II spurred further enthusiasm for the crop, and United Fruit began replanting large areas of disease-stricken banana fields with oil palm (Bozzi & Jaramillo, 1998).

Human entities: actors, markets, institutions, and social constraints

In Colombia, the first commercial oil palm plantation was established in 1945 on the Caribbean coastal plain near Sevilla, Magdalena, using seeds developed in

Honduras (Bozzi & Jaramillo, 1998). Early plantations were established mostly by families of rural elites, and today the production and processing of palm oil in the country continue to be dominated by Colombian companies. Colombian oil palm plantations are, on average, considerably smaller than those in Southeast Asia. In 2011, 74 percent of the total area planted was on plantations smaller than 2,000 hectares, and smallholder production (<50 hectares) represented 13 percent of the total area planted (Potter, 2020).

The major challenges facing Colombian palm oil producers are related to high production costs, marketing, and phytosanitary conditions. In 2010, producing a tonne of palm oil in Colombia cost roughly USD600 compared to less than USD500 in Malaysia, and only USD400 in Indonesia (Rueda-Zárate & Pacheco, 2015). Labour costs are the primary difference, reaching as high as 30 percent of total production costs in Colombia. Imported fertiliser inputs are also expensive and can range from 25 to 40 percent of the total cost of production (Ruiz et al., 2017). The high production costs make Colombian palm oil less competitive on the international market than Asian palm oil. These price issues have consequent impacts on marketing, which also presents a major challenge for Colombian palm oil producers. To avoid less profitable exports, producers have historically looked to the domestic market to sell palm oil and relied on favourable government policies to increase demand in these markets. As a result, most Colombian palm oil is consumed domestically. While Colombian palm oil exports have traditionally remained in the Americas (Furumo & Aide, 2017), the Food and Agriculture Organisation (FAO) states that exports to Europe have increased in the last decade and now represent over 60 percent of total exports (FAO, 2020).

The Colombian palm oil sector was originally established by the government to reduce reliance on foreign vegetable oil imports and has always been heavily influenced by the central government and other parastatal bodies. The sector is highly organised under the leadership of Fedepalma, the national federation of oil palm growers. Fedepalma has been instrumental in intervening on behalf of the sector in public policy forums. Together with its marketing (Acepalma) and technical research (Cenipalma) branches, Fedepalma has procured technical support and financial assistance for Colombia's palm oil producers to enhance the production and commercialisation of palm oil. At several junctures in history, the government of Colombia has promoted oil palm as a rural development strategy for peacebuilding (see below). Colombia is currently emerging from a decades-long armed political conflict largely centred around land inequality. In 2016, the government signed a peace accord with the largest armed rebel group in the country, the Revolutionary Armed Forces of Colombia (FARC). Demobilisation created a power vacuum in many rural areas, leading to large-scale land grabbing and deforestation (Furumo & Lambin, 2020). In the current post-conflict period, the government seeks to stabilise the countryside through illicit crop substitution programmes and rural development opportunities that include oil palm.

Non-human entities: Oil palm landscapes, ecosystems, and biophysical constraints

Compared to Southeast Asia, Colombian oil palm landscapes are highly transformed, given the long history of cattle ranching. Few large, undisturbed areas of lowland habitat remain in these landscapes. Oil palm is grown in four primary geographic production zones across Colombia (Figure 6.1), each featuring different biophysical, socioeconomic, and cultural conditions (Furumo & Aide, 2019; Castiblanco et al., 2015; Bozzi, 1998). The North zone sits on the coastal Caribbean plain with tropical dry broadleaf forests and xeric shrubland ecosystems. It is the oldest region of production with the most degraded oil palm landscapes. The core of the Central zone is located in the Middle Magdalena River region, with moist tropical forests. Early development was geographically concentrated in several adjacent large-scale operations (5–10 thousand hectares each) but has since become more diffused with growing numbers of small- and medium-scale producers setting up in more forested fringe areas. Located in the tropical savannah and grasslands of the Orinoco region, the East zone is the largest producer, with 230,000 hectares planted including the largest plantations in the country. The majority of plantations are found closer to the Andes foothills in areas long-since transformed by cattle ranching (Etter et al., 2008), but recent expansion is occurring in the more remote natural savannahs. Given the low population density of the region and strong cattle culture, much of the labour in the East zone is provided by Colombian migrant workers. Finally, the Southwest zone is the smallest producer, with 22,000 hectares of plantings concentrated in the Pacific Chocó region of Tumaco (Nariño department) bordering Ecuador. The area possesses one of the wettest ecosystems in the country. Each production zone has a primary urban market where most of its palm oil is sold, coinciding with the largest cities of Colombia—Bogotá (East Zone), Barranquilla (North Zone), Cali (Southwest Zone), Medellín, and Bucaramanga (Central Zone).

Phytosanitary issues have been detrimental at different stages of oil palm development in these production zones. Lethal bud-rot disease, known colloquially as PC (*Pudrición de Cogollo*), is the most pressing threat. PC is a plant pathogen that enters the palm through insect vectors (i.e., *Rhynchophorus palmarum*) and is particularly prone to spreading in conditions of flooding (Silva & Martínez, 2009), resulting in high palm mortality. The East Zone faced the earliest outbreaks in the late 1980s. The Southwest zone is still recovering from a 2006 to 2009 epidemic that caused over 35,000 hectares of loss (Ayala & Romero, 2019) and forced many workers to migrate to the plantations of the East zone. The Central zone lost nearly 38,000 hectares in a separate outbreak during this same period. The seasonal dry conditions of the North zone have historically insulated it from PC spread, but the region is currently in the midst of a serious outbreak. As of 2019, the total economic losses from PC have been estimated at nearly USD2.5 billion (Ayala & Romero, 2019). In response, the sector has developed a hybrid oil palm variety by crossing the American oil palm (*Elaeis oleifera*) with the African oil palm (*Elaeis guineensis*). This OxG hybrid has lower yields and higher production costs as it requires assisted pollination, but is more resistant to disease, produces a higher-quality oil, and can be harvested for longer due to its stunted physiology (Alvarado et al., 2013). In areas affected by PC, the hybrid palm has become an important strategy for oil palm companies to stay in business.

Historical stages of oil palm development: Enabling and regulating policies

This section describes the historical role of the state and public sector as enabling institutions of oil palm expansion in Colombia. Following Rueda-Zárate and Pacheco (2015), key enabling policies will be traced across three stages of oil palm development with unique political economies: (1) *Protectionist* (1950s–1990), (2) *Market liberalisation* (1991–2000), and (3) *Biodiesel boom and Consolidation* (2001–2012). A fourth stage of development that is currently underway will then be proposed, *Sustainability adoption* (2013–present), in which the Colombian oil palm sector prioritises sustainability commitments to access new markets abroad. This current stage marks the shift in the enabling role of the state and public sector to regulatory institutions. The chronic marketing and commercialisation challenges facing Colombian palm oil producers are reflected in the policy strategies advanced throughout these stages of development.

Early stage—Protectionist (1950s–1990)

In the post-war period, the political economy of the Colombian government prioritised the strengthening of national industries, giving rise to the modern commercial agriculture sector. The government created a new Ministry of Agriculture and implemented protectionist measures that increased production and exports. To increase cotton production for the national textile industry, an entity was created in 1950 under the Ministry of Agriculture called the *Instituto de Fomento Algodonero* (IFA). Although the IFA was initially focused on cotton fibre, its mandate expanded to include the development of other oilseed crops to satisfy a rapidly growing national demand for vegetable oil. Palm oil entered this discourse, and in 1958, it was estimated that 33,000 hectares would be needed to cover the national deficit of edible fats (Bozzi & Jaramillo, 1998).

The first national development strategy for oil palm was introduced in 1957 and consisted of financial incentives to encourage both production and uptake of palm oil in the national market (Rueda-Zárate & Pacheco, 2015; Bozzi & Jaramillo, 1998). The IFA financed plantation development for associations of producers in rural areas. The colonisation strategy largely failed due to poor infrastructure in frontier areas and a lack of technical capacity and experience in managing the crop, but it was successful in introducing oil palm into what would become the primary

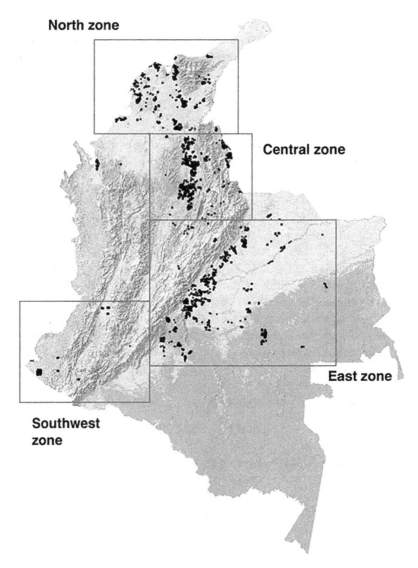

FIGURE 6.1 Map of four major geographic oil palm production zones in Colombia (Furumo & Aide, 2017).

geographic production zones (Figure 6.1). Under the national development plan, the government also offered favourable credit that reflected the long payback time of the perennial crop, and a system of quotas and tariffs on imports. The permanence of these benefits was inconsistent over the next several decades and became a source of intense lobbying by the sector. Adopting a successful model from the coffee sector, the national association of oil palm growers, Fedepalma, was established in 1962 to create a direct channel to the government and would eventually gain influential political weight towards securing favourable conditions for producers. Nearly 20,000 hectares had been planted by the end of the 1960s, but the 1970s were a decade of relatively little growth due to an increase in vegetable oil imports entering Colombia.

By the 1980s, the sector experienced its first "boom" in expansion, closing the decade with nearly 110,000 hectares planted (Bozzi & Jaramillo, 1998). The surge was largely fuelled by economic factors, including renewed protectionist tariffs, favourable tax legislation, the devaluation of the Colombian peso, and expanded government credit for perennial crops such as cocoa, rubber, and oil palm. With quota limits on vegetable oil imports, palm oil prices remained high. To circumvent the high costs of importing primary materials, domestic food processors and manufacturers had begun investing in plantations and controlled 40 percent of the national palm oil supply (Bozzi & Jaramillo, 1998), initiating the high degree of vertical integration observed in the sector today. There were dozens of palm oil producers at the time, but only a handful of processors and manufacturers, resulting in oligopolist conditions that created tension in the sector. By the end of the decade, the industry faced a crisis as increased production outpaced demand from the national market, exacerbated by contraband palm oil imports. With a crash in price and calls for fairer market conditions, the sector was propelled into the next stage of development under market liberalisation.

Intermediate stage—Market liberalisation (1991–2000)

Under the new globalised political economy and saturation of the domestic market, stabilising the internal commercialisation of palm oil and expanding exports became a major priority for the sector. Fedepalma created a marketing entity in 1991, Acepalma, that specialised in selling Colombian palm oil on the international market. The *Fondo de Fomento Palmero* (FFP), or Oil Palm Development Fund, was established by national law in 1994 and internalised an existing government price stabilisation mechanism applied across the agricultural sector (Law 101, 1993). The fund helped support Fedepalma's mission of increasing exports by directing 1.5 percent of palm oil sales from the domestic market to offset export costs (i.e., shipping). From 1991 to 1995, exports of palm oil products increased from 2,768 to 22,465 tonnes, directed mainly to Venezuela, Mexico, Honduras, El Salvador, Jamaica, and Great Britain (Bozzi & Jaramillo, 1998). By the end of the 1990s, 22 percent of palm oil produced in Colombia was exported (Rueda-Zárate & Pacheco, 2015).

This intermediate stage of development also focused on increasing competitiveness through improved yields and professional management, heralding large-scale institutional changes that would come to define the modern era of the Colombian oil palm industry. The FFP financed programmes to increase productivity and efficiency. Research and development expanded with the creation of Cenipalma, a research institution focused on improving yields through breeding and genetics programmes, pest and disease control, and improved agronomic management. Experimental field stations have now been established in each of the four major production zones. Although the opening of the Colombian palm oil market created alternative outlets, the oligopolist conditions of the national industry persisted with just four national companies behind 47 percent of palm oil sales in the country during this stage (Rueda-Zárate & Pacheco, 2015). Despite increased exports, there remained a large surplus, and production costs remained stubbornly high. This hampered the competitiveness of Colombian palm oil on the international market, creating a market barrier for excess national supply. The solution would be the creation of an entirely new domestic market for Colombian palm oil.

Late stage—Biodiesel boom and consolidation (2001–2012)

Oil palm expansion boomed during this stage of development under strong political support from the Uribe administration (2002–2010). The government introduced a law to provide financial incentives for biodiesel production in 2004; an official B5 blending mandate was passed in 2008 and increased to B10 in 2010. These policies aimed to support a government target of three million hectares of oil palm by 2020 (Castiblanco et al., 2013). The area planted grew from 169,564 hectares in 2001 to 419,870 hectares by 2012 (Fedepalma, 2002, 2013). Biodiesel policies practically doubled domestic demand for palm oil, giving rise to an entirely new national market in a short time (Figure 6.2). The first palm oil sales for biodiesel began in 2008, and at its peak in 2012, the biodiesel market represented 46 percent of domestic palm oil sales. Biodiesel production is concentrated among a few highly integrated firms, and these companies are among the largest producers of palm oil in Colombia. Five companies represented the entire domestic biodiesel market in 2012, and three of these each accounted for more than 10 percent of the total market share of domestic palm oil sales that year (Fedepalma, 2013). Most of the exported surplus palm oil was redirected to the biodiesel market during this stage (Figure 6.2).

This stage of development also saw the entry of a large number of smallholders into production through direct government intervention. As a peacebuilding strategy, the government promoted rural economic development through a model of "strategic production alliances" (*Alianzas Productivas Estratégicas—APE*) that would incentivise agro-industrial companies to incorporate smallholders into their supply bases (Rueda-Zárate & Pacheco, 2015). This built on previous policies introduced in the 1990s seeking to capitalise and modernise the countryside through agriculture. In the oil palm sector, subsidies and tax exemptions were offered to companies establishing smallholder alliances, and in turn, companies would provide low-interest credit for smallholders to establish their plantations as well as technical assistance to improve fruit yields and quality. This nucleus-outgrower arrangement created groups of producers similar to the "schemed" smallholders of Southeast Asia. Between 1998 and 2006, over 4,500 smallholders representing more than 80 alliances accounted for at least 52,000 hectares of expansion;

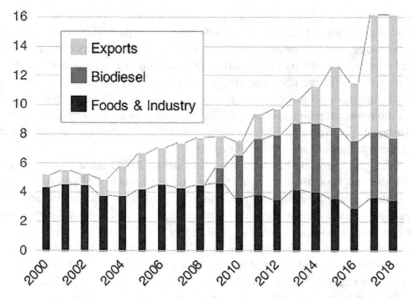

FIGURE 6.2   Annual sales of Colombian-produced palm oil to the three major markets from 2000 to 2018; food processing and other industries, biodiesel production, and exports. Values represent the amount of palm oil sold to each market (hundreds of thousands of tonnes).

by 2013, roughly 6,000 smallholders in 116 alliances represented approximately 15 percent of the total area planted in Colombia (Rueda-Zárate & Pacheco, 2015).

Within the framework of the armed conflict, smallholder expansion was further supported by illicit crop substitution programmes promoted under Plan Colombia—a bilateral effort between the United States and Colombia to stop cocaine production and drug trafficking during this period. Forced displacement from the armed conflict reached its height in 2002, with 400,000 people displaced that year (Núñez & Hurtado, 2014). By 2007, the More Investment for Sustainable Alternative Development (MIDAS) programme funded by the United States Agency for International Development (USAID) had established 51,300 hectares of oil palm among 4,800 families in conflict areas (Molano-Aponte, 2008). The area of oil palm cultivated by smallholders on farms 5–20 hectares in size increased tenfold from 1998 to 2011 (Potter, 2020). Biodiesel policies thus fuelled expansion during this stage and helped cement the current Colombian oil palm model: supply bases composed of an anchor company or mill that rely on production from smaller outgrowers upstream, and processing and manufacturing from vertically integrated actors downstream (Furumo et al., 2019).

Current stage—Sustainability adoption (2013–present)

The biodiesel boom rapidly accelerated domestic demand for Colombian palm oil, but with a corresponding increase in supply, the domestic market was again facing saturation by 2013. At this point, the proportion of palm oil being sold to the bio-diesel market began to decline relative to palm oil exports (Figure 6.2). At this current stage of development, producers were left with no choice but to export surplus palm oil. Contrary to the market liberalisation stage of the 1990s, sustain-ability criteria like the RSPO certification have increasingly become a barrier to entry to international markets (i.e., European Union/EU). The current strategy of the Colombian palm oil sector has thus become one of demonstrating commitments to sustainability to distinguish Colombian palm oil in the international marketplace and increase the competitiveness of exports.

Although Fedepalma has been a member of the RSPO since its inception in 2004, the first Colombian oil palm grower did not become RSPO-certified until 2010 and remained the only certified company until 2014. Certification adoption lagged in the early years of the RSPO since domestic market demand—particularly for palm oil in biodiesel—still outpaced supply, and there was virtually no con-sumer pressure for certified palm oil in Colombia. However, facing the looming saturation of the domestic market, a rush of companies began pursuing certification to access markets abroad. By 2016, six additional oil palm growers had become RSPO-certified, and in 2017, there were ten RSPO-certified producers in total. In addition to RSPO, other voluntary certifications in the Colombian palm oil sector include IFOAM—Organics International, the Rainforest Alliance, and the ISCC. By 2018, Colombia had 88,497 hectares certified by at least one of these standards, corresponding to 16.4 percent of the total area (Fedepalma, 2019). As plantations have acquired certification, exports have increased dramatically, from 16 percent of national production in 2013 to 52 percent in 2018. Over this same period, palm oil exports to Europe increased from 46 to 63 percent of Colombia's total exports (FAO, 2020), and the area planted in oil palm expanded by 15 percent with the fast-est growth rate recorded in the East Zone (Fedepalma, 2019).

To elevate the reputation of Colombian palm oil and increase exports, Fedepalma began an international campaign in 2017 to promote Colombian palm oil as "unique and differentiated" compared to other producing countries (i.e., Southeast Asia), emphasising the sector's low rates of deforestation. A national campaign, *La Palma es Vida* (Oil Palm is Life), was also launched by Fedepalma to increase dietary consumption of palm oil among Colombians (Fedepalma, n.d.). The Colombian government has also become more involved in regulating negative trade-offs of oil palm expansion. In 2017, a voluntary multi-stakeholder zero deforestation agree-ment was signed in the Colombian oil palm sector by growers, processors, civil society, and the government. This is the first national-scale zero deforestation agree-ment in a commodity sector and aims for all palm oil produced in Colombia to be free of deforestation by 2020 (Furumo & Lambin, 2020). Currently, the agreement includes the supply bases of 18 palm oil producers and traders, representing 33 per-cent of the national area planted. Most are already RSPO-certified or in the process of gaining certification. Fedepalma is focused on committing more companies to the national zero deforestation agreement to support its campaign of differentiated

Colombian palm oil. The zero deforestation agreement has roots in peacebuilding and a government pledge to eliminate deforestation in the Colombian Amazon. It is supported by international funding for Reducing Emissions from Deforestation and Forest Degradation (REDD+). A number of other climate-related REDD+ initiatives with a focus on oil palm development have been advanced at the subnational level, including the Orinoco Sustainable Integrated Landscapes Programme. The initiative is supported by the BioCarbon Fund and targets zero deforestation palm oil production in the East Zone (Furumo & Lambin, 2020).

Political economy narratives, regional adoption, and localised outcomes

The development of the Colombian oil palm industry is distinct from Southeast Asia in that demand was not driven by distal urban consumers. The industry was established to serve the national market, leading to the creation of domestic institutional structures that have ensured the continued growth of the sector despite a history of economic, social, and biophysical challenges. Due to the relatively high costs of producing palm oil in Colombia, international markets were never a strong driver of the sector, but rather a second-tier release valve when the national market became too saturated. Government policies have supported the industry in two main ways: (1) stimulating national demand for palm oil (e.g., import quotas, biodiesel mandates), and (2) facilitating expansion through new models of production (e.g., smallholder alliances, zero deforestation palm oil). The oil palm sector itself, led by Fedepalma, has played an important role in securing these opportunities from the state and shaping the policy narrative at different stages.

Biodiesel demand

The biodiesel policies introduced in the mid-2000s contributed to massive industry expansion and transformed the economic and geographic landscape of Colombian oil palm production. Colombian biodiesel policies were embedded in a larger global political economy narrative of expanding biofuel programmes in response to rising oil prices and concerns over greenhouse gas emissions (Dauvergne & Neville, 2010). Most influential was the 2008 EU biofuel policy that stimulated increases in palm oil imports for biodiesel production, particularly from Latin America (Gerasimchuk & Koh, 2013). While Colombia does not export large quantities of refined biodiesel due to its strong domestic demand, Europe has become a major importer of Colombian crude palm oil, much of which is used for biodiesel.

Biodiesel policies have produced their greatest effect on oil palm expansion in the East zone, where 58 percent of the country's biodiesel is refined (Fedepalma, 2019). The proximity of these plantations to the largest market in the country, the capital city of Bogotá, has driven this demand. Of the palm oil that the East zone sells domestically, 98 percent ends up in Bogotá. The Central and North Zones account for 22 and 20 percent of biodiesel production, respectively. Several refineries in the North Zone produce biodiesel for export, given their proximity to ports. Of the three primary markets for Colombian palm oil—domestic foods industry, biodiesel, and exports—the biodiesel market pays the lowest price to producers. The Ministry of Mines and Energy sets both the blending mandate and price of biodiesel in Colombia, whereas the price of palm oil sold to the food industry and export markets is regulated by the national price stabilisation mechanism. Nonetheless, Colombian biodiesel companies are highly integrated conglomerates, and thus a certain quota of the palm oil produced on their plantations is allocated to their biodiesel refinery.

The current biodiesel blending mandate in Colombia remains at 10 percent (B10), but there has been intense lobbying by the oil palm sector to increase the blend to 20 percent (B20). If achieved, this policy would double the demand for Colombian palm oil in the domestic biodiesel market and most certainly lead to a contraction of exports. Future exports may also decline if the EU continues phasing out palm oil-based biodiesel, which has been under consideration since a 2017 resolution (European Parliament, 2017). The EU resolution aims to prevent deforestation from its biofuel feedstocks, particularly palm oil from Southeast Asia. Given the more favourable land use history of the sector in Colombia, there is an environmental potential for Colombian palm oil-based biodiesel if social trade-offs can be minimised.

Peacebuilding through smallholder alliances

Leading up to the boom in palm oil demand for biodiesel, government policies had begun promoting new models of oil palm production. These policies were rooted within a peacebuilding framework and sought to stabilise the countryside by bringing rural economic development through more capital-intensive agriculture. In 1998, an Investment Fund for Peace was established (Law 487) that helped finance the production alliances (*APE*) for smallholders (Rueda-Zárate & Pacheco, 2015). The APE model was designed to be a "win-win" arrangement between palm oil mills and smallholders: mills would receive financial incentives to diversify their supply base while smallholders would receive capital and technical assistance to produce higher-quality fruits. Critics of the productive alliances model claim that it is merely a strategy for large rural elites to minimise risk while accessing more land to expand their supply base (Potter, 2020; Marin-Burgos & Clancy, 2017). For instance, a former Colombian minister of agriculture established one of the largest oil palm supply bases in the North zone largely through the productive alliance model, consolidating over 45,000 hectares of oil palm managed by nearly 3,000 smallholders.

The APE model is formalised through smallholder contracts that typically apply to the productive lifespan (20–30 years) of the palms that mills help finance (Rueda-Zárate & Pacheco, 2015). The extent to which these contracts benefit farmers has been contested. Farmers risk getting locked into oil palm as their live-lihood for a long period of time, losing a degree of autonomy over their land and the flexibility of planting other crops, especially for subsistence. Contracts also require farmers to sell their fruit only to the contracted mill, foregoing negotia-tion with other mills when prices fluctuate. This particularly benefits mills in years of drought or disease outbreaks when regional production is suppressed and fruit prices increase. The advantages for smallholders are that they have a guaranteed purchaser of their fruit over a long period of time, and they receive technical assis-tance to improve yields and quality. In a survey of smallholder oil palm farmers in the department of Magdalena in the North zone, many farmers regarded the model of production alliances positively, stating they were only able to plant oil palm through the credit provided by the anchor company as part of the alliance (Furumo et al., 2019). Studies have shown that farming oil palm is more profitable than cultivating other commodity crops grown in these landscapes or working as wage labourers (Potter, 2020) as oil palm smallholders across production zones gener-ate an income ranging from 3–8 times greater than the national minimum wage in Colombia (Rueda-Zárate & Pacheco, 2015).

Although oil palm is associated with higher incomes in the Colombian coun-tryside, the extent to which this has improved social indicators is more mixed. Castiblanco et al. (2015) found that oil palm municipalities in Colombia had higher fiscal incomes and lower unmet basic needs, but also greater inequality related to land and its ownership. This pattern varied by region, showing the greatest levels of inequality in the East Zone, where large agro-industrial planta-tions dominate, and the least disparity in the Central Zone, where smallholder productive alliances are more numerous. Today, Colombia remains one of the most unequal countries in terms of land distribution, and this has been the source of persistent conflict throughout the country's decades-long civil war. The most recent national agricultural census found that 70 percent of farmers and ranch-ers operate on less than five hectares of land, representing only 2 percent of the national rural land area, while 75 percent of the rural area is managed under large estates (>1,000 hectares) by just 0.2 hectares of producers (Oxfam International, 2017).

Episodes of violence related to the armed conflict have historically contributed to this pattern of inequality and have been more prevalent in oil palm-cultivating municipalities (Marin-Burgos & Clancy, 2017; Sabogal, 2013). In this context, armed actors—particularly paramilitary groups—have used oil palm plantations as a mechanism to appropriate land in strategic corridors of the country (Potter, 2020; Marin-Burgos & Clancy, 2017; Castiblanco et al., 2015; Oslender, 2008). There is evidence of this occurring in each of the major oil palm production zones of Colombia, including the regions of Montes de Maria and Urabá (North Zone), Magdalena Medio and Catatumbo (Central Zone), the eastern Andes foothills and Guaviare River (East Zone), and Tumaco (Southwest Zone) (Centro Nacional de Memoria Histórica, 2016). These incidents have been most widely reported during the late stage of oil palm development (early/mid-2000s) at the peak of forced displacement and rapid oil palm expansion (Potter, 2020). More equitable development has thus been obstructed in areas affected by the armed conflict despite the higher incomes brought about by the oil palm sector. On the other hand, less well documented are the cases of the positive impact the oil palm sector has brought to conflict areas. Anecdotal evidence from the Magdalena Medio region of the Central Zone suggests the important economic opportunity the sector can provide as an alternative to armed conflict. The town of Yarima in the department of Santander was severely affected by violence between guerilla and paramilitary groups. The arrival of an oil palm company in the region in the late 1980s provided an employment alternative and helped stabilise the social situation by the mid-1990s (Bozzi, 1998). Similarly, oil palm farmers in the North Zone described how the crop lowered risk during times of conflict as plantations do not require daily visits, and there is less concern over fruit theft.

Given the armed conflict and history of land inequality, land acquisition is one of the biggest barriers to oil palm expansion in Colombia. Without a national cadastre system, it is difficult to know if there are existing claims to land and if previous owners have been displaced. It is thus a risky investment for companies to purchase land and invest capital in plantations, only to later find out that there is an existing claim and the land must be returned. Several such incidents have occurred in the Colombian oil palm sector. For example, in the case of Las Pavas in the department of Bolívar, a consortium of oil palm growers from the North zone purchased land from which local residents had been forcefully displaced by paramilitaries (Marin-Burgos et al., 2015). Failed land reform attempts by the Colombian government have also contributed to these conflicts. In the Tumaco region of the Southwest production zone, communal Afro-Colombian lands were titled and sold to large oil palm companies (Martínez, 2016). These communities are still fighting for titles to 70,000 hectares of ancestral lands and the restitution of thousands of hectares of communal lands planted in commercial oil palm (Arenas, 2018). The case of oil palm in Tumaco demonstrates how multiple factors—political, socio-cultural, biophysical—can come together to create volatile conditions in the regional assemblage that result in conflict. Historical land inequality in the region resulting from government policies had left many smallholder oil palm farmers marginalised and confined to a small area of cultivation. In the early 2000s, the arrival of bud-rot disease devastated the local oil palm sector. The government-sponsored credit for replanting with the disease-resistant hybrid OxG palm was not made available to smallholders, citing the sophistication and labour-intensive requirements of the new variety (Arenas, 2018). As a result, many smallholders began planting coca to supply local guerilla groups, and the region remains a hotspot of coca cultivation and persistent land conflict today.

Government-sponsored crop substitution programmes have continued in the current stage of development. In addition to the USAID-MIDAS programme introduced under Plan Colombia, another illicit crop substitution programme was included in the peace accords signed with the FARC in 2016—the *Plan Nacional Integral de Sustitución* (PNIS). The programme promotes perennial crop alternatives such as oil palm, cocoa, and rubber, and according to the United Nations Office on Drugs and Crime (UNODC), the programme had enrolled 99,097 families in 56 municipalities of Colombia as of October 2019 (UNODC, 2019). Crop substitution programmes can lead to unintended, perverse environmental outcomes. In the Catatumbo region of the Central Zone, for example, crop substitution programmes increased smallholder oil palm expansion in the lowlands of Tibú and displaced coca growing to the adjacent forested highlands (Granados-Cabrera et al., 2020). These are nuanced, indirect land use dynamics that play out differently across regional assemblages.

Contradictory policies

More contemporary government zoning policies have contradicted smallholder support policies by facilitating large-scale oil palm expansion. An oil palm suitability analysis was conducted by the National Rural Planning Office (UPRA) in conjunction with Fedepalma which determined that over 16 million hectares of land were suitable for future oil palm expansion, equating to 14 percent of Colombia's national territory (UPRA, 2017). Much of these lands are situated in the vast eastern savannahs of the Orinoco region, the country's largest oil palm production zone. The region is considered one of the last agricultural frontiers on the planet and is being targeted by the government for further agro-industrial development.

In 2016, a law was introduced that targeted poverty-stricken areas with poor infrastructure as priority zones for rural, economic, and social development (*ZIDRES* in Spanish). Similar to the rural capitalisation efforts that have come to define previous periods of agrarian public policy, the rationale for the ZIDRES programme is to stimulate private sector investment in rural areas that create economic opportunities for smallholders. Soils in the Orinoco region, for instance, are highly acidic and require large-scale, capital-intensive chemical transformations. This is one reason why the East zone has the fewest smallholder alliances among the oil palm production zones (Rueda-Zárate & Pacheco, 2015). The ZIDRES policy provides financial incentives for companies to establish operations in these remote areas and waives existing legal limits on property sizes intended to prevent land accumulation. Critics argue that the ZIDRES are contradictory to the government's stated goals of integral rural land reform and will effectively exacerbate the concentration of land, particularly in favour of the large multinational companies that the law supports. The implications for the oil palm sector are vast. The ZIDRES cover 7.2 million hectares of land across Colombia, of which 5.5 million hectares (76 percent) coincide with areas deemed suitable for oil palm expansion, mostly in the Orinoco (Colombia Plural, 2018). Smallholders also have fewer opportunities to access capital. Of the total agricultural loans granted to the oil palm sector by the government in 2018, smallholders received just 3 percent of the financing, the rest going to medium and large-scale producers (Fedepalma, 2019).

Sustainability commitments to increase exports

With domestic markets saturated, the Colombian palm oil sector began adopting sustainability commitments to increase exports. The sector has historically been decoupled from international markets, facing little demand for sustainably produced palm oil. Instead, European importers of Colombian palm oil have largely initiated this shift and the first companies to become RSPO-certified in Colombia were those with the tightest links to international markets. Producers of the North zone in greater proximity to ports export a greater share of their palm oil and began pursuing certification earlier than those in the landlocked East zone, where most palm oil is destined for the domestic foods and biodiesel industries. Of the ten RSPO-certified companies in 2018, six were companies in the North zone, and only two were companies in the East zone (Fedepalma, 2019). The high transportation costs of shipping palm oil to ports have made internal markets more attractive to producers in the East zone. However, given increased domestic market saturation, East zone producers are left little choice but to pursue certification to increase exports. In 2013, the region exported only 8 percent of its palm oil, which increased to 35 percent in 2018 (Fedepalma, 2019). The ISCC, a European standard focusing on greenhouse gas emissions from biofuel feedstocks, is also becoming an important certification for Colombian producers selling palm oil to the EU for biodiesel.

The institutional response by the Colombian oil palm sector and the state has been to embrace these transnational certification standards in order to increase exports. Fedepalma has promoted the RSPO, intending to get 50 percent of the sector certified by 2021. Furthermore, the national zero deforestation agreement led by the Colombian government permits RSPO certification to demonstrate compliance. A promising development for the agreement has been its adoption by the largest biodiesel producer in Colombia—a vertically integrated refinery supplied by palm oil from 11 mills in the East zone that processes fruit from over 20 plantations covering roughly 40,000 hectares. The company is supporting its entire supply base to become RSPO-certified and zero deforestation compliant, showing the potential of the vertically integrated production model when committed to the adoption of sustainability standards.

The RSPO often dominates the discussion on the sustainability of the oil palm sector, but evidence of its ability to generate positive social and environmental outcomes is mixed, particularly for smallholders. In the North zone of Colombia, certified smallholders (RSPO and IFOAM—Organics International) used fewer chemical inputs, maintained more natural habitat on their farms, and received a higher price for certified fruits, but had lower yields due to organic management practices and did not show significant livelihood improvements compared to non-certified farmers (Furumo et al., 2019). The RSPO has been contested in Colombia and elsewhere, with opponents concerned that it legitimises a model of large-scale, agro-industrial expansion that puts the values of local communities at risk (Marin-Burgos et al., 2015; Ruysschaert & Salles, 2014). These risks are compounded by power asymmetries in the national interpretation process that favour industry interests and exclude smallholders (Brandi et al., 2015; Lee et al., 2011). The regional context and baseline socioeconomic conditions under which RSPO certification is adopted are increasingly understood as a determinant of its success in mitigating the socio-ecological trade-offs of oil palm cultivation. A national evaluation of RSPO certification on village well-being across Indonesia shows modest or even negative effects of RSPO on livelihoods; villages with commodity markets oriented towards subsistence rather than commercial production showed particularly negative livelihood impacts associated with oil palm expansion that RSPO certification was not able to overcome (Santika et al., 2020).

Conclusion

This chapter has demonstrated the importance of scale when considering the environmental and livelihood outcomes of enabling and regulatory policies in the oil palm sector. Colombia illustrates the different socio-ecological trajectories that oil palm expansion can take, not only between different regions or countries (i.e., Southeast Asia vs. Latin America) but also between different production zones *within* a single country. The major geographic oil palm production zones of Colombia represent trans-local governance assemblages of different human and non-human entities that shape the outcome of state-led enabling policies. The economic geography of palm oil production in Colombia influences where palm oil is marketed, with important implications for the adoption of sustainability commitments. Effective sustainability interventions, therefore, cannot simply be copied from one region and implemented in another. They must be tailored to local settings, highlighting the place-based context of policymaking and transfer. In settings with strong regionalisation like Colombia, a jurisdictional approach to sustainability governance might be particularly relevant. Jurisdictional approaches strive to create sustainability havens of standard-compliant actors along policy-relevant boundaries that are enforceable by local governments. These multi-stakeholder approaches have gained prominence in subnational REDD+ and other zero deforestation initiatives. They have also been applied to the sustainable production of palm oil (e.g., state-wide RSPO certification in Sabah, Malaysia and Central Kalimantan, Indonesia).

The assemblage of sustainability governance in the Colombian oil palm sector is a valuable case study in policy mobility. It considers the uptake of transnational regulatory mechanisms in an industry that has been historically decoupled from international markets. The state and sector have been important enablers of oil palm expansion in Colombia under the pretence of peacebuilding, rural development, and biodiesel expansion. More recently, the saturation of domestic markets and the growing need to export palm oil has led these institutions to adopt a more regulatory role by advancing a sustainability narrative for Colombian palm oil. To this end, RSPO certification has been embraced but also extended with the formation of novel, public–private governance arrangements that link oil palm sustainability with rural development, zero deforestation, and climate change domains. The multiplicity of actors and scales incites recombination and the emergence of increasingly complex policy assemblages. These assemblages are presented with opportunities to leverage and engage new resources and actors. However, they are also faced with challenges relating to the additionality, interaction, and coordination among interventions at different scales.

In Colombia, oil palm cultivation has created employment opportunities and brought greater income for farmers. However, institutional land conflicts have led to trade-offs that have kept the industry from reaching a greater potential for rural development. These boil down to issues of land access and rights, where the interests of large-scale landholders have often been prioritised over smallholders. In the post-conflict period, there is a major opportunity and imperative to address these inequalities through integrated rural land reform. The window for improved governance created by the peace deal—alongside a more favourable land use history—creates an opportunity for Colombia to lead sustainability in the oil palm sector. It will not be enough to simply prevent deforestation and avoid conflicts with communities. The industry needs to proactively seek ways to enhance smallholder inclusion and value-capture, ensure social safeguards, and undertake ecosystem restoration activities to replenish forests and biodiversity that have already been lost in oil palm landscapes.

References

Allen, J., & Cochrane, A. (2007). Beyond the territorial fix: Regional assemblages, politics and power. *Regional Studies*, *41*(9), 1161–1175. https://doi.org/10.1080/00343400701543348

Alvarado, A., Escobar, R., & Henry, J. (2013). The hybrid OxG Amazon: An alternative for regions affected by bud rot in oil palm. *Palmas*, *34*(1), 305–314. https://publicaciones.fedepalma.org/index.php/palmas/article/download/10689/10674

Andonova, L. B. (2013). Boomerangs to partnerships? Explaining state participation in transnational partnerships for sustainability. *Comparative Political Studies*, *47*(3), 481–515. https://doi.org/10.1177/0010414013509579

Apontes, D. M. (2008). Oil palm cultivation with USAID-MIDAS support: An example of alternative development in Colombia. *Palmas*, *29*. https://publicaciones.fedepalma.org/index.php/palmas/article/download/1377/1377/

Arenas, N. (2018, December 3). *Colombia: Oil palm in the midst of land conflicts in Tumaco*. Mongabay. https://es.mongabay.com/2018/12/restitucion-de-tierras-palma-de-aceite-colombia/

Astari, A. J., & Lovett, J. C. (2019). Does the rise of transnational governance "hollow-out" the state? Discourse analysis of the mandatory Indonesian sustainable palm oil policy. *World Development*, *117*, 1–12. https://doi.org/10.1016/j.worlddev.2018.12.012

Ayala Díaz, I. M., & Romero, H. M. (2019). OxG hybrid cultivars and the productive reactivation of problem areas with PC. *XV Reunión Técnica Nacional de Palma de Aceite*.

https://www.cenipalma.org/wp-content/uploads/2019/09/1.Ivan-Ayala-OxG-RTN-2019_compressed.pdf

Bozzi, M. L. O. (1998). *The African palm in Colombia: Notes and memories* (Vol. 2). Fedepalma.

Bozzi, M. L. O., & Jaramillo, D. O. (1998). *The African palm in Colombia: Notes and memories* (Vol. 1). Fedepalma.

Brandi, C., Cabani, T., Hosang, C., Schirmbeck, S., Westermann, L., & Wiese, H. (2015). Sustainability standards for palm oil: Challenges for smallholder certification under the RSPO. *The Journal of Environment & Development, 24*(3), 292–314. https://doi.org/10.1177/1070496515593775

Briassoulis, H. (2019). Governance as multiplicity: The assemblage thinking perspective. *Policy Sciences, 52*(3), 419–450. https://doi.org/10.1007/s11077-018-09345-9

Carreño, Á. S., & López, G. M. (2009). National plan for the management of bud rot. *Palmas, 30*(3), 97–121. https://publicaciones.fedepalma.org/index.php/palmas/article/view/1457

Castiblanco, C., Etter, A., & Aide, T. M. (2013). Oil palm plantations in Colombia: A model of future expansion. *Environmental Science & Policy, 27*, 172–183. https://doi.org/10.1016/j.envsci.2013.01.003

Castiblanco, C., Etter, A., & Ramirez, A. (2015). Impacts of oil palm expansion in Colombia: What do socioeconomic indicators show? *Land Use Policy, 44*, 31–43. https://doi.org/10.1016/j.landusepol.2014.10.007

Centro Nacional de Memoria Histórica. (2016). *Land and rural conflicts. History, agrarian policies, and protagonists*. CNMH.

Colombia Plural. (2018, March 24). *ZIDRES: In the shadow of the African palm*. Colombia Plural. https://colombiaplural.com/zidres-la-sombra-la-palma-africana/

Cramb, R. A., & McCarthy, J. F. (2016). *The oil palm complex: Smallholders, agribusiness, and the state in Indonesia and Malaysia*. NUS Press.

Curtis, P. G., Slay, C. M., Harris, N. L., Tyukavina, A., & Hansen, M. C. (2018). Classifying drivers of global forest loss. *Science, 361*(6407), 1108–1111. https://doi.org/10.1126/science.aau3445

Dauvergne, P., & Neville, K. J. (2010). Forests, food, and fuel in the tropics: The uneven social and ecological consequences of the emerging political economy of biofuels. *The Journal of Peasant Studies, 37*(4), 631–660. https://doi.org/10.1080/03066150.2010.512451

Etter, A., McAlpine, C., & Possingham, H. (2008). Historical patterns and drivers of landscape change in Colombia since 1500: A regionalised spatial approach. *Annals of the Association of American Geographers, 98*(1), 2–23. https://doi.org/10.1080/00045600701733911

European Parliament. (2017, April 4). *Palm oil and deforestation of rainforests*. https://www.europarl.europa.eu/doceo/document/TA-8-2017-0098_EN.html

FAO. (2020). *FAOSTAT*. Food and Agriculture Organisation of the United Nations. https://www.fao.org/faostat/en/#home

Fedepalma. (2002). *Statistical yearbook 2002: The oil palm agroindustry in Colombia and the world 1997–2001*. Centro de Información y Documentación Palmero. https://publicaciones.fedepalma.org/index.php/anuario/article/view/10480/10470

Fedepalma. (2013). Statistical yearbook 2013: The oil palm agroindustry in Colombia and the world 2008–2012. In *Centro de Información y Documentación Palmero*. https://publicaciones.fedepalma.org/index.php/anuario/article/view/11016/11001

Fedepalma. (2019). *Statistical yearbook 2019: The oil palm agroindustry in Colombia and the world*. Centro de Información y Documentación Palmero. https://publicaciones.

fedepalma.org/index.php/anuario/issue/view/1452/La%20agroindustria%20de%20
la%20palma%20de%20aceite%20en%20Colombia%20y%20en%20el%20mundo

Fedepalma. (n.d.). La Palma Es Vida; Cenipalma. https://lapalmaesvida.com/

Furumo, P. R., & Aide, T. M. (2017). Characterising commercial oil palm expansion in Latin America: Land use change and trade. *Environmental Research Letters, 12*(2), 024008. https://doi.org/10.1088/1748-9326/aa5892

Furumo, P. R., & Lambin, E. F. (2020). Scaling up zero deforestation initiatives through public-private partnerships: A look inside post-conflict Colombia. *Global Environmental Change, 62*, 102055. https://doi.org/10.1016/j.gloenvcha.2020.102055

Furumo, P. R., & Mitchell Aide, T. (2019). Correction to: Using soundscapes to assess biodiversity in Neotropical oil palm landscapes. *Landscape Ecology, 34*(5), 1195–1195. https://doi.org/10.1007/s10980-019-00833-8

Furumo, P. R., Rueda, X., Rodríguez, J. S., & Parés Ramos, I. K. (2020). Field evidence for positive certification outcomes on oil palm smallholder management practices in Colombia. *Journal of Cleaner Production, 245*, 118891. https://doi.org/10.1016/j.jclepro.2019.118891

Garcia-Ulloa, J., Sloan, S., Pacheco, P., Ghazoul, J., & Koh, L. P. (2012). Lowering environmental costs of oil-palm expansion in Colombia. *Conservation Letters, 5*(5), 366–375. https://doi.org/10.1111/j.1755-263x.2012.00254.x

Garrett, R. D., Levy, S., Carlson, K. M., Gardner, T. A., Godar, J., Clapp, J., Dauvergne, P., Heilmayr, R., le Polain de Waroux, Y., Ayre, B., Barr, R., Døvre, B., Gibbs, H. K., Hall, S., Lake, S., Milder, J. C., Rausch, L. L., Rivero, R., Rueda, X., & Sarsfield, R. (2019). Criteria for effective zero deforestation commitments. *Global Environmental Change, 54*, 135–147. https://doi.org/10.1016/j.gloenvcha.2018.11.003

Gerasimchuk, I., & Koh, P. Y. (2013). *The EU biofuel policy and palm oil: Cutting subsidies or cutting rainforest?* (pp. 1–19). The International Institute for Sustainable Development. https://www.foeeurope.org/sites/default/files/publications/iisd_eu_biofuel_policy_palm_oil_september2013_0.pdf

Granados-Cabrera, O. A., Rincón-Romero, V. O., Arango, E., & Arias, N. (2020). Oil palm in Puerto Wilches: Actors and transformation processes (1960–2016). *Anuario de Historia Regional Y de Las Fronteras, 26*(1). https://doi.org/10.18273/revanu.v26n1-2020004

Guereña, A. (2017). *X-ray of inequality: What the latest agricultural census tells us about the distribution of land in Colombia.* Oxfam International. https://oi-files-d8-prod.s3.eu-west-2.amazonaws.com/s3fs-public/file_attachments/radiografia_de_la_desigualdad.pdf

Hamilton-Hart, N. (2014). Multilevel (mis)governance of palm oil production. *Australian Journal of International Affairs, 69*(2), 164–184. https://doi.org/10.1080/10357718.2014.978738

Heilmayr, R., & Lambin, E. F. (2016). Impacts of nonstate, market-driven governance on Chilean forests. *Proceedings of the National Academy of Sciences, 113*(11), 2910–2915. https://doi.org/10.1073/pnas.1600394113

Hospes, O. (2014). Marking the success or end of global multi-stakeholder governance? The rise of national sustainability standards in Indonesia and Brazil for palm oil and soy. *Agriculture and Human Values, 31*(3), 425–437. https://doi.org/10.1007/s10460-014-9511-9

Jessop, B. (2013). Hollowing out the "nation-state" and multi-level governance. In P. Kennett (Ed.), *A Handbook of Comparative Social Policy* (pp. 11–26). Edward Elgar Publishing. https://doi.org/10.4337/9781782546535.00008

Lambin, E. F., Gibbs, H. K., Heilmayr, R., Carlson, K. M., Fleck, L. C., Garrett, R. D., le Polain de Waroux, Y., McDermott, C. L., McLaughlin, D., Newton, P., Nolte, C., Pacheco, P., Rausch, L. L., Streck, C., Thorlakson, T., & Walker, N. F. (2018). The role of supply-chain initiatives in reducing deforestation. *Nature Climate Change, 8*(2), 109–116. https://doi.org/10.1038/s41558-017-0061-1

Lambin, E. F., Kim, H., Leape, J., & Lee, K. (2020). Scaling up solutions for a sustainability transition. *One Earth, 3*(1), 89–96. https://doi.org/10.1016/j.oneear.2020.06.010

Law 101. (1993). *General Law of Agricultural and Fisheries Development.* The Congress of Colombia.

Lee, J. S. H., Rist, L., Obidzinski, K., Ghazoul, J., & Koh, L. P. (2011). No farmer left behind in sustainable biofuel production. *Biological Conservation, 144*(10), 2512–2516. https://doi.org/10.1016/j.biocon.2011.07.006

Leibold, M. A., Holyoak, M., Mouquet, N., Amarasekare, P., Chase, J. M., Hoopes, M. F., Holt, R. D., Shurin, J. B., Law, R., Tilman, D., Loreau, M., & Gonzalez, A. (2004). The metacommunity concept: A framework for multi-scale community ecology. *Ecology Letters, 7*(7), 601–613. https://doi.org/10.1111/j.1461-0248.2004.00608.x

Marin-Burgos, V., & Clancy, J. S. (2017). Understanding the expansion of energy crops beyond the global biofuel boom: Evidence from oil palm expansion in Colombia. *Energy, Sustainability and Society, 7*(1). https://doi.org/10.1186/s13705-017-0123-2

Marin-Burgos, V., Clancy, J. S., & Lovett, J. C. (2015). Contesting legitimacy of voluntary sustainability certification schemes: Valuation languages and power asymmetries in the Roundtable on Sustainable Palm Oil in Colombia. *Ecological Economics, 117*, 303–313. https://doi.org/10.1016/j.ecolecon.2014.04.011

Martínez, A. G. P. (2016). *Business colonisation and land concentration: The palm oil cultivation in Tumaco.* Instituto Colombiano de Antropología e Historia. https://repositorio.uniandes.edu.co/server/api/core/bitstreams/0b7d72b4-b7f6-417e-86c8-2ea8c92bc29e/content

McCann, E., & Ward, K. (2013). A multi-disciplinary approach to policy transfer research: Geographies, assemblages, mobilities and mutations. *Policy Studies, 34*(1), 2–18. https://doi.org/10.1080/01442872.2012.748563

McFarlane, C. (2009). Translocal assemblages: Space, power and social movements. *Geoforum, 40*(4), 561–567. https://doi.org/10.1016/j.geoforum.2009.05.003

Montoya, M. M., Villabona, M. V., Álvarez, E. R., Alfonso, D. L., Zamudio, L. E. C., Díaz, C. A. F., & Arenas, M. A. G. (2017). Production costs for oil palm fruit and palm oil in 2015: Estimation in a group of Colombian producers. *Palmas, 38*(2), 10–26. https://publicaciones.fedepalma.org/index.php/palmas/article/view/12122

Núñez, C. E., & Hurtado, I. P. (2014). *Forced displacement in Colombia: The footprint of the conflict.* CODHES.

Ocampo-Peñuela, N., Garcia-Ulloa, J., Ghazoul, J., & Etter, A. (2018). Quantifying impacts of oil palm expansion on Colombia's threatened biodiversity. *Biological Conservation, 224*, 117–121. https://doi.org/10.1016/j.biocon.2018.05.024

Oslender, U. (2008). Another history of violence: The production of "geographies of terror" in Colombia's Pacific Coast region. *Latin American Perspectives, 35*(5), 77–102. https://www.jstor.org/stable/27648121

Pacheco, P., Schoneveld, G., Dermawan, A., Komarudin, H., & Djama, M. (2018). Governing sustainable palm oil supply: Disconnects, complementarities, and antagonisms between state regulations and private standards. *Regulation & Governance, 14*(3). https://doi.org/10.1111/rego.12220

Pirker, J., Mosnier, A., Kraxner, F., Havlík, P., & Obersteiner, M. (2016). What are the limits to oil palm expansion? *Global Environmental Change*, *40*, 73–81. https://doi.org/10.1016/j.gloenvcha.2016.06.007

Potter, L. (2020). Colombia's oil palm development in times of war and "peace": Myths, enablers, and the disparate realities of land control. *Journal of Rural Studies*, *78*. https://doi.org/10.1016/j.jrurstud.2019.10.035

Prince, R. (2016). Local or global policy? Thinking about policy mobility with assemblage and topology. *Area*, *49*(3), 335–341. https://doi.org/10.1111/area.12319

Richardson, D. L. (1995). The history of oil palm breeding in the United Fruit Company. *ASD Oil Palm Papers (Costa Rica)*, *11*(1). https://asd-cr.com/wp-content/uploads/2022/10/ASD-OPP-No11-1995.pdf

Rueda-Zárate, A., & Pacheco, P. (2015). *Policies, markets and production models: An analysis of the situation and challenges of the Colombian palm sector*. Centre for International Forestry Research. https://doi.org/10.17528/cifor/005658

Ruysschaert, D., & Salles, D. (2014). Towards global voluntary standards: Questioning the effectiveness in attaining conservation goals. *Ecological Economics*, *107*, 438–446. https://doi.org/10.1016/j.ecolecon.2014.09.016

Sabogal, C. R. (2013). Special analysis of the correlation between oil palm cultivation and forced displacement in Colombia. *Cuadernos de Economía*, *32*(61), 683–719. https://www.redalyc.org/articulo.oa?id=282130076003

Santika, T., Wilson, K. A., Law, E. A., St John, F. A. V., Carlson, K. M., Gibbs, H., Morgans, C. L., Ancrenaz, M., Meijaard, E., & Struebig, M. J. (2020). Impact of palm oil sustainability certification on village well-being and poverty in Indonesia. *Nature Sustainability*, *4*. https://doi.org/10.1038/s41893-020-00630-1

Santo, R., & Moragues-Faus, A. (2019). Towards a trans-local food governance: Exploring the transformative capacity of food policy assemblages in the US and UK. *Geoforum*, *98*, 75–87. https://doi.org/10.1016/j.geoforum.2018.10.002

Savage, G. C. (2019). What is policy assemblage? *Territory, Politics, Governance*, *8*(3), 1–17. https://doi.org/10.1080/21622671.2018.1559760

Silva, A., & Martínez, G. (2009). National plan for the management of bud rot: Fedepalma - Cenipalma. *Palmas*, *30*(3), 97–121.

Stroud, J. T., Bush, M. R., Ladd, M. C., Nowicki, R. J., Shantz, A. A., & Sweatman, J. (2015). Is a community still a community? Reviewing definitions of key terms in community ecology. *Ecology and Evolution*, *5*(21), 4757–4765. https://doi.org/10.1002/ece3.1651

UNODC. (2019). *Report no. 19: Comprehensive national programme for the substitution of illicit crops—PNIS* (pp. 1–61). United Nations Office of Drugs and Crime. https://www.unodc.org/documents/colombia/2020/Febrero/INFORME_EJECUTIVO_PNIS_No._19.pdf

UPRA. (2017). *Colombia: 16 million hectares suitable for oil palm*. Unidad de Planificación Rural Agropecuaria. https://www.upra.gov.co/sala-de-prensa/noticias/-/asset_publisher/GEKyUuxHYSXZ/content/colombia-16-millones-de-hectareas-aptas-para-palma-de-aceite

Wilson, D. S. (1992). Complex interactions in metacommunities, with implications for biodiversity and higher levels of selection. *Ecology*, *73*(6), 1984–2000.s https://doi.org/10.2307/1941449

7

MAKING SUSTAINABLE PALM OIL? DEVELOPMENTALIST AND ENVIRONMENTAL ASSEMBLAGES IN THE BRAZILIAN AMAZON

Diana Córdoba, Renata Moreno and Daniel Sombra

Introduction

In February 2008, the Brazilian federal government launched Operation *Arco de Fogo* (Arch of Fire) in the northern state of Pará as part of its strategy to combat deforestation in the Amazon—a subject garnering strong national and international concern since the 1990s (Josenaldo, 2008). More than 1,000 federal and state soldiers and policemen destroyed 1,326 illegal charcoal kilns, seized 245 tonnes of illegally logged timber, closed about 50 unlicensed sawmills, and imposed R$31.8 million in fines (Josenaldo, 2008). The deforestation occurred alongside acute social conflict and alarming violence rooted in land disputes and precarious working conditions in the sawmills and charcoal ovens (Nepstad et al., 2001). The Brazilian state saw this crisis as an opportunity to address environmental concerns over deforestation, advance a developmentalist agenda via agribusiness development, and meet social justice objectives through the economic integration of small farmers.

In 2010, Brazil introduced the *Programa de Produção Sustentável do Óleo da Palma* (Sustainable Oil Palm Production Programme/SPOPP), which aimed to facilitate the expansion of large-scale plantations for biodiesel production throughout the region. Then-President Lula da Silva of the *Partido dos Trabalhadores* (PT or Workers' Party) discursively connected oil palm, socio-economic development, conservation, and the social and environmental crises in the Amazon region. Speaking to an audience of peasants, agribusinessmen, and local authorities at SPOPP's launch, Lula hailed it as a sustainable initiative for the Amazon:

> The programme we are launching today opens a new horizon of possibilities for Brazil and the Amazon region. It *represents the marriage between environmental*

DOI: 10.4324/9781003459606-9

protection and generation of income and decent employment for thousands of people who live in the Amazon. Today, Brazilians can proudly say that we protect the Amazon, one of the greatest natural patrimonies of the Planet.
(Biblioteca Presidencia da Republica, 2010, p. 4, emphasis added)

SPOPP promoted a new kind of palm oil governance that led with sustainability. Its policies aimed to prevent deforestation and recover degraded lands; facilitate the economic integration of family farmers through access to financial support, knowledge and technology transfer, and adequate training; in addition to developing the biodiesel sector through blending targets with diesel and specific support incentives (Brandão et al., 2021; da Mota et al., 2019). The SPOPP radically transformed the landscape in the Northeast Amazon, more than quadrupling the number of palm oil hectares from around 50,000 in 2009 to 226,835 in 2020 (Abrapalma, 2021).

Critical scholarship often frames the rapid expansion of palm oil in the Brazilian Amazon since 2010 as a top-down state-agribusiness alliance that dispossessed peasants and Afro-Brazilians from their land (e.g. Backhouse, 2013; Acevedo, 2010). SPOPP allowed agribusiness to incorporate new land stocks into the international commodities market, while farmers linked to palm oil cultivation did not reap the benefits (Cordoba et al., 2018; Backhouse, 2013). More mainstream scholars examine the performance and efficiency of combining conservation and developmentalist aims in regional development. They assess the impacts on job generation and the rehabilitation of "degraded" areas (Becker, 2010; Santos, 2008), the livelihoods of family farmers (Brandão et al., 2021, 2019), food security (dos Santos, 2013), and deforestation (Brandão et al., 2019; Benami, 2018).

However, little work has examined how SPOPP fits within longer histories of multiscalar policy interventions aimed at balancing environmental and developmental management and sustainability in the Amazon. This chapter aims to fill this gap by challenging cause-and-effect understandings of policy implementation, impact, and regional transformation. Drawing on assemblage thinking, we critically examine how human and non-human actors (including previous and new environmental and developmentalist policies) came together in SPOPP's implementation to tackle the ongoing socio-environmental crisis in the Brazilian Amazon.

For De-Landa (2006, p. 5), assemblages "are wholes whose properties emerge from the interactions between parts". They do not work through unidirectional domination and exploitation relations between the parts, but in myriad, complex ways through both internal relations and in connection with external entities. We argue that palm oil governance is not unified, but rather an assemblage of heterogeneous elements shaped by the actions of different actors over time. Within this oil palm assemblage, different actors try to define relationships to reflect their perceived interests.

The chapter is organised as follows: The first section introduces assemblage thinking and clarifies our approach to this study of palm oil expansion and its related policies. The second section describes the study methodologies and data collection. The third section offers a historical perspective to explain how palm oil is entangled with past environmental and development policies in the Amazon. The fourth section presents our results in three parts. The first and second parts explore the process of state formation since 2003, when PT won federal control and deployed the SPOPP. These sections analyse how social inclusion policies and environmental strategies towards sustainability were assembled. The third part explores how both bud rot (*Amarelecimento fatal*), a disease affecting oil palms, and the specific materiality of palm oil biofuel hindered SPOPP's success. We conclude by reflecting on how SPOPP actors, within and beyond the state, tried to garner support for their developmentalist interests and exclude or depoliticise environmental and social concerns, thus limiting change.

By covering these aspects, this chapter seeks to engage the book's three main research objectives with regard to the Brazilian context: how do different governance arrangements facilitate and legitimise the expansion of palm oil industry; how do power differentials impact oil palm governance; and how does governance enable the accumulation of wealth?

Assemblage thinking and the multiplicity of palm oil governance

Assemblage thinking helps us understand how seemingly contradictory environmentalist and developmentalist sustainable governance objectives were merged in the SPOPP. Murray Li (2007, p. 266) defines an assemblage as a "gathering of heterogeneous elements consistently drawn together as an identifiable terrain of action and debate". She adds that the process of assembling can include forging alignments, rendering technical, authorising knowledge, managing failures and contradictions, anti-politics, and reassembling (Murray Li, 2007). An assemblage approach decentres any claim that policy outcomes are the result of rational measures underpinned exclusively by state capacities, capital, technology, or coherently-designed institutional arrangements. According to Deleuze and Guattari (1988), assemblages are not about an outcome, but rather the process of reconstitution itself. Assemblage thinking centres policies, technologies, social relations, materiality, and relations to nature that come together in the change, maintenance, extension, and operation of actors' agency (Delanda, 2016; Carolan, 2013).

Multiple assemblages with the same elements may coexist and overlap. Change in one assemblage is related to change in another; they are influenced by "relations of exteriority" (Deleuze & Guattari, 1988). This means that an assemblage emerging around SPOPP implementation in the state of Pará could emerge through the interaction of actors located in different space-temporal locations. Transformations in an assemblage are produced through processes of "territorialisation" and "deterritorialisation". Territorialisation stabilises the assemblage's identity, increasing its degree of internal homogenisation and sharpens its boundaries. Deterritorialisation changes the assemblage, sometimes so much that a new one is formed (Deleuze & Guattari, 1988). Territories are constantly changing—"assembling", "disassembling", and "reassembling" complex networks of relationships that interact and overlap with discourses, practices, and materiality (Fox & Alldred, 2013; Anderson & McFarlane, 2011; Massey, 2004).

Following Murray Li (2007) and Deleuze and Guatarri (1988), this chapter operationalises assemblage thinking in two ways. First, we map how policies and human entities' diverse agendas come together to address environmental and social inclusion issues via oil palm governance and "what they can bring into being" in policy implementation (Deleuze & Guattari, 1988, p. 4). Palm oil assemblages are provisional territories where "relations may change, new elements may enter, alliances may be broken and new conjunctions may be fostered" (Anderson & McFarlane, 2011, p. 126). Territorialisation may refer to the various components of the SPOPP as an assemblage of developmentalist and environmentalist policies and actors to enact palm oil expansion in Northeast Pará.

Second, we use assemblage thinking to analyse the non-human elements of the palm oil assemblage and how they co-constitute with human entities through interactions. In assemblage processes, nature and material objects are equal participants in social relations and the construction of knowledge. This blurs conventional ideas of politics and institutional practices as a purely human affair. Rather, actors can be natural and social (socio-natural), and their agency is conditioned through their relationships with other actors over time. Human and non-human entities are arranged into socio-material relationships for certain strategic ends to change the very nature of policy implementation. Entities such as "persons, artefacts, plants, organisations, documents, beliefs, [and] technology" (Briassoulis, 2019, p. 425) can affect other entities linked in assemblage relations.

These entities are "relatively autonomous, have multiple memberships, variable spatiotemporal reach and importance, and play material and symbolic/expressive roles" (Briassoulis, 2021, p. 5). A crop or a disease, for example, is a non-human agent that shapes how policies are enacted. We examine how entities (components of an assemblage) are linked in relationships that determine how different components affect one another. For example, palm oil's physical characteristics and growing conditions shape other entities within the chain (i.e. production, labour, location of processing plants, and distribution networks).

Palm oil assemblages encompass historical conditions and future expectations that will shape the power struggle between territorialising and deterritorialising forces (Pløger, 2008). New territorialisations invoke new elements and the reduction and/or removal of others. They also reconfigure existing entities, practices, and discursive meanings to construct new relationships with places. An assemblage "establishes connections between certain multiplicities" involved in socio-natural relations (Deleuze & Guttari, 1988, p. 21); every socio-natural actor can influence the policy implementation process. However, not all actors have the same influence or power—their influence depends on the scope, size, and resources of their networks. Within assemblages, actors seek to re-define relationships to better reflect their perceived interests, with more powerful actors having a greater capacity to actually do so. In palm oil assemblages, certain entities attempt to incorporate others into stable and coherent relationships (territorialities) that serve their strategic interests. This theory of power within the assemblage helps explain how inclusion, dominance, and exclusion of developmentalist and environmentalist views are justified and achieved.

Methods

This chapter draws on fieldwork carried out in three phases between 2014 and 2021. In October 2014, we conducted 19 interviews with oil palm family farmers and 17 interviews with non-governmental organisation (NGO) representatives, local government officials, and policymakers in the municipalities of Concordia do Pará, Tailândia, and Moju in Northeast Pará. Interview questions focused on the interviewees' perceptions of changing job opportunities, their quality of life and land use, contract farming relationships, as well as the impact of palm plantations on the community and access to land.

The next phases investigated the diverse actors, scenarios, dynamics, scales, and processes involved in SPOPP. In October 2014, as well as May and October 2015, we conducted 35 interviews with academics, researchers, social movement leaders, and government officials in the municipalities of Tailândia, Moju, and the state capitals of Belém and Brasilia. Follow-up interviews were carried out in October 2016 with one agribusiness representative, four policymakers, and one palm oil industry representative in May 2021. Following Saldaña (2015), the qualitative data were coded using NVivo 11 based on emerging key themes. The interviews were supplemented by a document analysis of government policies and programmes and notes from ongoing fieldwork in Pará. Our fieldwork took place at a conjunctural moment (2014–2016), during which SPOPP was hailed by policymakers, agribusiness, and participating family farmers as a government initiative that supported both agribusiness and family farming while preserving the environment.

The territorialisation of palm oil in the Brazilian Amazon

The development of palm oil in the Brazilian Amazon is rooted in projects designed to incorporate the territory into the rest of Brazil during the military dictatorship (1964–1985). To advance their territorial occupation, the military government created a new developmental economic model for the "empty" region (Velho, 2009). The 1964 *Estatuto da Terra* increased the migration of landless peasants, responding to a promise of socio-economic assistance (Ianni, 1979). The 1971 national development plan re-organised the Amazonian territory around "integration corridors" of development that disregarded local history, people, and physical characteristics (Velho, 2009). Brazilian legislation accelerated the rapid deforestation of the

Amazon through the construction of roads through primary forests, the nationalisation of millions of square kilometres of land, and tax incentives for agribusinesses. The state also incentivised large-scale agriculture, while simultaneously introducing policies to deliver socio-economic development (Martins, 1981). The low taxes on farm incomes incentivised large-scale ranchers to convert forests into pastures.

The military regime believed the palm oil agroindustry would generate domestic agro-industrial modernisation and development. Seeds were collected from oil palms (*Elaeis guineensis*) introduced to Bahía by enslaved peoples from West Africa in the sixteenth century (Watkins, 2018). The Ministry of Agriculture first introduced these seeds to Pará in 1942. The variety's high productivity, the region's adequate infrastructure and favourable tax incentives from the military government led to the first agro-industrial palm oil projects in Northwest Pará in the 1960s and 1970s (Homma et al., 2000).

In the 1990s, a "new phase" for the Amazonian frontier marked further integration into neoliberal international markets (Loureiro, 2009). In search of pest-free land, agribusinesses opened palm oil plantations in the Tomé-Açu region. Stimulated by strong state support, the planted area reached 75,000 hectares, legitimising the potential of palm oil agribusiness (Homma et al., 2000) and intensifying the conversion of primary forests into palm oil plantations. The Agropalma Group, a Brazilian national business conglomerate, built a territorial complex in the Tome-Açu region in partnership with the state that included its plantations, mills and refinery. Agropalma, which produced for both the national food industry and export, practically monopolised palm oil production chains. New infrastructure (roads, ports and airports) and the creation of the first palm oil refining company in Brazil connected the industry's command centres and production areas (da Rocha, 2015). This vertical integration of agribusiness was made possible by a surplus labour supply from the municipalities of Acará, Moju, and Tailândia.

This growth deepened inequalities among producers. For example, growth in the palm sector increased gaps in access to capital, seeds, and production technology between small and large producers. Agribusiness actors also successfully lobbied the Brazilian government for credit allocations and price levels that would only benefit large-scale producers. Large palm oil companies, together with local authorities and public institutions, only began to actively promote small palm oil production in the late 1990s (Nahum & Malcher, 2012). The Pará government organised an official visit to Malaysia in 1995 to learn about palm oil contract farming between agroindustry and peasant communities. Upon returning, the state government and the palm oil industry devised a contract farming pilot model for small producers (Vieira & Magalhães, 2013). These contract farming arrangements increased the agribusiness actors' ability to territorialise relations between themselves and other linked entities.

Environmental groups and social movements proliferated during the military dictatorship in response to the geopolitical development projects in the Amazon.

This growth gained momentum from 1985 to 1996 as NGOs, religious organisations and political parties all hoped to territorialise rainforest development based on global environmental discourse (Acselrad, 2010). These responses influenced regional dynamics by spurring projects for alternative development within the state (Gabinete de Segurança Institucional, 2004). Environmental movements advocated for the creation of a Ministry of the Environment and for state-led initiatives to demarcate indigenous territories and conservation areas (e.g. the Pilot Programme for the Protection of the Brazilian Tropical Forests) (Serrao & Thompson, 2005). The "professionalised" environmental movement (Alonso & Maciel, 2010) that dominated the Amazon in the late 1990s both resisted and cooperated with the dominant developmentalist approach, which intensified during the formation of the neoliberal state. Ultimately, some social movement members achieved positions of power to subvert neoliberal objectives from within (Alonso & Maciel, 2010).

Developmentalism and environmentalism in palm oil policy assemblages

The first palm oil boom in Pará resulted from diverse arrangements and relationships between developmentalist and environmentalist goals to expand or resist large-scale palm oil production. The following sections analyse how the second palm oil boom, beginning in the 2000s, configured a new palm oil assemblage that merged social inclusion and environmental protection through the SPOPP.

Palm oil governance as social inclusion: SSPOP's contract farming for biodiesel production

The election of Luiz Inácio Lula da Silva and the PT in 2003 gave a new direction in state support for the expansion of palm oil in the Amazon. After 21 years of military rule, followed by 17 years of centre-right governments, Lula took power amid intense social conflicts in rural areas prompted by widespread discontent with the neoliberal project. Between 2004 and 2005, the PT government undertook several initiatives to adopt biofuels. First, the PT instituted an Inter-Ministerial Executive Commission to investigate the production and use of vegetable oil for energy purposes. The commission also researched the viability of biodiesel production and consumption, as well as the legal frameworks needed to introduce it into the country's energy matrix. Second, the PT created the National Biodiesel Production and Usage Programme (PNPB) (BRASIL, 2004). The Federal Government defined lines of finance and restructured the productive chain and the technological basis of biodiesel and its regulatory framework (Córdoba et al., 2018).

Third, the PNPB framed biodiesel as a "social fuel" with supposed benefits for the environment and the social inclusion of family farmers through contract farming schemes and rural proletarianization. The PNPB, in addition to requiring a mix of 5 percent of biodiesel to diesel, assigned the management of the biodiesel industry to the National Petroleum Agency (ANP). The government also promoted biodiesel market access for farmer unions and farmer cooperatives by creating the so-called "social fuel seal" (Leite et al., 2014). Both new and established biodiesel companies needed to involve family farmers in their productive arrangements to participate in ANP-promoted auctions and to take advantage of federal economic and tax incentives.[1]

The PNPB forged new relationships within the palm oil assemblage through a framework that articulated large-scale plantation, climate change mitigation, and social inclusion aims. Palm oil was the biodiesel feedstock chosen for Pará due to the state's favourable soil and climate conditions for growing the plant (Santos, 2008). Government documents speculated that the region could become a major global producer of palm oil biodiesel (Ferreira et al., 2010; Müller & Furlan-Júnior, 2001). The PT government launched the SPOPP in 2010, providing a solid discursive and material base for new capitalist formations in the Amazon territory (Presidência da República, 2010). Resources and people were mobilised in different modes of production, ranging from smallholder plantations under contract farming schemes to joint ventures with medium-scale farmers and large plantations managed by multinational corporations.

The assemblage of developmentalist SPOPP policy goals aimed to achieve a more diversified and technologically advanced economic structure under the premise that a more socially inclusive economy required agribusiness to gain wealth before wealth redistribution could occur. The state helped legitimise agribusiness accumulation by promoting the inclusion of small contract farmers. The SPOPP, through *Programa Nacional de Fortalecimento da Agricultura Familiar* (PRONAF) credits, supported a maximum of 10 hectares for each family under the contract farming scheme to supply palm fruit to agribusinesses (Brandão et al., 2021). As Córdoba et al. (2018) argue, SPOPP assumed that family farming and agribusiness could coexist and even create synergistic alliances to advance both sectors. Contract farming, however, contributed simultaneously to capitalist accumulation and further commoditisation of the crops and farmers' livelihoods (Little & Watts, 1994). SPOPP promotional materials promised family farmers a future with regular access to larger incomes so that they could overcome the limits of slash-and-burn agriculture. This "win-win" scheme, the government argued, would also protect biodiversity and enhance environmental sustainability (Presidência da República, 2010).

When describing their decision to opt for palm oil, most small producers began with the precarious conditions at the edge of the agrarian frontier. Basic services such as sewage, electricity, piped water, roads, and schools were largely unavailable. Contract farmers understood the project as a desirable state intervention that could support these basic services. When asked about improvements to quality of life, interviewees connected the palm oil expansion with increased investment in local infrastructure (water, energy, and roads) carried out by the PT government (2003–2016). One palm oil grower explains this relationship:

> With the entry of palm, the situation has improved because we did not have electricity. Nor did we have these houses. We managed to build this house; although it is still unfinished, we did it. We have drinking water that we did not have, which is for everyone in the community, electricity, [and] roads too.
>
> *(Interview 223)*

This farmer's interaction with the SPOPP assemblage suggests that these small-scale producers are knowledgeable and motivated, not slavishly incorporated into contract farming by agribusiness.

Another producer compared palm oil's materiality and advantages with other local cash monocrops. Since crops like black pepper can be sold almost anywhere, they are subject to widespread theft. In contrast, contracted palm is only sold to the agribusiness sector under the contract farming model and is never stolen. He explained:

> We live in a region where you cannot produce anything if you are not looking after the crop 24 hours because people will steal it. With palm, it is different because you can leave it without security, nobody will be interested in it. Nobody can steal your product and take it to sell to companies.
>
> *(Interview 224)*

However, state interventions to support contract farming and agribusiness technology have never met the assemblage's social inclusion goals. The lack of alternative markets for palm oil made farmers reliant on agribusiness and subject to price inequalities. Until 2018, the only palm processing mills belonged to Agropalma and Dendê do Pará S/A (DENPASA). Therefore, farmers had no alternative markets for their product and could not establish a mill for themselves. This situation may differ in countries where there are multiple mills and buyers (Cramb & McCarthy, 2016; Khor and Tamilwanan, this volume).

The initial SPOPP ambition of incorporating family farmers into contract farming schemes also fell short. By 2021, only eight companies had family farmers under contract. In 2018, Biovale (now BBF) intended to sign contracts with 2,000 families, Belem Bioenergy Brazil with 1,000 families and Archer Daniel Midland (ADM) with 600 families (de Sousa, 2019). However, these numbers never materialised: of the 3,600 estimated families, only 1,313 were signed (Brandão et al., 2019). According to our interviews, family farmers decided not to participate in these contracts due to several reasons. These included the lack of access to credit lines to start the plantation, the unavailability of family members and the inability to hire labour to work in the plantation, the risk of reducing food crops and the desire to diversify crop production and livelihood strategies (Interviews 350,

353, 358 and 360). Moreover, Brandão et al. (2021) also suggest that declining corporate extension and government support since 2015 contributed to the lack of contract farming arrangements.

Assembling environmentalist agendas in palm oil governance: Agroecological zoning of palm oil (ZAE-Dendê) and the 2012- Forest Code

The environmentalist agenda for palm oil governance seeks to stop the expansion of the agrarian frontier. In 2010, the *Empresa Brasileira de Pesquisa Agropecuária* (Embrapa)—Brazilian Agricultural Research Corporation—outlined the Agroecological Zoning of Palm Oil (ZAE-Dendê) plan for deforested or degraded lands in the Legal Amazon (Presidência da República, 2010) to address environmental concerns over deforestation (see Figure 7.1). The ZAE's territorial delimitation document clearly distinguishes so-called degraded areas that can be used for palm from the ongoing deforestation. In most cases, these degraded lands were previously cleared for livestock or agricultural production. While Embrapa identified 29,655,133 hectares suitable for palm cultivation, under new ZAE-Dendê rules, it was illegal to plant palm in 96.3 percent of the Brazilian territory (Glass, 2013). Land use models like ZAE-Dendê "co-produce a new sociotechnical imaginary for the Amazon" with supposed "neutrality" while providing a practical and "rational" framework through which the Brazilian government can orient and legitimate palm oil expansion (Hecht & Rajão, 2020, p. 10).

The ZAE, as a governance tool, assembled the interests of agribusiness and their representatives in the state, environmental activists and NGOs, as well as government officials. Since the mid-1990s, the state and municipalities have used territorial planning projects established through agreements between the Inter-American Development Bank and the Secretariat of Strategic Affairs of the Presidency of the Republic. Their objective was to reconcile developmental and new environmentalist visions promoted by social movements while making sustainable use of natural resources (Becker & Egler, 1992, p. 4). The Brazilian geographer, Bertha Koiffmann Becker, one of the intellectual proponents of agroecological zoning for palm and other agricultural commodities, argued that:

> The welfare of the population in the region cannot be achieved without economic growth, without production growth. There is no social development without economic performance. In the Amazon specifically, it must be economic growth based on the use of regional natural potential, without destroying it.
>
> *(Gusan, 2007)*

The longstanding confrontation between the developmentalist and environmentalist visions within the PT informed the creation of the ZAE. The *Aceleração do Crescimento Program* (Accelerated Growth Programme) of the Lula and Dilma

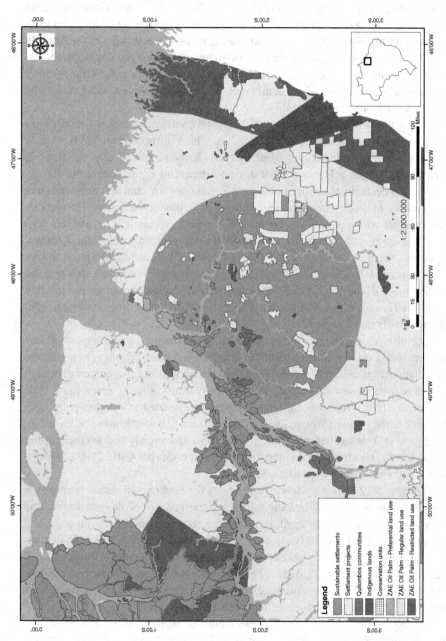

FIGURE 7.1   Agroecological zoning of palm oil and traditional territories in northeast Pará.

Roussef administrations (2008–2016) continued the developmentalist measures promoted by previous governments but also sought to green the Accelerated Growth Programme using the 2007 *Amazonia Sustentável Plane* (Programme for a Sustainable Amazon). This plan tried to integrate conservation goals with Lula's anti-poverty mission while also developing the extractive resources of the Amazon (Baletti, 2012). It articulated both pro-environmentalist and pro-developmentalist policy positions under the slogan "developing without devastating" (Villela, 2009). This revealed the hegemonic dominance of developmentalist ideas about the economy-environment relationship.

During the first PT government (2003–2006), environmental activism began to engage with state environmental agencies, which influenced the ZAE. In June 2007, a new Strategic Affairs Ministry was designed specifically for Roberto Mangabeira Unger, a Harvard professor and important figure within the PT. By presidential decision, the Programme for a Sustainable Amazon was removed from the Ministry of the Environment and transferred to this new ministry. Mangabeira was charged with considering far-reaching issues, especially Amazon development. He expressed interest in large-scale development projects in the Amazon, especially the development of palm oil for biodiesel (Filho & de Souza, 2008). In an interview, he championed agroecological zoning: "The whole axis and the starting point of the project is ecological-economic zoning, based on the solution of agrarian issues, which will allow a definition of different economic strategies for different parts of the region."

Assemblages and the territories they create are dynamic and contingent. At any moment, other entities can destabilise existing arrangements, forming new patterns and relationships within an assemblage (Murray Li, 2007). Environmental organisations opposed a new forest code bill that the "ruralist block" (representing large landowners in the National Congress) argued was needed to develop palm agribusiness in the region (Moran, 2011). Environment Minister Marina Silva, a central figure in Amazon deforestation protests who had already had several clashes within the PT, resisted the ZAE and the new forest code (BRASIL, 2003), aligning with the environmental organisations.

The new 2012-Forest Code eased the rules for land regularisation purposes in the following cases: (1) When the state has approved agroecological zoning and more than 65 percent of its territory is occupied by nature conservation units, and by homologated indigenous lands; and (2) When more than 50 percent of a municipality's area is occupied by conservation areas and by approved indigenous lands (Embrapa, n.d.; BRASIL, 2012). The environmental movement dubbed the new code "Forest Zero", as it encouraged actors to plant monocrops in degraded areas and incentivised further deforestation of legal reserve areas.

The 2012-Forest Code changed the palm oil assemblage by deterritorialising environmental aims. The developmentalist view gained the support of the central government and Silva resigned from her position five days after launching the Accelerated Growth Programme. The agreement signed between the newly-appointed Minister of Environment and Minister of Agriculture allowed the planting of palm oil and other exotic species in native forest recovery areas in the Legal Amazon and reduced the legal reserve requirement from 80 percent to 50 percent in areas demarcated by the ZAE (Nunes et al., 2016).

In our interviews, agribusiness representatives and government representatives relied on techno-scientific discourse when explaining that the ZAE contributes to sustainable palm oil cultivation by avoiding further expansion of the agrarian frontier and by recovering already deforested areas (Interviews 373 and 384). They considered the ZAE to be a non-contentious technical, environmental governance tool based on agroecological variables.

While Embrapa's expert knowledge gives the impression of sustainability, the ZAE obscures the fact that the state facilitated agribusiness' disproportionate advance over *Quilombola* (Afro-Brazilian) collective lands, riparian, and family farming communities. While many of these traditional territories are in the process of being recognised by the state, they do not appear on official maps. The ZAE no longer recognises 111 specially protected areas, including indigenous lands and remaining Quilombo communities, 45 federal settlement projects and sustainable state settlements (da Silva, 2015) (see Figure 7.1). Therefore, traditional communities and peoples have strongly contested palm oil expansion in the region and questioned ZAE's environmental aims (Acevedo, 2010). A Quilombola leader who we interviewed added:

> In São Domingo do Capim, for example, Quilombolos are under pressure from ADM. In Concórdia, it is Biopalma [BBF]. There is a lot of pressure for land. In Moju, Petrobrás entered to plant palm. So, the Northeast of Pará was dreamed of for large companies to be there. Petrobrás is in Tocantins, in the Cametá region. Agropalma in Tome Açu with a long history in this territory. After the announcement of the palm programme, right, for biodiesel. And in these areas, there is no guarantee for the Quilombolas, even for their already titled land. It is like in Moju, there are Quilombolas who have titled land, but they are surrounded by palm, and inside their territory there is an area of the Marborges company. They should leave but the plantation has not yet been removed.
>
> *(Interview 372)*

Institutional incentives to advance palm oil promoted a race to appropriate agricultural lands in the preferential ZAE zones (Nahum & Santos, 2014, 2013). National and transnational agribusiness quickly accumulated large tracts of land, forming a new agrarian landscape in the territory. The effects of this transformation are witnessed in the changing ratio of palm oil plantations to subsistence crops (da Silva, 2015), a rapid urbanisation of surrounding areas and the "proletarianization" of family farmers (Nahum, 2014). Palm oil companies deploy three complementary strategies for land use appropriation: (1) the acquisition of rural properties, especially farms and family farming areas, although SPOPP expressly prohibits purchasing lots from farmers; (2) leasing of rural properties, mainly farms with abandoned pastures; and (3) contracting family farming (da Silva, 2015; Interviews 369 and 370).

Environmentalist framings of palm oil enhanced through ZAE and sustainability were weakened as the developmentalist views shared by agribusiness and the PT government aligned. In the next section, we analyse how material entities affect the expansion of palm oil in Northeast Pará and the use of the crop as feedstock for biodiesel.

The deterritorialisation of palm oil in Pará: Bud rot and biodiesel production

Bud rot (*Amarelecimento fatal*, AF), a disease caused by an unknown pathogen and characterised by leaf yellowing and drying, is considered the most serious phytosanitary problem for palm oil in Brazil. The disease has repeatedly limited palm expansion and shaped the spatialisation of the crop, as well as the implementation of palm oil policies in Brazil, because effective control measures are impossible without eradicating the crop (Kastelein et al., 1990). For example, despite the favourable political-economic conditions offered by the Brazilian government during the 1980s, the presence of bud rot delayed the advancement of oil palm (van Slobbe, 1988). The disease spread so ferociously from 1988 to 1990 that DEN-PASA had to destroy 5,300 hectares of palm oil (Mariau et al., 1992), leading to the company's closure in late 2000.

The public sector first responded to AF in the 1980s (de Franqueville, 2003). Embrapa developed an oil palm hybrid called HIE (OxG) from the American (*Elaeis oleifera*) (Kunth) Cortés and African (*Elaeis guineensis*) varieties. Although HIE (OxG) is less productive than the African variety, it has promising genetic characteristics for further commoditisation. Its smaller palm fruit facilitates easier harvest, produces good quality oil, and resists diseases and pests to which the African palm is susceptible (Homma et al., 2016). Embrapa proposed replacing all AF-susceptible varieties with the new resistant hybrid to establish a phytosanitary barrier and prevent the spread of the disease. As an agricultural technician explained, "Using hybrids can allow us to isolate the disease. We made a barrier with hybrid varieties or forests that prevent the disease from spreading to other parts of our plantation" (Interview 367).

The hybrid is already being cultivated by small and medium producers. By 2018, Embrapa had distributed between 4.5–5 million seeds and another 5 million seeds in partnership with palm oil agribusiness in Pará (Monteiro & Homma, 2017). However, its limited production capacity is insufficient to meet SPOPP-involved farmers' or large-scale plantations' demands (interview 373). For example, Biopalma (now BBF) alone requires more than 45 million resistant variety seeds to achieve their near-term expansion goals (Monteiro & Homma, 2017).

Embrapa's Genetic Improvement Programme began multiplying the resistant varieties in its laboratories. In 2018, Embrapa's oil palm portfolio included

43 research projects (under execution and/or completed), with a total budget of R$3.7 million from the wider Embrapa research budget and R$2.9 million from partnerships with the private sector. Researchers believed that effective control of AF in Brazil requires greater investment in genetic improvement research, seed propagation and distribution, in addition to technical and scientific recommendations for the management of the new variety in areas already affected by AF (Interviews 382 and 383).

Several studies have shown how the materiality of plant diseases can alter capital accumulation (de la Cruz & Jansen, 2017). Political and economic action plans are created to respond to the disease; these strategies and regulation efforts shape the spatialisation of the accumulation processes. To establish new plantations, agribusinesses must follow environmental and agroecological regulations within the preferential and regular ZAE areas *and also* find AF-free land with lower crop management risks and higher profits. A former palm oil expansionist explained how he identified the best agricultural land:

> In this region, we have the knowledge and experience to define where you can find the disease. We know there is bud rot disease in Moju and that there is no disease in Castanhal. When planting in Castanhal, we may choose the African species because you will not have problems, while in Moju, you may plant hybrids to avoid the disease. Production costs are higher, though.
>
> *(Interview 376)*

Land in AF-free communities is highly sought after by agribusiness companies

If AF presence and the use of hybrid varieties as a crop management strategy affected the outcomes of SPOPP and palm oil expansion in Pará, the materiality of palm oil biodiesel as a non-human actor hindered the opportunities to produce and commercialise biodiesel—a main objective of the PNPB, the large federal programme in which SPOPP was inserted. The PNPB was intended to be "scalable" (without significant changes in design or framework assumptions) throughout different biofuel feedstocks and Brazil's vastly different socio-ecological contexts (cf., Tsing, 2015a, p. 38). However, despite increasing technological and capital intensiveness in the biodiesel chain, the unpredictability of nature persisted, affecting the possibilities for the commoditisation of palm oil biodiesel.

SPOPP's main objective was biodiesel production. Yet, by 2021, no companies in Pará produced biodiesel or even had the Social Fuel Seal required to sell to ANP. Only Agropalma produced biodiesel from oil palm in 2010. This biodiesel, known as palm diesel, was produced from a byproduct of the palm oil refining process using technology developed with local universities but has not been used for commercial purposes. However, the oil did not meet PNPB's or ANP's requirements (which were meant to regulate biodiesel produced from oil palm, not from its residue) (da Silva César et al., 2013). Today, corporations focus on supplying palm oil for other sectors such as food, the chemical-industrial sector, cosmetics, and animal food. So long as the Brazilian food sector demands oil for domestic consumption, there is little hope for palm oil biodiesel (Interview 385).

Additionally, oil palm biodiesel's biophysical properties make it what Bakker (2004) calls an "uncooperative commodity", that is, difficult to commodify. Most of Brazil's biodiesel comes from soybeans, which are mainly grown for animal feed. Unlike soy oil, a by-product relatively easily converted to biodiesel, palm oil's acidity makes it deteriorate rapidly (it is 50 percent saturated fatty acids and 50 percent unsaturated fatty acids) (Interview 382). One Embrapa researcher explained that this was the main problem in complying with biodiesel demands:

> The oil cannot be acidic, there is a specification that says the acidity must be low. ANP has strict regulations for oil parameters. The problem is during the production, if you produce acid oil you will have problems in its transformation into biodiesel. When the oil is not acidic like soy, there is only one production step; when it is acidic like palm oil, there is one extra step in the biodiesel process.
>
> *(Interview 382)*

Since 2007, Embrapa has tried to develop a one-step system for acid oils like palm. Embrapa-Agroenergy was created in 2006 as part of the 2006–2011 Brazil agroenergy plans to develop a crop-specific medium-term Research and Development programme to identify supply chain bottlenecks and solutions.

Other factors limiting palm oil biodiesel in Brazil include poor infrastructure and lengthy transportation times between mills, biodiesel units, and gas stations. Despite Brazil's abundant productive feedstock capacity, biodiesel mills are concentrated in the Central-West and South regions. The Southeast and North regions are net importers of biodiesel, with the deficit set to increase in the coming years. Another problem with biodiesel is its susceptibility to water absorption, for example, from water residues in tanks and trucks that also transport or store other fuels. A researcher working for Embrapa-Agroenergy stressed the presence of water residues and transportation as important obstacles to producing good quality palm oil biodiesel:

> Sometimes our recommended best practices of production are not followed. The presence of water is a big problem because microorganisms cause chemical degradation and deterioration. There is a long way from the mills to biodiesel factories, to the gas station and overall, it can create problems in the final product.
>
> *(Interview 383)*

Non-human actors, bud rot disease and biodiesel dissociated and transformed SPOPP's aims. As these rhizomes (Deleuze & Guattari, 1988) connect with SPOPP policy, they create new configurations affecting the expansion of oil palm in Brazil, despite and against government and agribusiness objectives.

Conclusion

The question of how to generate development while preserving the environment is central to the history of the Brazilian Amazon. Decades of top-down state interventions driven by developmentalism have resulted in an ongoing socio-environmental crisis. This chapter, therefore, investigated how SPOPP policy attempted to overcome developmentalist views and implement environmentalist policies for agricultural development. We analysed SPOPP as a governance assemblage formed by heterogeneous human and non-human elements (including technical measures for environmental governance, crop management strategies, and social justice measures to include smallholders and influence their farming practices).

The first research question was answered through our analysis of SPOPP policy assemblages, which revealed how the state has enabled and legitimised palm oil expansion through narratives of social inclusion and climate change mitigation via biodiesel production. The government materialised and legitimised palm oil expansion using relations between state institutions. This included bringing new actors into the region (oil palm agribusiness and state institutions), creating legislation and programmes to "open up" the Amazon for large-scale agriculture and economic opportunities, and shaping agribusiness and smallholder spaces using new territorial governance tools such as palm oil agroecological zoning or policy incentives for social inclusion. Previous developmentalist policy assemblages also helped legitimise palm oil expansion in the Brazilian Amazon.

We also showed how actors in the assemblage act and react to dominant developmentalist views and examined the new relationships established with non-humans and materialities, thereby addressing the second research question, which touches on power differentials and their effects on oil palm governance. Plant diseases and oil palm hybrids or expressions of relationships such as technological artefacts and knowledge, were a destabilising obstacle in the spatiality acquired by palm oil assemblages in the Amazon region. Legislative changes to facilitate palm oil plantations in the Amazon region deterritorialised the assemblage, allowing new alignments between governance entities, Embrapa, and agribusiness. They pushed the state to produce new knowledge about oil palm crop materiality to facilitate crop disciplining and expansion. Actors like Embrapa were charged with fixing diseases and managing failures in biofuel materiality. One result of this process was the propagation of hybrid bud rot-resistant palm varieties. Our analysis revealed that the palm oil governance assemblage cannot be considered exclusively local, regional, or global, and should not be analysed solely through micro, macro, or meso frames (Delanda, 2006).

Studies on the impact and efficiency of SPOPP (Brandão et al., 2019) have not accounted for its reliance on developmentalist views. In contrast, this chapter explores how SPOPP policy assemblage actors established specific relationships to enrol other actors in their perceived developmentalist interests. Although environmental representatives advocated for stricter forest protection laws, developmentalists managed to facilitate large-scale agriculture, including palm oil, in the Amazon. As previous studies have noted, environmental interests, such as agroecological zoning for palm oil, were integrated into the SPOPP and led to a reduction in deforestation (Brandão et al., 2021; Benami et al., 2018).

However, the zoning had unintended social injustice outcomes, including the appropriation of family farming agricultural land in preferred agroecological zones (Backhouse, 2013; Nahum & Santos, 2013). This unintended outcome speaks to the consequences of depoliticising or reposing political questions (e.g. sustainability and long-term large-scale palm oil in the agricultural frontier) as problems needing spatial technical solutions. As Lysgård (2019, p. 12) notes, "policy becomes real and valid only when the ideational and the material work together in a specific context". The policy goal to link family farming and agribusiness portrayed this articulation as a synergistic alliance. However, the answer to the third research question—how does governance enable the accumulation of wealth—shows that the actuality of events is a far cry from this idealised portrayal. Material realities on the ground, such as limited access to credit and availability of labour (Brandão et al., 2021; Córdoba et al., 2018), have resulted in small farmers being excluded from contract farming arrangements. In the context of alarming deforestation and violence, a multiplicity of institutional mechanisms continues to reassemble a developmentalist vision for the Amazon, including stronger support for large-scale palm oil expansion.

Note

1 For the northern region of the country, the minimum participation percentage was 15 percent of the total production volume of agribusiness companies (BRASIL, 2004, 2019).

References

Abrapalma. (2021). *Palm in Brazil and the world.* http://www.abrapalma.org/pt/a-palma-no-brasil-e-no-mundo/

Acevedo, R. (2010). Quilombola territories faced with the expansion of palm oil in Pará. In S. M. F. Buenafuente (Ed.), *Amazon: Carbon Dynamics and Socioeconomic and Environmental Impacts* (pp. 165–184). Universidade Federal De Roraima.

Acselrad, H. (2010). Environmentalisation of social struggles—The case of the environmental justice movement. *Estudos Avançados, 24*(68), 103–119. https://doi.org/10.1590/s0103-40142010000100010

Alonso, A., & Maciel, D. (2010). From protest to professionalisation: Brazilian environmental activism after Rio-92. *The Journal of Environment & Development, 19*(3), 300–317. https://doi.org/10.1177/1070496510378101

Anderson, B., & McFarlane, C. (2011). Assemblage and geography. *Area, 43*(2), 124–127. https://doi.org/10.1111/j.1475-4762.2011.01004.x

Backhouse, M. (2013). The sustainable dispossession of the Amazon. The case for investments in oil palm in Pará. *Fair Fuels?* Working Paper 6. https://d-nb.info/1276600720/34

Bakker, K. J. (2004). *An uncooperative commodity: Privatising water in England and Wales.* Oxford University Press.

Baletti, B. (2012). Ordenamento territorial: Neo-developmentalism and the struggle for territory in the lower Brazilian Amazon. *Journal of Peasant Studies, 39*(2), 573–598. https://doi.org/10.1080/03066150.2012.664139

Becker, B. K. (2010). Recovery of deforested areas in the Amazon: Would oil palm cultivation be a solution? *Confins, 10.* https://doi.org/10.4000/confins.6609

Becker, B. K., & Egler, C. A. G. (1992). *Brazil: A new regional power in the world economy.* Cambridge University Press.

Benami, E., Curran, L. M., Cochrane, M., Venturieri, A., Franco, R., Kneipp, J., & Swartos, A. (2018). Oil palm land conversion in Pará, Brazil, from 2006–2014: Evaluating the 2010 Brazilian Sustainable Palm Oil Production Programme. *Environmental Research Letters, 13*(3), 1–12. https://doi.org/10.1088/1748-9326/aaa270

Biblioteca Presidencia da Republica. (2010). *Speech by the President of the Republic, Luiz Inácio Lula da Silva, at the launch ceremony of the National Programme to stimulate palm oil production.* http://www.biblioteca.presidencia.gov.br/presidencia/ex-presidentes/luiz-inacio-lula-da-silva/audios/2010-audios-lula/06-05-2010-discurso-do-president e-da-republica-luiz-inacio-lula-da-silva-na-cerimonia-de-lancamento-do-programa-nacional-de-estimulo-a-producao-de-oleo-de-palma-tome-acu-pa-41min26s/view

Brandão, F., de Castro, F., & Futemma, C. (2019). Between structural change and local agency in the palm oil sector: Interactions, heterogeneities and landscape transformations in the Brazilian Amazon. *Journal of Rural Studies, 71,* 156–168. https://doi.org/10.1016/j.jrurstud.2018.09.007

Brandão, F., Schoneveld, G., Pacheco, P., Vieira, I., Piraux, M., & Mota, D. (2021). The challenge of reconciling conservation and development in the tropics: Lessons from Brazil's oil palm governance model. *World Development, 139.* https://doi.org/10.1016/j.worlddev.2020.105268

BRASIL. (2003). *The Sustainable Amazon Plan.* Ministério Do Meio Ambiente. https://antigo.mma.gov.br/florestas/controle-e-preven%C3%A7%C3%A3o-do-desmatamento/plano-amaz%C3%B4nia-sustent%C3%A1vel-pas.html

BRASIL. (2004). *Decree No. 5, 297 of December 6, 2004.* Presidência Da República Casa Civil. http://www.planalto.gov.br/ccivil_03/_Ato2004-2006/2004/Decreto/D5297.htm

BRASIL. (2012). *Law No. 12, 651 of May 25, 2012.* Presidência Da República Casa Civil. https://www.planalto.gov.br/ccivil_03/_ato2011-2014/2012/lei/L12651compilado.htm

BRASIL. (2019). *Ordinance No. 144, of July 22, 2019.* Ministério Da Agricultura, Pecuária E Abastecimento. https://www.in.gov.br/web/dou/-/portaria-n-144-de-22-de-julho-de-2019-203419910

Briassoulis, H. (2019). Governance as multiplicity: The assemblage thinking perspective. *Policy Sciences, 52*(3), 419–450. https://doi.org/10.1007/s11077-018-09345-9

Briassoulis, H. (2021). Becoming e-petition: An assemblage-based framework for analysis and research. *SAGE Open, 11*(1), 1–17. https://doi.org/10.1177/21582440211001354

Carolan, M. (2013). Doing and enacting economies of value: Thinking through the assemblage. *New Zealand Geographer, 69*(3), 176–179. https://doi.org/10.1111/nzg.12022

Córdoba, D., Selfa, T., Abrams, J. B., & Sombra, D. (2018). Family farming, agribusiness and the state: Building consent around oil palm expansion in post-neoliberal Brazil. *Journal of Rural Studies, 57,* 147–156. https://doi.org/10.1016/j.jrurstud.2017.12.013

Cramb, R. A., & McCarthy, J. F. (2016). *The oil palm complex: Smallholders, agribusiness and the state in Indonesia and Malaysia.* NUS Press.

da Mota, D. M., Ribeiro, L., & Schmitz, H. (2019). Family working arrangements in the production of palm oil in Tomé-Açu, Pará. *Boletim Do Museu Paraense Emílio Goeldi. Ciências Humanas, 14*(2), 531–552. https://doi.org/10.1590/1981.81222019000200014

da Rocha, J. F. (2015). *The formation of new territorialities: The "revision" of municipal limits promoted by oil palm farming and public actions: A study on the territorial limits of Moju, Acará and Tailândia/PA* [Dissertation]. http://repositorio.ufpa.br/jspui/handle/2011/7638

da Silva César, A., Batalha, M. O., & Zopelari, A. L. M. S. (2013). Oil palm biodiesel: Brazil's main challenges. *Energy, 60,* 485–491. https://doi.org/10.1016/j.energy.2013.08.014

da Silva, E. P. (2015). *Agrostrategies and palm oil monocultures: The silent transfer of land from agrarian reform to big capital in Pará, the Amazon* [Dissertation]. http://repositorio.ufpa.br/jspui/handle/2011/7630

de Franqueville, H. (2003). Oil palm bud rot in Latin America. *Experimental Agriculture, 39*(3), 225–240. https://doi.org/10.1017/s0014479703001315

de la Cruz, J., & Jansen, K. (2017). Panama disease and contract farming in the Philippines: Towards a political ecology of risk. *Journal of Agrarian Change, 18*(2), 249–266. https://doi.org/10.1111/joac.12226

de Sousa, C. (2019). *The firm arrived, weighed it, took it, then it will take its billions (...), but the farmer does not get a cent out of those: Expropriation and resistance of peasant women in Para, the Amazon* [PhD Thesis].

DeLanda, M. (2006). Deleuzian social ontology and assemblage theory. In M. Fuglsang (Ed.), *Deleuze and the Social* (pp. 250–266). Edinburgh University Press.

DeLanda, M. (2016). *Assemblage theory.* Edinburgh University Press.

Deleuze, G., & Guattari, F. (1988). *A thousand plateaus: Capitalism and schizophrenia.* Athlone Press.

dos Santos, C. B. (2013). Oil palm farming, traditional communities and food security in Pará, the Amazon. *Anais Das Semanas de Geografia Da Unicamp, 9,* 240–244.

Embrapa. (n.d.). *Legal reserve area.* https://www.embrapa.br/codigo-florestal/area-de-reserva-legal-arl

Ferreira, D., Strpasson, A. B., de Andrade, P. M., & da Silva, S. F. S. (2019). The Oil Palm Sustainable Production Programme in Brazil: Advances and challenges. *7th Congress of the Brazilian Biodiesel Technology and Innovation Network.*

Filho, R. M., & de Souza, J. L. (2008, November 3). *Roberto Mangabeira Unger—Minister of Strategic Affairs explains long-term projects for Brazil.* Ipea; Instituto de Pesquisa EconômicaAplicada.https://www.ipea.gov.br/desafios/index.php?option=com_content&view=article&id=1350:entrevistas-materias&Itemid=41

Fox, N. J., & Alldred, P. (2013). The sexuality-assemblage: Desire, affect, anti-humanism. *The Sociological Review, 61*(4), 769–789. https://doi.org/10.1111/1467-954x.12075

Gabinete de Segurança Institucional. (2004). Cycle of studies on the Amazon. In *Dados Abertos PR.* http://dadosabertos.presidencia.gov.br/dataset/a20f69e0-2807-4770-94d3-5463b5bceed0/resource/54bad2df-d827-4975-8a9a-a21c36123389/download/ciclo_de_estudos_da_amazonia2004.pdf

Glass, V. (2013). *Expansion of oil palm in the Brazilian Amazon: Elements for an analysis of the impacts on family farming in the northeast of Pará* (pp. 1–15). Reporter Brasil. https://reporterbrasil.org.br/documentos/Dende2013.pdf

Gusan, M. (2007, April 16). *Global warming.* Xapuri Agora! https://raimari9.blogspot.com/2007/04/aquecimento-global.html

Hecht, S., & Rajão, R. (2020). From "Green Hell" to "Amazonia Legal": Land use models and the re-imagination of the rainforest as a new development frontier. *Land Use Policy*, *96*, 1–12. https://doi.org/10.1016/j.landusepol.2019.02.030

Homma, A. K. O., de Amorim Carvalho, R., Ferreira, C. A. P., & de Deus Barbosa Nascimento Jr, J. (2000). *The destruction of natural resources: The case of Brazil nuts in southeastern Pará*. Embrapa Amazônia Oriental.

Homma, A. K. O., de Menezes, A. J. E. A., dos Santos, J. C., Gomes Jr, R. A., da Silva, R. P., Monteiro, K. F. G., & dos Santos Sena, A. L. (2016). *Commercial producers of hybrid oil palm (HIE—oleifera x guineensis) integrated into Denpasa, in the northeast of Pará* (pp. 1–38). Embrapa Amazônia Oriental.

Ianni, O. (1979). *Colonisation and agrarian counter-reform in the Amazon* (Vol. 11). Editora Vozes.

Josenaldo Jr. (2008, February 26). *Operation "Arch of Fire" begins in Tailândia*. Portal Tailândia. https://portaltailandia.com/tailandia-pa/operacao-arco-de-fogo-tem-inicio-em-tailandia/

Kastelein, P., van Slobbe, W. G., & De Leeuw, G. T. N. (1990). Symptomatological and histopathological observations on oil palms from Brazil and Ecuador affected by fatal yellowing. *Netherlands Journal of Plant Pathology*, *96*(2), 113–117. https://doi.org/10.1007/bf02005135

Leite, J. G. D. B., Silva, J. V., & van Ittersum, M. K. (2014). Integrated assessment of biodiesel policies aimed at family farms in Brazil. *Agricultural Systems*, *131*, 64–76. https://doi.org/10.1016/j.agsy.2014.08.004

Little, P. D., & Watts, M. (Eds.). (1994). *Living under contract: Contract farming and agrarian transformation in Sub-Saharan Africa*. University of Wisconsin Press.

Loureiro, V. R. (2009). *The Amazon in the 21st century: New dilemmas and their implications on the international scene*. Empório do Livro.

Lysgård, H. K. (2019). The assemblage of culture-led policies in small towns and rural communities. *Geoforum*, *101*, 10–17. https://doi.org/10.1016/j.geoforum.2019.02.019

Mariau, D., van de Lande, H. L., Renard, J. L., Dollet, M., de Souza, L. R., Rios, R., Orellana, F., & Corrado, F. (1992). Heart rot diseases on oil palm in Latin America: Symptomatology, epidemiology, incidence. *Oléagineux*, *47*(11), 605–618.

Martins, E. (1981). *Amazon, the last frontier*. CODECRI.

Massey, D. (2004). Geographies of responsibility. *Geografiska Annaler: Series B, Human Geography*, *86*(1), 5–18. https://doi.org/10.1111/j.0435-3684.2004.00150.x

Monteiro, K. F. G., & Homma, A. K. O. (2017). Current scenery of the market seeds in the chain of palm oil in Brazil. *Revista Observatorio de La Economía Latinoamericana*, 1–21. https://www.alice.cnptia.embrapa.br/alice/bitstream/doc/1073791/1/mercadosementesbrasil.pdf

Moran, S. (2011). Brazilian bill weakens Amazon protection. *Nature*. https://doi.org/10.1038/nature.2011.9584

Müller, A. A., & Furlan Júnior J. (2001). *Palm oil agribusiness: A social, economic and environmental alternative for the sustainable development of the Amazon*. Embrapa Amazônia Oriental.

Murray Li, T. (2007). Practices of assemblage and community forest management. *Economy and Society*, *36*(2), 263–293. https://doi.org/10.1080/03085140701254308

Nahum, J. S. (2014). *Oil palm cultivation and territorial dynamics of the agrarian space in the Amazon of Pará*. GAPTA/UFPA.

Nahum, J. S., & Malcher, A. T. C. (2012). Territorial dynamics of the agrarian space in the Amazon: The palm culture of palm oil in micro region of Tome-Açu (PA). *Confins*, *16*. https://doi.org/10.4000/confins.7947

Nahum, J. S., & Santos, C. B. (2013). Socio-environmental impacts of palm oil plantations on traditional communities in the paraense Amazon. *Revista ACTA Geográfica*, 63–80. https://doi.org/10.5654/actageo2013.0003.0004

Nahum, J. S., & Santos, C. B. (2014). Oil palm farming and depeasantisation in Pará, Amazon. *Campo – Território*, *9*(17), 469–485. https://doi.org/10.14393/rct91723628

Nepstad, D., Carvalho, G., Barros, A. C., Alencar, A., Capobianco, J. P., Bishop, J., Moutinho, P., Lefebvre, P., Silva, U. L., & Prins, E. (2001). Road paving, fire regime feedbacks, and the future of Amazon forests. *Forest Ecology and Management*, *154*(3), 395–407. https://doi.org/10.1016/s0378-1127(01)00511-4

Nunes, S., Gardner, T., Barlow, J., Martins, H., Salomão, R., Monteiro, D., & Souza, C. (2016). Compensating for past deforestation: Assessing the legal forest surplus and deficit of the state of Pará, eastern Amazonia. *Land Use Policy*, *57*, 749–758. https://doi.org/10.1016/j.landusepol.2016.04.022

Pløger, J. (2008). Foucault's dispositif and the city. *Planning Theory*, *7*(1), 51–70. https://doi.org/10.1177/1473095207085665

Presidência da República. (2010). *Decree no. 7, 172, Approves the agroecological zoning of oil palm cultivation and provides for the establishment by the National Monetary Council of standards relating to financing operations for the oil palm segment, under the terms of the zoning.* http://www.planalto.gov.br/ccivil_03/_ato2007-2010/2010/decreto/D7172.htm

Santos, A. M. (2008). *Analysis of the potential of palm biodiesel for electricity generation in isolated systems in the Amazon* [Dissertation].

Serrão, E. A., & Thompson, I. S. (2005). Impacts of land cover change in the Brazilian Amazon from a resource manager's perspective. In M. Bonell & L. A. Bruijnzeel (Eds.), *Forests, Water and People in the Humid Tropics: Past, Present and Future Hydrological Research for Integrated Land and Water Management* (pp. 59–65). Cambridge University Press.

Tsing, A. L. (2015a). Feral biologies. *Anthropological Visions of Sustainable Futures*.

Tsing, A. L. (2015b). *Friction: An ethnography of global connection.* Princeton University Press.

van Slobbe, W. G. (1988). Amarelecimento fatal at the oil palm estate Denpasa, Brazil. *International Seminar about the Spearrot Syndrome in Oil Palm.*

Velho, O. G. (2009). *Authoritarian capitalism and peasantry: A comparative study from the moving frontier.* Centro Edelstein.

Vieira, A., & Magalhães, S. B. (2013). Transformations in the Amazonian rural space: Planting of palm oil in peasant communities in lower Tocantins, municipality of Moju/PA. *International Seminar on Ruralities, Work, and Environment.*

Villela, A. A. (2009). *Palm oil as a sustainable energy alternative in degraded areas in the Amazon* [Dissertation].

Watkins, C. (2018). Landscapes and resistance in the African diaspora: Five centuries of palm oil on Bahia's Dendê Coast. *Journal of Rural Studies*, *61*, 137–154. https://doi.org/10.1016/j.jrurstud.2018.04.009

8

(DE)CO$_2$LONIAL STRUGGLES WITHIN "GREEN" OIL PALM ASSEMBLAGES

Shady monoculture entanglements and fissures of hope in Ecuador and its Chocó borderlands

Julianne A. Hazlewood, Geovanna Lasso, María Moreno Parra and Iñigo Arrazola Aranzabal

Introduction

In this chapter, we investigate three principal questions: (1) to what extent does the Ecuadorian government pass bundles of laws and disseminate discourses based on the oil palm industry's promise of greater environmental sustainability (rather than a fossil fuel-dependency)—what we call "green" oil palm assemblages—and instead, allow the industry to violate constitutional and international law; (2) what are the different forms in which the oil palm industry expands its reach and power; and, (3) what are the communities negatively affected by the oil palm industry doing to activate hope? These questions are contextualised within an ongoing and ever-increasing bifurcation between the deepening social and environmental costs of oil palm expansion across the country and the rights guaranteed in the 2008 Ecuadorian constitution. Essentially, they tie back to all four of the book's main research questions of how the different governance arrangements facilitate and legitimise oil palm expansion; how power differentials impact palm oil governance and how this, in turn, enables the accumulation of wealth for certain actors in the assemblage; and finally, how conflict is governed and moderated in the palm oil sector.

In 2008, a new Ecuadorian constitution was approved. The document represented a turning point in the country's history and was globally applauded for the progressive, transformative ways in which it conceptualised well-being or "Living Well" (Hazlewood, 2012, 2010a, 2010b; Walsh, 2011; Acosta, 2009b; Acosta & Martinez, 2009b; Gudynas, 2009), human-nature relations (Acosta, 2009a; Acosta & Martinez, 2009a; Galeano, 2009), and interrelations between the state and diverse cultures/nationalities (Acosta & Martinez, 2009c). The document emphasised planning and implementing a "post-extractive development" strategy that claimed

DOI: 10.4324/9781003459606-10

to simultaneously improve the economy and mitigate climate change, an approach that was referred to as "post-development" (Escobar, 2012, 2008; Esteva, 2011; Gudynas, 2011).

Ecuadorian articulations of post-development frameworks spring from the Kichwa concept of *Sumak Kawsay* meaning *"Vivir Bien"* or "Living Well". Applying *Sumak Kawsay* entails supporting communities' ability to build and maintain their self-determinative livelihoods based in specific and diverse ways of self-determination, cosmovisions, cultural practices, and knowledges to cultivate "harmonious relationships" (Asamblea Constituyente del Ecuador, 2008, Article 275)—i.e. "the right to difference" (Escobar, 2012, 2008; Mignolo, 2005). In the constitution, the latter is referred to as "plurinationality" (Articles 1, 6, 257, and 380, Asamblea Constituyente del Ecuador, 2008) and demands equality, as well as true democratic and intercultural representation of the 13 different Indigenous nationalities and 20 peoples in Ecuador, both in policy making and implementation (Acosta & Martinez, 2009c).

Access to ancestral territories, healthy forests and lands, clean rivers, and water are essential to *Sumak Kawsay* and go hand-in-hand with securing the "Rights of Nature" (Hazlewood, 2012, 2010b; Acosta, 2009a, 2009b; Acosta & Martinez, 2009a, 2009b; Carrere, 2009; Houtart, 2009; Quintero, 2009; Wray, 2009; Asamblea Constituyente del Ecuador, 2008). The 2008 Constitution recognised nature as a social actor (Galeano, 2009) with constitutional rights "to have its existence and the maintenance and regeneration of its life cycles, structure, and functions and processes of evolution integrally respected" (Acosta, 2009a; Article 71; Lander, 2009; Asamblea Constituyente del Ecuador, 2008).

Walsh (2011), however, argues that the Ecuadorian state co-opted *Sumak Kawsay*, and consequently, is not post-extractive. Indigenous and Afro-descendant peoples' articulations of *Sumak Kawsay* are based on specific, place- and cultural-based versions of "development otherwise"—ways of living life distinct from Western capitalist, growth-centred models of development. Walsh highlights that the resulting constitutional interpretations of *Sumak* Kawsay mask the continuing colonial development as business-as-usual.

Olindo Nastacuaz—President of the Federation of Awá Centers of Ecuador when the constitution was approved, and later President of the Confederation of Coastal Indigenous Nationalities of Ecuador (CONAICE)—concurs. Nastacuaz claimed that the government's ongoing support for mining, oil, and agrofuel development shows that the laws are merely words on paper: "The actual government is only a discourse. In the end, the new laws are pure politics" (Nastacuaz, interview, 26 May, 2009).

For instance, even in the context of the newly approved constitution, the Ecuadorian state, together with oil palm companies, promoted biofuels as a "sustainable solution" to climate change, claiming that plantations produce oxygen, provide employment, and develop the agricultural sector (Benalcázar, 2009; Gonzalez, 2007; Armendáriz, 2002; Carrere, 2002). This "solution", however is linked to considerable social and environmental costs. Proponents argue for expanding and investing in the oil palm industry to mitigate climate change and enhance well-being, while critics state that the industry in fact, deepens and expands detrimental colonial and extractive socio-ecological relations and suffering. The Indigenous Environmental Network (2007) refers to these processes as rolling out of CO$_2$lonialism. Since the turn of the millennium, multiple studies have demonstrated the detrimental socio-environmental effects of the expansion of oil palm (*Elaeis guineensis*) experienced by forest protectors and *campesinos'*—meaning rural producers, a political conceptualisation of family farmers that is often translated to peasants in English—specifically in Ecuador (Hazlewood, 2023, 2012, 2010a, 2010b; Minda Batallas, 2020, 2013, 2002; Lasso, 2019, 2012; Moreno Parra, 2019; Naizot, 2011; Mideros Zamora, 2010; Editor PCN, 2009; Tuinstra, 2008; Bravo, 2007; García Salazar, 2007). Other studies (Oslender, 2016, 2007; Marin-Burgos, 2014; Ballvé, 2013; Cardenas, 2012; Escobar, 2008; Tauli-Corpuz & Lynge, 2008; Tauli-Corpuz & Tamang, 2007; Carrere, 2002) provide ample evidence of the negative consequences of oil palm expansion for people and nature in other countries where oil palm production is prominent.

Lawsuits (*Isaha Ezequiel Valencia Cuero v Palmeras de los Andes*, 2017) reports, manifestos, campaigns (Roots & Routes IC, 2020; Benalcázar, 2009; Editor PCN, 2009; Misión de Verificación al Pacífico Sur de Colombia, 2009; Rogers, 2009; Santa Barbara, 2007), and letters document that, as opposed to mitigating climate change, the expansion of large oil palm plantations, and related palm oil extraction and processing aggravate deforestation (Minda Batallas, 2013; Hazlewood, 2012, 2010a, 2010b), promote soil erosion, produce adverse carbon emission effects, and contribute to the loss of biodiversity in forest and agricultural lands (Ortega-Pacheco & Jiang, 2009). Replacing diversified cropping meant for national consumption with oil palm monoculture also negatively impacts national food sovereignty (Lasso, 2019, 2012). This reflects a wider body of concerns about the greenhouse gas emission impacts of palm oil (Dominguez, 2008; Tauli-Corpuz & Lynge, 2008; Pachauri & Reisinger, 2007). Oil palm plantations also contaminate rivers and put Ecuadorian citizens' access to clean water at risk (Núñez Torres, 2004, 1998). Additionally, they reorganise people's relations to one another, which leads to conflicts between people experiencing greater pressures on their ancestral territories (Lasso, 2019; Carlet & Ferreira, 2018; Hazlewood, 2012). Yet, despite growing concerns about the detrimental socio-environmental impacts of palm oil and its violation of international and national laws in Ecuador, the state continues to prioritise the sector as a viable way forward (Public Letter to President Moreno, 2020). Amidst the COVID-19 pandemic and human crisis in Ecuador, on 7 June 2020, the Ecuadorian National Assembly unanimously approved the "Law for the strengthening and development of production, commercialisation, extraction, export, and industrialisation of oil palm and its derivatives" (Asamblea Nacional República del Ecuador, 2020; Armendáriz, 2002).

Building upon the research and social movements that underscore the CO_2lonising processes of oil palms within Ecuador and beyond, our chapter demonstrates how "green" sustainability discourses and legislation incentivise and reward the oil palm industry—owned and operated by economic elites—to accelerate the spread of oil palm monoculture across the Ecuadorian tropical lowland landscapes. We show how shady entanglements of oil palm create inroads towards racialised land dispossession and ethnocide throughout Ecuador. In its wake, it leaves fallen primary and secondary rainforests, contaminated rivers, and uprooted and/or diseased individual and collective landowners. In the face of such acts (that we suggest must be considered crimes against nature and humanity), fissures of hope nevertheless make their way through these entanglements. Our research with Esmeraldas communities demonstrates that when everyday survival is at stake, people united and collaboratively mobilised with local, regional, national, and international allies to defend their ancestral lands and demand the dignity and justice guaranteed in the 2008 Constitution.

In what follows, we first discuss how we employ the concept of assemblage. The authors show that on all scales—international, national, regional, and local—and all dimensions—political, economic, ecological, and psychological—oil palms act as subjects that impose social processes that become interlaced with other people, other-than-humans, objects (means of production and otherwise), and processes in ways that render profit as the priority and people they encounter as both obstacles and disposable. In effect, under the pretext of advancing economies while reducing emissions, expanding CO_2lonial oil palm assemblages sow seeds of *la muerte lenta* (slow death) that deepen the chasm between the aspirations of the 2008 Constitution and the realities that human and non-human communities face.

Lasso (2019) has identified three principal ways in which the oil palm industry infiltrates and occupies human and environmental landscapes: (1) through the total transformation of rural people's livelihoods; (2) the dispossession of rural people and communities through indirect violence; and (3) dispossession through direct violence. In this chapter, we unpack and build upon these.

We conclude by delving especially deeply into the circumstances complicit in dispossessing rural people and communities through direct violence. Within the coastal rainforests of Northern Esmeraldas, the Ecuadorian Chocó Borderlands, we focus on two historical and present-day environmental-with-racial justice and hope assemblages. One is a legal struggle led by the Black community of Wimbí, and the other is an intercultural effort by the Afro-descendant community of La Chiquita and the Awá Indigenous community of Guadualito. Despite decades of historical abandonment by the state, a lack of access to basic human services, and outright corruption, these communities carve out *geographies of hope* emerging from decolonising the day-to-day realities that unfold due to the arbitrary implementation of false solutions to climate change. These Chocó rainforest-based hopeful geographies aim to protect collective territorial rights, defend rivers and the living forests they inhabit, and demand environmental-with-racial justice for their communities.

Assemblages that execute ecological racism via CO$_2$lonial slow death processes

Assemblage theories have been employed to understand the forms of socio-material organisation as consisting of multiple, heterogeneous entities that constitute a whole for a certain time (Müller, 2015). Heterogeneity is a central aspect of assemblages, which also underscores emergent relationalities. Indeed, assemblage frameworks are not based on organic unity but are instead emergent, contingent, relational, co-existing, and multiple. They tend to reject *a priori* assumptions as to what can be related and how: "they are socio-material, eschewing the nature-culture divide" (Müller, 2015, p. 29). The concept illuminates how human and non-human elements are meaningfully related in contingent ways through particular times and spaces. In our case, assemblages allow us to explore the relations that arrange in diverse heterogeneous arrays of the following:

- Human agents—local communities, companies, labourers, the oil palm industry, government officials, illegal armed groups, special forces employed for security, scientists, alongside local, regional, national, and international activists;
- Other-than-humans—water buffalos used to transport the harvest, as well as plants and animals in rivers, forests, plantations, community lands, and ancestral territories;
- Non-living elements—territory, bulldozers, trucks, monocultures, fences and gates, road closures, guns, drug trafficking, machetes, agrochemicals, palm oil waste/blackwaters, agrochemicals, and other artefacts.

Assemblage framings bring together all these subjects, drawing particular attention to their interrelations. They are "defined solely by their external relations of composition, mixture, and aggregation" (Nail, 2017, p. 23). The relationships between the elements of the assemblages are neither essential nor nascent to organic unity. Instead, assemblage relations constitute and are constituted by socio-material processes, while (re)generating assemblages necessitates "…agency, the hard work required to draw heterogeneous elements together…, [in the] face of tension" (Murray Li, 2007, p. 264). In these ways, (Rocheleau & Roth, 2007, p. 435) claim that assemblages are at once productive and unpredictable, and can be conceptualised as "…creative entanglements with agency on all sides, a relational web shot through with power". New territorial organisations, behaviours, expressions, actors, realities, and other possible worlds are novel potentialities resting on the horizon of still unthinkable collisions and encounters (Müller, 2015).

To better understand how assemblages are constructed and rolled out, Nail (2017) suggests that we need to ask empirical questions relative to what they do in specific contexts. In other words, why bother with assembling assemblages? The assemblages discussed in relation to Ecuador underscore the importance of specificities in conceptualising events within specific places in time and space.

Yet, a question also emerges: do assemblage frameworks struggle to account for long-term structural forces? This next subsection alludes to such long-term racialising in environmental assemblages and within circumstances fostering *la muerte lenta* (slow death).

Environmental racism, environmental suffering, and la muerte lenta (slow death)

Assemblages have a multiplicity of effects, distinct for all agents involved. Here, we want to share some reflections about environmental assemblages that are intentionally (re)produced to racialise. Pulido (2017) defines environmental racism as both the structural dimensions of racism and the processes that create territories and devalue certain lives. Global capitalism sustains its insatiable appetite for ever-increasing surplus by producing relations of severe inequality among human groups (Pulido, 2017); it accumulates and produces profit by devaluing, appropriating, and exploiting the labour and bodies of Black and other non-white peoples. Thus, capitalism on all scales can be considered to be racial capitalism, as sustaining its metabolic intake and profit-based output requires segregating people and exploiting labour in relation to already existing power differentials based on class, race, region, and nationality.

Earlier studies of race and environmental relations show that unequal exposure to environmental risks and negative impacts are known to affect communities of colour disproportionately (White, 1998; Bullard, 1993). Later, as race was recognised as "…a powerful factor in the selective distribution of people in their physical environment…influencing land use, housing patterns, and infrastructure development" (Bullard, 2004, p. 1), scholars further conceptualised ecological racism. They demonstrated that extractive frontiers are configured racially, rendering certain peoples and spaces vulnerable to disposability, the loss of lives, and the unequal differentiation of human and natural value. Ever diverse forms of environmental suffering are intimately linked to the (re)production of social inequality.

Racism marginalises, enabling and deepening the inequalities required by capitalism. Some communities "…are treated as non-citizens, as usable and disposable beings, to the extent that capital can dispense with them or consider them an obstacle to the development of some new project" (Pacheco, 2007, p. 2). Extractivist projects and agribusiness in Latin America block social and environmental justice. Environmental degradation is linked to and reproduces social inequality: daily forms of environmental suffering generated from the toxicity of water and soil, loss of forest, and land use change are linked to the (re)production of poverty in direct relation to race.

In places such as Esmeraldas—where most of the population of the Ecuadorian Chocó borderlands is Black—historical and ongoing ecological racism results from, and contributes to extractivist and productive processes that funnel benefits to actors external to the province. Hazlewood (2023) underscores how the concept of "long-wave disasters" that Barnett and Blaikie (1994) introduce is helpful here. Distinct from environmental catastrophes such as floods or earthquakes, these slowly ripple through and reorganise communities, gradually eroding multiple scales of resilience and perpetuating socio-environmental crises.

Applicable to the Ecuadorian Chocó, Black scholar Cuero Campaz (2022) conceptualises state planning in the Colombian Pacific Region (also forming part of the Chocó lowland rainforest) as an ongoing racist territorial ordering that includes deliberately setting aside "ethnic enclaves" for extraction, demonstrating a predetermined intent to mine Black spaces via scarce state presence and non-intervention in today's drug-trafficking-related violence. Waldmueller (2020, p. 1) describes this ongoing phenomenon as "…the broader permanently neglected disaster in the Pacific border area." While Harvey (1989, p. 303) pinpoints "organised abandonment"—a set of intentional strategies which sets up organised political-economic chaos, Gilmore (2008) emphasises the racialisation processes inherent to them.

Moreno Parra (2019, p. 20) adopts Zaragocín's (2018) concept of *la muerte lenta* to describe the environmentally racialising wave that degrades, territorially dispossesses, and causes violence in Esmeraldas as "slow motion genocide". Moreno Parra demonstrates that slow death takes place through the state's actions and omissions. Afro-descendant and Indigenous peoples suffer the consequences and harm, and despite their denouncement at all scales, it continues to take place. Hazlewood's (2023, p. 1473) theoretical proposal of "long-wave extractive coloniality" contributes to Moreno Parra's conceptualisation of slow death taking place in Esmeraldas, emphasising the multiple waves within the greater centuries-long racialised slow-motion disaster and on-going colonisation through planned state abandonment.

(De)CO$_2$loniality: an everyday reality based in resisting systematised slow death

Where there is colonising, there is likely resistance. Tuhiwai Smith (1999) describes decolonising as Indigenous peoples taking back and transforming the colonising powers of scientific research by researching, recreating, and re-presenting their own versions of history. Walsh (2011) adds the parenthesis—*(de)*colonising—to emphasise the constant tension in everyday life in which the colonising-and-*de*colonising processes take place. Furthermore, she and other decolonial scholar-activists from the South (Mignolo & Walsh, 2018; Escobar, 2008; Mignolo, 2005) transform the verb (de)colonise to (de)coloni*ality*—an *ontology*, or way of being, that permeates psychologies, emotions, economies, politics, *et cetera*.

Grossman et al. (2012) pinpoint a "triple whammy" of colonial waves: (1) settler colonisation, (2) climate change, in addition to (3) violations of sovereignty/self-determination and land grabbing that come with climate change mitigation development. CO$_2$lonialising processes become CO$_2$loni*ality* when such slow death processes draped in guises of "green" development are reconceptualised as conditioning everyday reality, filtering into every dimension of life. In this article, (de)CO$_2$loniality emerges from day-to-day uprisings and resurgences in relation to slow death and ethnocide within geographies of still-expanding oil palms, especially within the Coastal Province of Esmeraldas.

Spatial consequences of Ecuadorian legislative inconsistencies

Different power strategies—contingent upon different political links between the government and the economic elites that manipulate governance and regulatory policies, ordinances, and laws—have accelerated oil palm territorialisation between the 1970s and 1990s (Lasso, 2019). Land concessions and/or changes in the category of land use from forest to agricultural use, for example, allowed big industries to access territory otherwise off limits in the Ecuadorian Amazon in the east and the coastal Chocó Rainforest of Northern Esmeraldas in the west (Buitrón, 2001; Ashley, 1987). In Northern Esmeraldas, for example, a mix of national policies with fraudulent local mechanisms—what Massey (1993) refers to as a power geometry—worked to illegally appropriate indivisible ancestral collective territories, encourage deforestation, and foster misery.

Lasso and Clark (2016) show how neo-development power-knowledge-discourse power geometries benefit the oil palm sector and some of Ecuador's economic elites. Since 2004, with the Ecuadorian Congress passing the National Programme for Biofuels (República del Ecuador, 2004), several legislative acts encouraging continued expansion of biofuel production followed. The National Programme for Biofuels consisted of plans to mix biofuels with petroleum-based fuels up to 10 percent in Guayaquil and Quito, the principal metropolitan areas of Ecuador. Recognising the opportunity, Ecuador's Ministry of Mines and Petroleum's rhetoric in 2006 also suggested that the country's diverse microclimates, quality of soils, and availability of labour were suited to ethanol and biodiesel production (elEconomista.es, 2006).

Following the inauguration of Rafael Correa in January 2007, Ecuador created a National Biofuels Council (Jull et al., 2007). The council consists of the President of Ecuador, five Ecuadorian Ministries (Agriculture, Environment, Industry, Electricity, and that of Energy, Mines and Oil), two sugarcane producer associations, the National Association of African Oil Palm Cultivators (ANCUPA), fuel sale and distribution organisations, and PETROECUADOR (Gonzalez, 2007; República del Ecuador, 2007). With the passing of the Promotion of the Biofuels Law, also in 2007, the government formalised plans outlined in the 2004 National Programme for Biofuels, mandating increased domestic use of biofuels, diversification of energy sources, and reductions in carbon emissions (República del Ecuador, 2007).

The World Bank Group, the International Finance Corporation, the State of Ecuador, and the private and public sectors which support oil palm expansion, production, and commercialisation promoted cultivating oil palms with discourses related to how the sector holds potential to do the following:

1 Eradicate poverty and violence.
2 Generate employment and integrate small farmers into a newly flourishing economy.
3 Hold potential for environmental sustainability, climate change mitigation, and food security (World Bank & International Finance Corporation, 2011).
4 Oil palm as just generally "good business" for small farmers (Lasso, 2019).

The National Biofuels Council initiated the National Programme of *Negocios Rurales Inclusivos* (Inclusive Rural Business, IRB) that was based on growing oil palm. For Correa, integrating small farmers into palm oil supply chains was presented as a redistributive approach to aid historically excluded people to become active in the economic system (Correa, 2007). Therefore, the government prioritised commodity chains that had both international demand and would include many small and medium producers (Lasso, 2019). One of the principal policies included the Oil Palm Competitive Improvement Plan (PMC-PA), built with the participation of the private sector. Additionally, several ministerial decrees promoted labour flexibility in the production phase (No. 0060. Ministry of Labor Relations) and biofuels, alongside underscoring social, environmental, and economic well-being (Interministerial Agreement No. 030).

The RSPO especially influenced the green and environmental sustainability discourses. The private sector began to adapt to and adopt national and international "green" certification requirements. The role of the two national oil palm unions—FEDEPAL (the oil palm exporters union) and ANCUPA (the palm oil producers union)—was critical in promoting several environmental benefits of oil palm, such as how oil palms prevent soils from eroding and compacting, are efficient in capturing CO$_2$, have the capacity to auto-generate energy, and the possibility of having green reserves within the plantations themselves (Lasso, 2019). Oil palm corporations also highlighted how companies' high levels of innovation contribute to their businesses' economic growth while reducing the environmental impacts of production and industrial processes (Lasso, 2019). In this way, the oil palm industry claimed that it benefits social development, improves quality of life, and mitigates rural-urban migration, thereby contributing to small farmers' economic integration, as well as to towns and cities' overall sustainability and innovation (Danec Group, 2014).

The governments of Lenin Moreno (2017–2021) and Guillermo Lasso (2021–2023) continued to propagate these discourses. Additionally, they especially favour the neoliberal policies that encourage direct interrelations between economic and political elites, which benefit the oil palm sector not only through legal frameworks, but also through economic incentives such as tax exemptions.[1]

Oil palm assemblages: Three velocities and typologies of deterritorialising *campesinos*

The first commercial oil palm trees in Ecuador were planted in 1953. Today, Ecuador ranks fourth in Latin America and eleventh in the world in terms of oil palm area (US Department of Agriculture, 2023). Significant expansion began in the 1980s. Between 1980 and 1993 (the year in which exports of crude palm oil (CPO) began), the planted area increased by 467.11 percent. Between 1993 and 2016, the increase was 307.31 percent. As of 2016, there were 319,602 hectares nationwide producing 577,000 tonnes of CPO (INEC-ESPAC, 2016). By 2019, Ecuador had approximately 200,908 hectares of palm oil plantations and these crops were cultivated in 13 provinces: Esmeraldas is the province with the biggest area of oil palm (40 percent), followed by Los Ríos Provinces (18.5 percent), and Santo Domingo (10 percent) (Borja, 2020). In 2022, Ecuador exported USD193 million in palm oil (OEC World, 2023). Currently in Ecuador, there are four main areas of oil palm production: the north western block, the central western block, the south western block, and the Amazonian block (see Figure 8.1).

Profit accumulation in the oil palm sector is highly concentrated. Five companies control most of the palm oil production, processing, and export: La Fabril, the Danec Group, the Ales Group, the EPACEM Group, and the Cielcopalma Oleana Group. Currently, these companies occupy 21.57 percent (60,390 hectares) of the land cultivated for oil palm (while 69.76 percent of the 8,149 producers registered in 2017 occupied 18.36 percent of the total area), collection centres, oil extractors, as well as some refiners and processors (Lasso, 2019; Fierro, 1991).[2] These five groups concentrate 88 percent of sales (Centre for Economic and Social Rights, 2015) and lead exports (FEDEPAL's President, interview, 13 May 2016). These patterns of concentrated profit are also reflected in these entities' earnings and assets, which have been growing steadily. La Fabril, the Danec Group, and the Ales Group are among the 100 biggest economic groups in Ecuador, occupying positions 24, 34, and 53, respectively, competing with economic groups linked to oil, finance, cellular technology, and agroindustry. If their income increases, the percentage of sales destined to pay taxes does not exceed more than 0.75 percent, due to state tax benefits. The oil palm value chain has been prioritised in productive public policy and has been recognised as a strategic sector (Lasso, 2019).

Economic groups that lead the sector accumulate profits and control through territorialising oil palm assemblages—through discourses, laws, and implementation, as well as social, economic, and racial processes. At the same time, developing and expanding the industry has been made possible by deterritorialising *campesinos*, and oil palm supply chains replace traditional land use, ownership, and practices. The adverse effects documented in this chapter's introduction should be emphasised. Patterns demonstrate decreasing food sovereignty and sustainability as exploitation of labour flexibility increases (Lasso, 2019, 2012; Hazlewood, 2010; Cañas, 2009).

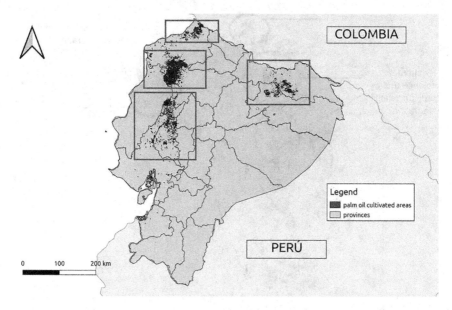

FIGURE 8.1   The four principal blocks of oil palm plantations in Ecuador, 2020. Created by Cartographer Iñigo Arazola Arranzabal, *Colectivo de Geografía Crítica de Ecuador*, 2023 (*El Ministerio de la Agricultura, Ganadería, Acuacultura y Pesca de la República del Ecuador* (MAGAP)—SIG Tierras).

In each of the four main blocks, the state and the private sector operationalise diverse oil palm assemblages. They utilise distinct strategies and mechanisms contingent upon the local environmental and social conditions, histories, events, and communities.

These oil palm assemblages have led to deterritorialising *campesinos* through three principal typologies: (1) dispossessing with indirect violence, (2) transforming *campesinos'* everyday lives and livelihoods, and (3) dispossessing with direct violence (Lasso, 2019). Figure 8.2 shows the spatial distribution of these processes; while Table 8.1 illustrates the different types of deterritorialisation tactics adopted by the state and palm oil companies in the Ecuadorian oil palm assemblage.

Deterritorialisation by dispossession using indirect violence

Within this typology, oil palm plantations expand via materially and immaterially deterritorialising *campesinos*, thereby increasing their precariousness. Left without access to land, they often migrate to new places in Ecuador to work on big farms.

In the central and south western oil palm blocks, the Ecuadorian government frequently works with the private sector to enable the process of indirect deterritorialisation. Their tactics involve economic violence and subsequent land dispossession. The state plays a double role: legislative inaction *and* action.

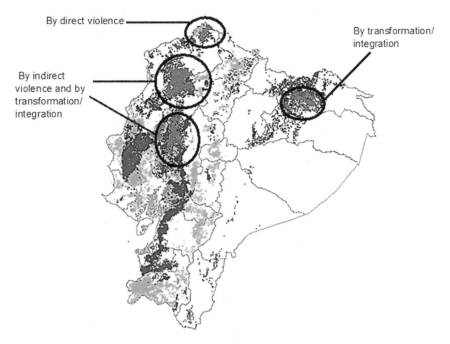

FIGURE 8.2 Typologies of deterritorialising *campesinos* by different oil palm assemblages (Lasso, 2019).

The state imposes violence through inaction by not supporting *campesinos'* livelihoods based on diversified polycultural production. Simultaneously, the state takes action by promoting a rural development model based on the boom commodity, creating preferential credit lines for those willing to produce monoculture crops. Small and medium producers who take out loans often end up in debt, resulting in increased vulnerability to production losses (due to price fluctuations or phytosanitary crises, for instance). To pay the acquired debts and cut their losses, producers then sell their land, frequently to big oil palm farms (Lasso, 2019).

In fact, this kind of economic incentive/debt and subsequent deterritorialisation process leads to *campesinos* who received loans to grow monoculture coffee in La Sexta or Las Golondrinas (located in the central western block) to hand over their lands to oil palm companies. These exchanges mark the transition from the coffee boom to the oil palm boom, underscoring the indirect violence used in the assemblage (Lasso, 2019).

Through these legislative inactions and actions, eventual land grabs take place, with the majority of the land area concentrated in the hands of big producers. Big producers end up choking out isolated small producers who manage to hold onto their lands by limiting the *campesinos'* road or water access. Throwing their hands up, they often also sell their land. The Danec Group expanded their land holdings in Quinindé by playing their part in deterritorialising by dispossessing using

TABLE 8.1 Three deterritorialisation tactics of Ecuadorian state-oil palm assemblages (Lasso, 2019)

Three deterritorialisation tactics of Ecuadorian state oil palm assemblages

Type of deterritorialisation	Main mechanisms and socioeconomic context	Role of the state	Impacts
By indirect violent dispossession	• Mainly through selling land for the following reasons: • Unfavorable economic situations, indebtedness, and crises • Compromised access to water and roads due to the dominance and presence of surrounding big farms. Territories dominated by big oil palm farms had lost community relations, and delinquency had increased.	• <u>Absence:</u> There is a lack of public policy that attends small and diversified *campesinos'* production • <u>Action:</u> Use of public policy to promote a model based on: • "The commodities in turn", linked to international booms • Monocrops and external chemical inputs, creating economic losses due to pests or international price fall.	• Land concentration by big oil palm farms • Small and mid-size producers migrating to cities, thereby losing their livelihoods • People are forced to sell their lands due to accumulating debts, yet remain in the territory as big farm workers • Loss of food sovereignty
By integration and transformation	• Unfavourable economic situations for small and medium *campesinos* • *Campesinos'* integration to the oil palm value chain • Lack of alternative markets for a diversified production	• <u>Action:</u> Definition of a public policy promoting the integration of small and mid-size producers into the oil palm value chain i.e. through an inclusive business programme • Generating preferential credit lines to grow oil palm	• Transforming producers' livelihoods and productive systems • Loss of autonomy and food sovereignty if oil palm replaces all the diversified production • High dependency on the industry conditions • Increased probability of small producers' expulsions

(Continued)

TABLE 8.1 (Continued)

Three deterritorialisation tactics of Ecuadorian state oil palm assemblages

Type of deterritorialisation	Main mechanisms and socioeconomic context	Role of the state	Impacts
By violent dispossession	• Illegal and legal extractive activities, contamination, destruction of local livelihoods • Socio-economic exclusion, historical patterns of violence and discrimination • Use of coercion, fraudulent legal tools and corruption linked to the state	• <u>Absence</u>: There are no public policies to take care of the local Indigenous and Afro- descendant peoples nor protect their rights. Not regulation to illegal activities, violence, etc. • <u>Action</u>: Legal tools, decrees, Institutions protecting oil palm enterprises	• Dispossession of local people's land • Local communities live in precarious conditions and violent environments. Lack of access to roads and water • Local livelihoods are depleted

indirect violence (Lasso, 2019). As a result, most *campesino* families evacuated the town of Las Golondrinas in Quinindé, and those who remained lived on small land holdings surrounded by a sea of oil palm (Lasso, 2012).

Deterritorialisation by transforming campesinos' everyday lives and livelihoods

In this oil palm assemblage typology, oil palm elites subordinate and incorporate *campesinos* and their mosaic of mixed practices and polycultures into their ever-expanding economic and territorial enterprise. Especially in the central, southwestern, and Amazonian blocks (see Figure 8.2), Ecuadorian small and medium producers with access to few market alternatives are persuaded/forced to grow oil palm; and when this happens, their everyday lives and livelihoods dramatically shift. Within this typology, rural producers at the local and national level experience a loss of economic and environmental sustainability, food sovereignty, and general well-being, which are contrary to the discourses put forth by the state and the private sector in promoting the oil palm commodity chain (Lasso, 2019).

This neoliberal assemblage, integrating small and medium producers into the oil palm industry, is very common in Colombia and other oil palm-producing countries (Castellanos-Navarrete & Jansen, 2018; Mingorría et al., 2014; Cardenas, 2012). In Ecuador, this is a more recent development for the industry, as access to territory has decreased.

Private enterprises integrate small and medium producers into their businesses to simultaneously expand and externalise costs. In fact, this strategy has been promoted globally through the World Bank Group Framework and International Finance Corporation Strategy for Engagement in the Palm Oil Sector (World Bank & International Finance Corporation, 2011), which offer 10-year credit lines to Ecuadorian enterprises. The World Bank and the International Monetary Fund claim that oil palms purportedly contribute to global-level sustainable commodity chains as well as corporate social responsibility (CSR) (Lasso, 2019).

Between 2001 and 2014, the state and private collaborative endeavours were able to increase the participation of rural producers in the oil palm sector from 2,159 to 6,097. In 2014, the government initiated the IRB programme to alleviate poverty and contribute to overall rural development. The goal was to integrate small and medium producers into strategic sectors, with palm oil as one of the commodity chains that the government targeted (Lasso, 2019). Participation and integration into the oil palm sector have soared since then.

The government created preferential credit lines through BAN Ecuador, a public financial institution, or through a local enterprise. Private enterprises also played a key role. They were appointed the *empresas anclas* (anchor companies), whose role, among others, was to be the middleman, purchasing small and medium producers' oil palm fruit. As owners of the palm oil extractors, they could pay the small and medium producers' credit and manage the services and tools that the latter needed as credit.

For example, Energy & Palma's Associative Project PAPA currently integrates 200 producers into oil palm production, with a projection of integrating 1,400 more in three cantons in the Province of Esmeraldas: Rio Verde, San Lorenzo, and Eloy Alfaro. Up until 2014, in the Province of Sucumbios, the Danec Group had 74 producers (among *mestizos*, Secopai, and the Chachi Indigenous peoples) and 840 hectares; through its extractor in the Amazon, the company integrated 502 producers, covering 5,097 hectares. During this period, the EPACEM Group started two new projects—in 2008, *El Desarrollo de Negocios Inclusivos en la Base Económica de la Pirámide* (Development of Inclusive Businesses at the Economic Base of the Economic Pyramid), and in 2011, *El Desarrollo de Producción Sostenible de Palma Aceitera en Fincas de Pequeños Productores en el Oriente Ecuatoriano* (The Development of Sustainable Oil Palm Production on Farms of Small Producers in Eastern Ecuador)—both financed by public and private banks (Lasso, 2019). In some localities, where collection centres and extractors were located, the groups competed with one another to harvest as much oil palm fruit as possible by reducing prices and using diverse mechanisms such as technical assistance, agrochemicals, tools, and credit facilities to lure producers (Lasso, 2019).

Consequent to their integration into oil palm assemblages, *campesinos'* ways of life drastically changed from subsistence to market dependency. They also began experiencing an increase in their vulnerability as monoculture implies less resilience to market fluctuations, losses caused by climatic variations, and other phytosanitary problems. Additionally, agroindustrial models that require monocultures' agrochemical-intensive methods cause the progressive degradation and loss of fertility of the soil, so that producers fall into a progressive spiral of dependency. For producers with greater access to capital, becoming dependent through market integration is less problematic than for small and medium producers, for whom increasing use of fertiliser on degrading soils is not possible. In the short- and medium-term, oil palm productivity loss leads to loss of income. Eventually, environmental degradation can force those *campesinos* who participate in oil palm market integration to sell their lands, making yet more land available for oil palm elites.

Deterritorialisation through dispossession using direct violence

The deterritorialisation by dispossession using direct violence is most prominent in the northwestern oil palm block, specifically in the Canton of San Lorenzo, in the northern Esmeraldas Province that borders Colombia. Awá and Afro-Ecuadorian communities have inhabited these territories for centuries; their ways of life have been based on living from and caring for, as well as protecting the coastal Chocó lowland rainforest (Antón, 2015). In the final section of this chapter, we unpack deterritorialisation through the usage of direct violence, which leads to landscapes of destruction and destitution, but also to communities coming together from within or across Black and Indigenous cultural differences to protect the living world around them and their self-determinative, culturally-specific ways of being. We will first discuss how this extractive approach of treating the people as disposable and violating their ancestral lands enables certain actors in the assemblage to eke out as much profit as possible.

(De)CO$_2$loniality and hope in the Ecuadorian Chocó borderlands: San Lorenzo Canton, Esmeraldas Province

The Ecuadorian Chocó's histories and present-day circumstances discussed in this section are not only little known to the world, but also to most of the country's population. The Afropacífico, Tsa'chila, Chachi, Épera, and Awá Indigenous peoples of the Esmeraldas Province have stewarded the Ecuadorian Chocó rainforests and rivers since the 1500s—a "periphery of the periphery" (Granda, 1977).

Historical colonial entanglements in Northern Esmeraldas

Reintegrating Moreno Parra's (2019) framing of Esmeraldas undergoing a slow death, the Ecuadorian state long ago set aside the province as a "sacrifice zone" (Silveira et al., 2017; Klein, 2014; Leon & Rosa, 2013), meaning a place that historically and present-day would endure the collateral damage for the progress and economic growth of the rest of the country. The Ecuadorian government continues to treat Esmeraldas as an exception. Hazlewood (2023) demonstrates that since the birth of the Republic, the Ecuadorian state precipitated and has perpetuated a long-wave crisis here in the country's Chocó borderlands by treating the forested, occupied lands as empty, open for the taking, and the people as less-than-human (merely labour, at the very best). A San Lorenzo collaborator (December 2009) explains how Northern Esmeraldas was officially demarcated as an extraction zone from 1859 to 1939:

> Ecuador could not pay the debt (to the British for providing economic backing in the Ecuadorian revolution against Spain). Then, the Esmeraldas Province was like a poker chip, where the British extracted gold in the province. It was considered almost like a reserve to protect Ecuador until General Rodríguez Lara's government started to export petroleum to pay the debt to the British. It was until then that it was considered a reserve zone.

Archival documents support this testimony. Due to forces led by Simón Bolívar calling for extra reinforcements from Great Britain (Alfaro, 1896) during battles for independence, Ecuador was born into a form of state-level external debt-peonage, needing to repay 1,824,000 pounds to Great Britain (Convenio acerca de la Deuda Extranjera, 1855). In 1856, the Ecuadorian President renounced state sovereignty

(Mensaje del Presidente de la República de Ecuador a las Cámaras Legislativas en 1856, 1856) over 200,000 square kilometres in Northern Esmeraldas (Ecuador Land Company, Limited Prospectus, 1859)—ancestral Black and Indigenous territories between Mataje River and La Tola (Terán, 1896)—permitting Great Britain (then Germany, followed by the United States) to plunder Black and Indigenous ancestral territories for 80 years. Minda Batallas (2020, 2013, 2002), a Black scholar from Esmeraldas, demonstrates that Northern Esmeraldas remains as much in the fringes today as ever and continues to socially and environmentally suffer from long-duration extractive pillage. It experiences some of the highest poverty rates in the country. In the canton of San Lorenzo, 84.6 percent of the population lives below the poverty line, according to the 2010 Population Census, a percentage well above 60 percent nationally and 51 percent in the province of Esmeraldas (INEC, 2010). Moreover, 47.3 percent of the people live in extreme poverty (Minda Batallas, 2020). The illiteracy levels in the cantons of San Lorenzo (15.3 percent) and Eloy Alfaro (17.2 percent) are the highest in the province, with percentages three times higher than the national rate. In the San Lorenzo canton, only 23 percent of households have basic services (INEC, 2010). It is especially worrisome that most of the communities that settled on the banks of contaminated rivers do not have drinking water systems (Lapierre Robles & Macías Marín, 2019). Many residents suffering from these social and environmental conditions remember and still await repayment for the gold, timber, rubber, vegetables, ivory, and other products extracted from their lands (Roots & Routes IC and the Communities of La Chiquita and Guadualito in collaboration with Selvas Producciones, 2024).

Marginalised San Lorenzo Black and Indigenous communities in the Ecuadorian borderlands refuse to be treated as disposable; they rise to make known their ancestors' struggles and *re-narrate* official versions of Ecuadorian history. Figure 8.3 shows San Lorenzo youth marching to "re-exist" (Cuero Campaz, 2022; Walsh, 2013; Zinn, 2006), what Walsh (2023) theorises as re-existing as pedagogical practices of "…rising up *and* living on in the cracks".

Present-day (de)CO$_2$loniality in the Ecuadorian Chocó Borderlands

Thirty years ago, much of Esmeraldas' coasts and banks of the Carchi, Mira, Santiago-Cayapas, Esmeraldas, and Muisne River basins were still covered by the lowland Ecuadorian Chocó rainforest. This area forms part of a contiguous ecological zone, designated in the late 1980s as the Chocó Biogeographic Region (Escobar, 2008; García Salazar, 2007; Myers et al., 2000; West, 1957), which then forms part of the greater Tumbes-Chocó Magdalena biodiversity hotspot (Conservation International, 2024). Encompassing 187,400 km² (Anwar, 2021), the Chocó extends through southern Panama (also known as the Darien Gap) and down the coasts of Colombia (also known as *El Pacífico*), and Ecuador. It includes the rainiest rainforests in the world as well as the western range of the Andes Mountains' cloud forests and highly diverse montane forests. Hundreds of resident Afro-descendant,

FIGURE 8.3 March for Independence Day. From left to right, the signs say, "In 1869, the zone of San Lorenzo was handed over to the English company, Ecuador Land. The English were the only ones with access to the zone's activities. They had their coin called *pailón* for their commercial transactions. The *pailón* circulated in San Lorenzo, Concepción, Eloy Alfaro, and Borbón. They converted us into colonists in our own lands. They deprived us of our constitutional rights. This colonialist situation lasted until 1939 when our territory was returned to Ecuador." (Photo by author, 10 August 2009).

Indigenous, and *mestizo* communities call the Chocó their home, catalysing the possibility of reconceptualising this forest/river home as an ethnic ancestral territory-region (GAIDEPAC, 2016; Escobar, 2008).

When people reside in the peripheries of society, the state often takes advantage of this invisibility by neglecting their obligations to provide services to these communities. Additionally, within these societal margins, outside the sight and interest of the rest of the nation, states can deliberately create legislative and economic assemblages that abandon these communities. Effectively, they can enable the activities of companies intent on the exploitation and dispossession of the marginalised, accomplices to plunder. This is characterised in the third typology discussed above, dispossessing with direct violence (Lasso, 2019).

Before the turn of the millennium, a drastic shift towards direct violence took place in the Chocó. Many local people report that oil palm companies began forcibly grabbing land from ancestral Afro-descendant communities with violent tactics. These included threats and harassment from bands of "hitmen", causing intercultural land conflicts (Hazlewood 2012, 2010a, 2010b; Lasso, 2019, 2012), which are facilitated through the collaboration of local state officials. Many Afro-Ecuadorian people sold their lands out of desperation, for laughable sums (Critical

Geography Collective of Ecuador, 2018; Cañas, 2009; Comité Internacional de Verificación, 2007; Buitrón, 2001). The result was that forest landscapes and food forests were converted into monocultural oil palm plantations.

Legal frameworks also serve to accelerate and embolden the commission of direct violence. Since Ecuador became a republic, legal mechanisms have been used to support industries that contribute to the national economy. For example, through Executive Decree No. 2961, 50,000 hectares of state land that formed part of the National Forest Reserve previously meant for conservation—including 30,000 hectares of tropical forests (Tuinstra, 2008; Armendáriz, 2002), 5,000 hectares of Afro-Ecuadorian and 1,000 hectares of Awá ancestral territories (Ramos, 2003)—were converted into a new agricultural frontier for "sustainable agricultural development" (Bravo, 2007, p. 111), allowing the "legal" expansion by big oil palm farms through coercion and illegal ways.

Expanding oil palm monocultures embodies the most recent disastrous wave of extractive CO_2loniality (Figure 8.4). Forest protector Aquilino Erazo speaks to the effects of deforestation: "[The oil palm companies] have brought us problems because they have destroyed many hectares of forest. They have extinguished many species of animals that no longer exist" (Roots & Routes IC-Selvas Producciones, 2020).

With such high degrees of violence and a limited ability to come and go, those Afro-Ecuadorian and Indigenous communities who have stood their ground against the encroaching violence of expanding oil palm companies literally and figuratively live like castaways, dispersed and isolated rainforest islands in a sea of oil palms (Hazlewood, 2012, 2010a, 2010b). To make matters worse, surrounding oil palm plantations contaminate the rivers with agrochemicals, and palm oil extractors dump their chemical-laden blackwater runoff directly into the rivers, as seen in Figure 8.5 (San Lorenzo TV, 2020; *Isaha Eszequiel Valencia Cuero v Palmeras de los Andes*, 2017; Núñez Torres, 2004; 1998).

Deforestation and water pollution impair every aspect of life for Black and Indigenous communities throughout San Lorenzo Canton (Public Letter to President Moreno, 2020; Indigenous Environmental Network, 2015). Water contamination leads to rashes and diseases, endangers the food security of local communities and their sovereignty, causes psychological stress, and disintegrates Afro-descendant and Indigenous peoples' ancestral ways of being (Lasso, 2019; Hazlewoood, 2012, 2010a, 2010b).

Local people testify that there is malicious intent, a kind of biological warfare. Oil palm companies have sent their workers to cut the banana, manioc, and sugar cane plants so that local communities do not have a way to sustain themselves on their land (Awá teacher, 2009, interview). La Chiquita resident Anaina Quintero Cortez states, "Now one lives with everything contaminated, with stomach aches all the time" (Roots & Routes IC-Selvas Producciones, 2020).

Even those who hope, may find it hard to see the light amidst contemporary circumstances in the Ecuadorian Chocó borderlands. Since 2008, Esmeraldas has

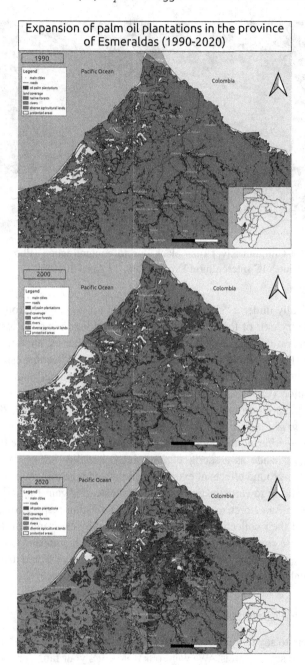

FIGURE 8.4 Comparative land use and cover change, 1990, 2000, and 2020: Oil palm expansion in San Lorenzo. Created by Cartographer Iñigo Arrazola Arranzabal, *Colectivo de Geografía Crítica de Ecuador*, 2023. Sources: MAGAP and *El Ministerio del Ambiente, Agua y Transición Ecológica de la República del Ecuador* (MAATE).

FIGURE 8.5   La Chiquita River runs black due to contamination (Photo by Roots & Routes IC Intercultural Drone Team, 17 April 2019).

been periodically under a "state of exception"—a no-go zone for tourism, having a patrolled entry/exit of the area due to drug trafficking, a special forces military presence, curfew, the presence of guerillas and paramilitaries, and money laundering (Green & Ferndández-Flores, 2023). The situation has become especially tense since 2018, when two newspaper journalists were kidnapped and killed in San Lorenzo Canton (Casey, 2018). Expanding oil palm plantations are inextricably tied to such socio-environmental landscape transformations (Ballvé, 2013; Misión de *Verificación al Pacífico Sur de Colombia*, 2009).

Yet, despite suffering the consequences of enduring life in a place that has been intentionally set aside as a sacrifice zone, deCO$_2$loniality insists on resisting the negative on-the-ground effects of extractive climate change mitigation-development strategies, working towards justice and dignity instead. Against all odds, ancestral communities in San Lorenzo strive to re-exist, requesting respect for the rights of nature, and for their collective rights to live in a healthy environment. Here we discuss two of these community-based instances in (1) La Chiquita and Guadualito, and (2) 5 de Junio (also known as Wimbí), where the people have united to activate hope to deCO$_2$lonialise and demand environmental-with-racial justice.

La Chiquita and Guadualito

If you were to say to me, 'Let's go see the President [of Ecuador],' the first thing that I would say to him is, 'We want our water! It's your fault for giving permission to the oil palm companies that the people are dying! I would like you to go and drink that water.' It's not fair what the government is doing to us.

—Anaina Cortez, La Chiquita resident (Roots & Routes IC-Selvas Producciones, 2020)

La Chiquita resident, Anaina Quintero Cortez, speaks to the injustice of the Ecuadorian state in allowing oil palm plantations to poison their biologically and culturally rich territories. Quintero Cortez is not alone. She speaks to what many people across the San Lorenzo Canton want: to make the invisible visible, deCO$_2$ lonise on-the-ground consequences of purported climate mitigation politics, and environmental-with-racial justice.

Many Chocó residents struggle to continue intra-island ways of being, doing, and knowing based on taking care of their ancestral island territories (Hazlewood, 2012, 2010a, 2010b). In La Chiquita and Guadualito, they cultivate their "territories of difference" (Escobar, 2008, p. 25). They challenge colonial state and international plans for a one-size-fits-all development, and instead, insist upon respect for "sustainabilities" based on specific pluricultural expressions of "reciprocal relationality" (Todd, 2017, p. 107). Amidst globalised forces that seemingly push towards conforming to a monocultural universalism of what is considered valuable, these place-based pluriversal ways of being in the world (Kothari et al., 2019; Oslender, 2019) root down as hopeful paradigms in praxis (Hazlewood et al., 2023; Haran, 2010; Wright, 2010, 2008). According to their own four-world cosmovision, the Awá nationality, for example, explain their territorial life plan, "*Inkal Awá Katsa Kual Wat Uzan*":

Our elders struggled to defend the life, territory, language, culture, and the unity of our people. For this reason, we have decided to recuperate and strengthen the historical memory of our ancestors. We have decided to continue *trochando* (carving out trails) and walking together in *minga* (work trade), for in following the footsteps of our elders, like the Big Awá Family (across Colombia and Ecuador), we remain strong like the big tree (*Katsa Tikana*, Ceibo tree, from where they originated).

(Gran Familia Awá Binacional, 2009, p. 3,
translated from Spanish by Hazlewood)

Similar to most Black and Indigenous communities throughout the San Lorenzo Canton, the people of La Chiquita and Guadualito use swidden farming techniques—*chacras* (farm plots) that rotate back to forest—horizontally organising their ancestral territories. They plant what is used closest to homes, leaving primary forests furthest away for hunting or, when needed, tree harvesting (Figure 8.6). Vertical zoning strategies include fostering diverse, multiple-tiered agroecological communities from forest floors into high canopies. Such culturally-based agroforestry approaches act as preventative food security methodologies because subsistence and market products are interwoven into their crop patterns to withstand economic booms and busts, as well as social crises (Hazlewood, 2023, 2012, 2010a, 2010b, 2004).

The Chocó peoples of La Chiquita and Guadualito also demonstrate that protecting their flourishing polycultural landscapes is not possible without fashioning

FIGURE 8.6   Horizontal and vertical zoned landscapes of the Awá community of Guadualito (Drone Shot by Roots & Routes IC Intercultural Team, 18 April 2019).

inter-island geographies. Simultaneously, they put ancestral versions of reality (Cagüeñas et al., 2020; Escobar, 2018; Oslender, 2016; Walsh, 2013; García Salazar, 2010; Grim & Marglin, 2001; Apffel-Marglin & Proyecto Andino de Tecnologías Campesinas, 1998; Esteva, 1987) front and centre, and decolonise false solutions to climate change. On 23 July 2010, those from the mostly isolated peripheries of Ecuador stepped to the frontlines of constitutional-shifts-in-action, initiating a case against palm companies that raised the bar for Earth Jurisprudence—an emerging field of law that is gaining popularity worldwide, confirming nature as a subject and humans as part of a wider "community of beings" (Kauffman, 2020; United Nations, 2019; Kauffman & Martin, 2017). La Chiquita, Guadualito, and their ancestral territories' diverse flora, fauna, and incarnate beings formed the plaintiff community. They initiated a landmark intercultural and interspecies lawsuit to guarantee ancestral rights to clean water and the rights of La Chiquita River. Calling for a reprioritisation of economic profit that often subjugates the intrinsic value of all sentient beings by applying national-level legislation, they filed the first constitutionally-based Rights of Nature (Article 71) and *Sumak Kawsay* (Living Well, Articles 12–34) civil lawsuit against two oil palm companies—Esteros EMA SA Palesema and Los Andes SA—to the Provincial Court of Esmeraldas. La Chiquita, Guadualito, and Nature demanded that the two oil palm companies compensate them for environmental and communal health damages and repair the La Chiquita River contamination (Indigenous Environmental Network, 2015).

After six and a half years, in January 2017, Ecuador's Esmeraldas Provincial Court handed down its decision on the world's first accepted-for-trial Rights of

Nature lawsuit based on a national-level constitution. It was partially favourable. Judge Morales Suarez confirmed that the oil palm companies were guilty of environmental and psychological damages. Guadualito, La Chiquita, and Nature thus broke through the confines of constricting politics (Alianza Periodística Tras las Huellas de la Palma, 2022; *Isaha Ezequiel Valencia Cuero v Palmeras de los Andes,* 2017). Incongruously, the judge's decision avoided forcing the oil palm companies to pay for and correct their violations. While charging that 12 state and provincial institutions should start providing the services they were designed to do, he merely ordered the companies to plant bamboo and teach a local folklore course to their workers (Hazlewood et al., 2017).

The guilty as-charged defendants have yet to address the rulings against them charged in 2017 (Alianza Periodística Tras las Huellas de la Palma, 2022; *Isaha Ezequiel Valencia Cuero v Palmeras de los Andes,* 2017). The Los Andes palm oil processing plant continues to dump their palm oil processing waste into the La Chiquita River (Roots & Routes IC, 2020; Roots & Routes IC-Selvas Producciones, 2020; San Lorenzo TV, 2020), and the Black and Awá plaintiffs still lack access to clean water to bathe, wash, cook, or drink. Having yet to receive adequate attention and support within Ecuador and beyond for taking a stand for their living territories, the greatest challenge has been to rise above the authorities' inaction and find the wherewithal to take the next step forward.

Since 2015, the communities, together with Hazlewood as communication advisor, have begun efforts to decolonise methodologies of environmental-with-racial justice through truth-telling, bridge-building, weaving intercultural Rights of Nature networks, indigenising technology, and re-storying their hope-with-justice struggles (Tuhiwai Smith, 1999) and culturally-based "life visions otherwise" (Walsh, 2015). An audiovisual school was set up in La Chiquita and Guadualito (see Figure 8.7), and the team produced a full-length documentary of the communities' struggles for clean water, *Queremos Nuestra Agua* (*Together for Water* in English), aimed for release in 2024 (see Figure 8.8).[3] A United States-based non-profit intercultural network, Roots & Routes IC, assists in their efforts and other initiatives related to what is called *corazostenabilidades* (sustainabilities with heart).[4]

5 de Junio (also known as Wimbí)

5 de Junio (also known as Wimbí) is a Black community in the San Lorenzo Canton. Their lands are under pressure from palm oil plantations and gold mining, which not only threaten the basis of local livelihoods but also cause ecological degradation and environmental suffering. These processes of environmental racism not only lower the quality of life for Black people in the region, but also expose them to the process of slow death.

The Wimbí consists of 416 inhabitants. Part of their territory is in the Río Santiago Cayapas Commune, inalienable communal land. The other part is located outside the commune, in territories of ancestral possession without titling. Here,

FIGURE 8.7   Eriberto Gualinga, an Amazonian film director of the Sarayaku Kichwa Original People, together with youth from La Chiquita and Guadualito in the Chocó-Amazon Film School. Gualinga is demonstrating how to use a camera in the background (Photo by Hazlewood, 25 February 2017).

Wimbí faced a land conflict with an oil palm company and was affected by processes of deterritorialisation (García Salazar & Walsh, 2017). This occurred in 2000, due to the adjudication and subsequent sale of lands of ancestral possession—an irregular process that ended up being recognised by titles endorsed by the courts and the Land Registry. This land sale led to processes of migration, intra- and extra-community social conflict, in addition to the loss of livelihoods and food sovereignty (Antón Sánchez, 2015).

The conflict with the company goes back to a process through which ancestral land, passed into the possession of a community member of Wimbí, was then sold to an investor from Quito. Subsequently, the lands were quickly transferred to the company, Energy & Palma (formerly Palmeras del Pacífico, La Fabril)– established in 1999 to acquire communal land in the communities of Carondelet, 5 de Junio (Wimbí), Urbina, and San Javier (Critical Geography Collective of Ecuador, 2018; Cañas, 2009). Energy & Palma currently has the title of ownership of the 1,200 hectares in dispute. Conflicts began shortly after adjudicating and illegally selling the land, with the Energy & Palma manager declaring that they had legally purchased the indivisible communal lands from community members.

In early 2016, Energy & Palma filed a lawsuit against four local leaders from Wimbí on charges of illegal use of land and land trafficking. This lawsuit was supported by the Ministry of Agriculture's Land Secretariat and the Judge of the San Lorenzo Judicial Unit, who affirmed that the company is the holder of legal property titles (Critical Geography Collective of Ecuador, 2018). According to the

FIGURE 8.8   Queremos Nuestra Agua: Unidos por el Bosque del Chocó film cover in English. Created by graphic designer Matias Canales and co-produced by Roots & Routes IC and the communities of La Chiquita and Guadualito in collaboration with Selvas Producciones, led by Eriberto Gualinga.

MAGAP report, the purchase of the properties took place through logging companies, who then passed legal possession to Energy & Palma (Defensoría del Pueblo, 2011).

The conflict escalated in November 2016, when the manager of Energy & Palma claimed that the residents of the ancestral territory were illegally invading his private property and demanded their eviction. Additionally, a police squad used a bulldozer to destroy crops and fences in the disputed area. Several members of the community came out to face the squad and stop the machinery. The eviction was stopped, and the communities confiscated the bulldozer as a measure of retaliation. Subsequently, the company sued community leaders for the theft of machinery.

In August 2017, the company and Wimbí leaders reached an agreement where the community returned the machinery and the company withdrew the charges of theft. However, the trial for illegal use of land continued. The community did not accept the terms proposed by the company, which intended to give them 100 hectares of land and offered them work on the plantation. "We are not going to go back to being slaves," commented one of the local leaders. The leaders of Wimbí have been criminalised as thieves and invaders of their own territory; they, however, continue to fight for their ancestral right to it.

We can analyse the case of Wimbí as an example of the intersection between the reproduction of poverty, racism, and environmental suffering. Structural forms of violence such as those described above that perpetuate conditions of poverty are amplified by the socioeconomic impacts of the extractive activities of the oil palm agro-industry and gold mining. The San Lorenzo Canton has the second highest number of hectares dedicated to palm, constituting a total of 26,641 hectares (Brown, 2018). In tandem, the impact of mining on land use removes vegetation and contaminates river water with heavy minerals. By 2011, the mining activity zone included 5,709 hectares of direct impact and a local influence zone amounting to 224,284 hectares (Lapierre Robles & Macías Marín, 2019).

The Catholic University of Esmeraldas investigated 15 mining fronts in 2012 and later expanded the investigation to state mining fronts, finding that levels of heavy minerals were well above permitted standards and higher contamination in state mining fronts. The investigation was contracted by the Environmental and Social Remediation Programme (PRAS) of the Ecuadorian Ministry of Environment (MAE). The MAE, nevertheless, did not provide this information to the participating communities. Additionally, the environmental remediation implemented by state institutions was "insignificant and deficient" (Lapierre Robles & Macías Marín, 2019, p. 256), intervening only in 60 hectares, or 1.2 percent of the total affected area, according to the PRAS.

These extractive activities have produced toxic environments, with water and soil being contaminated with heavy metals such as aluminium and iron (Lapierre Robles & Macías Marín, 2019). In the province, an estimated 4,800 pools in former mining fronts have been left open with contaminated water. Water contamination also causes short-term dermatological diseases, vaginal infections and problems, stomach diseases, and others that were reported by inhabitants of Wimbí. Heavy metals in water sources could likewise produce long-term health problems such as Alzheimer's (interview with Eduardo Rebolledo, researcher, at Pontificia Universidad Católica del Ecuador-SE, 31 October, 2017).

Similarly, in her study of the effects of oil palm production, Hazlewood (2012) found that between 1999 and 2003, there was a threefold increase in the number of cases of people affected by agrochemicals from the improper disposal of waste from the oil palm industry. It is estimated that the life and health of 93 percent of the population of the cantons of San Lorenzo and Eloy Alfaro, approximately 94,000 people, are at risk (Lapierre Robles & Macías Marín, 2019).

The seriousness of the situation caused by the effects of environmental pollution and the inaction of the Ecuadorian state leads to forms of slow death—the harmful

FIGURE 8.9   Children in Wimbí filling water up at hoses (Photo by Moreno Parra, October 2017).

short- and long-term effects of everyday actions in toxic environments or environments affected by environmental threats over which the inhabitants have little control. The case of Wimbí illustrates this process of slow death of Black communities, in contexts of sustained land grabbing, forest and biodiversity disappearance, environmental degradation, insecurity, and state neglect (Zaragocín, 2018). The state response has been to "let die", in a racialised regime of disposability that makes the workings of capital possible.

Nevertheless, communities of Esmeraldas are resourceful agents. They resist this process of slow death. In 2011, in a failed effort to resist and contest, communities of Esmeraldas, together with the Ombudsman's Office, were able to file precautionary measures in relation to the right to water and a healthy environment and a moratorium on mining activities. These measures were never applied, and the extractive activities continued. Wimbí, however, continued its search for a solution to water contamination and, in 2019, the community worked with the municipal government of San Lorenzo and secured funding from the State Bank (*Banco del Estado*) for the construction of a potable water system. The studies carried out determined that the best place for water intake was the estuary in the area in dispute with the company (see Figure 8.9). The water intake system was built there, and the provincial government declared it a conservation area. Due to Ecuadorian laws regarding water for human use, no oil palms can be planted in this area, "The company maintains ownership of the land in court, but in practice the community maintains the use of its territory." (Bonilla, 2022, p. 125). Despite the river still being contaminated, the community was successful in securing safe water for human consumption and, simultaneously, regaining control of the use of the lands in dispute with the company.

Conclusion

The Ecuadorian state sponsored "green" oil palm assemblages and shady mono-culture entanglements of CO_2loniality. Nonetheless, perhaps unknowingly, these geographies have spurred deCO_2lonial struggles and opened fissures of hope throughout Ecuador and its Chocó Borderlands.

Together, the state and the oil palm industry have neglected to recognise the ancestral Black and Indigenous communities' claims of companies' social and environmental disruption and harm. The state has granted the oil palm industry more power, and aggravated their processes of racialisation through very specific political and economic assemblages, as in the case of the passing of the oil palm law (Asamblea Nacional República del Ecuador, 2020), regardless of all the evidence of the negative consequences of biofuels/oil palm plantations as a supposed way to mitigate climate change. The "green" oil palm assemblages work in three principal typologies and velocities: (1) dispossession through using indirect violence, (2) by transforming the *campesinos'* everyday life and livelihoods, and (3) dispossession through using direct violence.

Despite the dispossession of land and living like refugees on ancestral territorial islands amidst a sea of oil palms in the San Lorenzo Canton of Northern Esmeraldas, Black and Indigenous communities in the most marginalised province of the country continue rising up and re-existing. They, together with their allies, (1) declare "We are still here! And our lives matter!", (2) root into and visualise the Ecuadorian Chocó borderlands, and (3) apply progressive national-level constitutional legislation to deCO_2lonise false solutions to climate change mitigation by demanding racial-with-environmental justice.

Acknowledgement

The authors would like to extend our deep gratitude to Leni Lindemann, MSc in Development Studies, for her help with the in-text citations, compiling the majority of the bibliography, and adding accessibility and aesthetics to the table. All errors and omissions are the authors' responsibility.

Notes

1 The types of legal and regulatory instruments, the year issued and by whom, the discourses, and the benefits for the oil palm sector that encouraged expansion and territorialisation between 2008 and 2020 in Ecuador is illustrated in Lasso (2019). In 2020, a new law (RO No. 255) was introduced, aiming to strengthen and encourage the development of production, commercialisation, extraction, exportation, and industrialisation of oil palm and its derivatives. This, in turn, would promote food security, sustainability, good living, and low productivity.

2 For more information on these main economic groups and their concentration of land, extractors, and collection centres, see Lasso (2019).

3 See also https://www.together4water.com/ for more information.

4 Visit https://www.rootsroutes.org/ for more information about Roots & Routes IC.

References

Acosta, A. (2009a). Big changes require bold efforts. In A. Acosta & E. Martínez (Eds.), *Rights of nature: The future is now* (pp. 15–23). Abya Yala.

Acosta, A. (2009b). Always more democracy, never less: As a prologue. In A. Acosta & E. Martínez (Eds.), *Good living: A path to development* (pp. 19–30). Abya Yala.

Acosta, A., & Martínez, E. (2009a). *Rights of nature: The future is now*. Editorial Abya-Yala.

Acosta, A., & Martínez, E. (2009b). Presentation. In A. Acosta & E. Martínez (Eds.), *Good living: A path to development* (pp. 7–18). Abya Yala.

Acosta, A., & Martínez, E. (2009c). *Plurinationality: Democracy in diversity*. Abya Yala.

Alfaro, E. (1896). *Gordian debt*. Imprenta Nacional.

Alianza Periodística Tras las Huellas de la Palma. (2022, October 11). *The bittersweet victory of La Chiquita against two palm giants in Ecuador.* Mongabay. https://es.mongabay.com/2022/10/la-agridulce-victoria-de-la-chiquita-contra-dos-gigantes-de-la-palma-en-ecuador/

Antón, J. (2015). Pressure on the right to the ancestral territory of the Afro-Ecuadorian people. The case of the Federation of Black Communities of Alto San Lorenzo. *Colombian Journal of Sociology*, 38(1), 107–144. http://bdigital.unal.edu.co/67248/

Antón Sánchez, J. (2015). The right to the ancestral territory of the Afro-Ecuadorian people: FECONA. In F. García & J. Antón Sánchez (Eds.), *Monitoring racism. Four cases of community observation of the right to non-discrimination in indigenous and Afro-Ecuadorian communities* (pp. 79–119). Instituto de Altos Estudios Nacionales (IAEN).

Anwar, A. (2021). Atrato River: History, features, routes, aids, flora, fauna. *Global Research Journal of Natural Science and Technology*, 1(1), 7–14.

Apffel-Marglin, F., & Proyecto Andino de Tecnologías Campesinas. (1998). *The spirit of regeneration: Andean culture confronting Western notions of development*. Zed Books.

Armendáriz, O. (2002). African palm sector. *Quito, Ecuador: Superintendency of banks and insurance, national directorate of studies and statistics, directorate of investigations.* Recuperado de http://www.flacsoandes.edu.ec/biblio/catalog/resGet.php.

Asamblea Nacional Constituyente. (2008). Constitution of the Republic of Ecuador. https://www.ambiente.gob.ec/wp-content/uploads/downloads/2018/09/Constitucion-de-la-Republica-del-Ecuador.pdf

Asamblea Nacional República del Ecuador. (2020). Subject: Law for the strengthening and development of production. Marketing, extraction. export and industrialiation of oil palm and its derivatives. AN-SG-2020-0425-0. Available at: https://vlex.ec/vid/ley-fortalecimiento-desarrollo-produccion846767045

Ashley, J. M. (1987). The social and environmental effects of the palm-oil industry in the oriente of Ecuador. Latin American Institute, University of New Mexico, 19. http://repository.unm.edu/handle/1928/7731

Ballvé, T. (2013). Grassroots masquerades: Development, paramilitaries, and land laundering in Colombia. *Geoforum*, 50, 62–75. Crossref.

Barnett, T., & Blaikie, P. (1994). AIDS as a long wave disaster. In A. Varley (Ed.), *Disasters, development and environment* (pp. 139–162). J. Wiley.

Benalcázar, W. (2009). *Esmeraldas loses its tropical forest*. El Comercio.

Bonilla, N. P. (2022). *The long journey of the Afro-Ecuadorian community of Uimbi, creating autonomies, in the midst of environmental racism*. [Unpublished Master's thesis]. FLACSO Ecuador.

Borja, S. (2020, December 1). *Ecuador's palm oil law a boon for producers, but not people and planet, groups say.* Mongabay. https://news.mongabay.com/2020/12/ecuadors-palm-oil-law-a-boon-for-producers-but-not-people-and-planet-groups-say/

Bravo, E. (2007). *Biofuels, energy crops, and food sovereignty in Latin America: Igniting the biofuels debate.* Acción Ecológica.

Brown, K. (2018). *Community vs. company: A tiny town in Ecuador battles a palm oil giant.* Mongabay. https://news.mongabay.com/2018/08/community-vs-company-a-tiny-town-in-ecuador-battles-a-palm-oil-giant/

Buitrón, R. (2001). *The case of Ecuador: Paradise in seven years? The bitter fruit of oil palm: Dispossession and deforestation.* World Rainforest Movement. http://wrm.org.uy/oldsite/plantations/material/OilPalm.pdf

Bullard, R. (Ed.) (1993). *Confronting environmental racism. Voices from the Grassroots.* South End Press.

Bullard, R. (2004). *Environment and morality: Confronting environmental racism in the United States* (pp. 1–31). United Nations Research Institute for Social Development. Identities, Conflict and Cohesion Programme Paper Number 8.

Cagüeñas, D., Galindo Orrego, M. I., & Rasmussen, S. (2020). Atrato and its guardians: Ecopolitical imagination to weave new rights. *Colombian Journal of Anthropology,* 56(2), 169–196.

Cañas, V. (2009). Socio-environmental and labour conflict between the Carondelet community and the palm grower Palmeras del Pacífico: Plural actors and diverse perspectives. [Master's thesis]. FLACSO Ecuador.

Cardenas, R. (2012). Green multiculturalism: Articulations of ethnic and environmental politics in a Colombian 'black community'. *The Journal of Peasant Studies,* 39(2), 309–333. https://doi.org/10.1080/03066150.2012.665892

Carlet, F., & Ferreira, F. (2018). For a postcolonial legal socio-anthropology: From Western ethos to Afro-Ecuadorian narratives of resistance. Report no. 3133055, SSRN Scholarly Paper. Social Science Research Network. https://papers.ssrn.com/abstract=3133055

Carrere, R. (2002). Oil palm: The expansion of another destructive monoculture. In R. Carrere & L. Lohmann (Eds.), *The bitter fruit of oil palm: Dispossession and deforestation* (pp. 8–12). World Rainforest Movement.

Carrere, R. (2009). Development and forests. In A. Acosta & E. Martínez (Eds.), *Good Living: A path to development* (pp. 93–101). Abya Yala.

Casey, N. (2018). Marxist rebels killed journalists and driver, Ecuador's president says. New York Times, April 13. https://www.nytimes.com/2018/04/13/world/americas/ecuador-journalists-killed.html

Castellanos-Navarrete, A., & Jansen, K. (2018). Is oil palm expansion a challenge to agro-ecology? Smallholders practicing industrial farming in Mexico. *Journal of Agrarian Change,* 18(1), 132–155. https://doi.org/10.1111/joac.12195

Center for Economic and Social Rights. (2015). *Participation in sales of main agricultural and agroindustrial companies: 2009–2013.* Center for Economic and Social Rights.

Comité Internacional de Verificación. (2007). *Verification report on the expansion of African palm monocultures in northern Esmeraldas.*

Conservation International. (2024). Ecuadorian Chocó. https://www.conservation.org/ecuador/nuestro-trabajo/programas/choco

Convenio acerca de la Deuda Extranjera. (1855). *Celebrated by Messrs. Marcos Espinel and Elías Mocatta.* Imprenta Del Gobierno.

Correa, R. (2007). Accountability of the social agenda. https://www.presidencia.gob.ec/wp-content/uploads/downloads/2012/11/2007-07-27-Discurso-Rendici¢n-de-Cuentas-de-la-Agenda-Social.pdf

Critical Geography Collective of Ecuador. (2018). *Cartographic and social expertise in relation to land conflicts in the community of Wimbí, San Lorenzo.*

Cuero Campaz, H. (2022). *Keys and logic for a new understanding of territorial planning in multi-ethnic and multicultural territories*. [PhD Thesis]. Universidad Politécnica de Cataluña, Barcelona.

Danec Group. (2014). *Social report 2014* (pp. 1–55). https://danec.com.ec/en/wp-content/uploads/2022/04/INFORME-DE-RESPONSABILIDAD-SOCIAL-2014.pdf

Defensoría del Pueblo. (2011). *Complaint by the Ombudsman's Office by the Santiago Cayapas community, Eloy Alfaro Canton, province of Esmeraldas*. Eloy Alfaro, Esmeraldas.

Dominguez, J. M. (2008). *The ethics of biofuel production*. PowerPoint presentation. FLACSO Ecuador.

Ecuador Land Company, Limited Prospectus. (1859). https://www.europa.clio-online.de/quelle/id/q63-28358

Editor PCN (2009). International verification mission to document five areas affected by monocultures for the production of agrofuels in Colombia: Oil palm and sugarcane. https://renacientes.net/blog/2009/07/21/mision-de-verificacion-internacional-para-documentar-5-zonas-afectadas-por-los-monocultivos-para-la-produccion-de-agrocombustibles-en-colombia-palma-aceitera-y-cana/

elEconomista.es (2006). CSR—The government of Ecuador promotes its biofuels programme to promote the use of bioethanol and biodiesel. https://www.eleconomista.es/mercados-cotizaciones/noticias/41747/07/06/RSC-El-Gobierno-de-Ecuador-impulsa-su-Programa-de-Biocombustibles-parapotenciar-el-uso-de-bioetanol-y-biodiesel-.html.

Escobar, A. (2008). *Territories of difference: Place, movements, life, redes*. Duke University Press. https://doi.org/10.1215/9780822389439.

Escobar, A. (2012). Beyond development: Postdevelopment and transitions towards the pluriverse. *Journal of Social Anthropology*, 21, 23–62.

Escobar, A. (2018). *Designs for the pluriverse: Radical interdependence, autonomy, and the making of worlds. New ecologies for the twenty-first century*. Duke University Press.

Esteva, G. (1987). Regenerating people's space. *Alternatives*, 12(1), 125–152.

Esteva, G. (2011). Beyond development: The good life. In *Andean Contributions*, No. 28. Simón Bolívar Andean University, Andean Human Rights Programme.

Fierro, L. (1991). *Financial groups in Ecuador*. Popular Education Centre, CEDEP.

GAIDEPAC. (2016). Another peaceful possibility. https://pazificopedia.blogspot.com/p/cartas-gaidepac.html

Galeano, E. (2009). Nature is not mute. In A. Acosta & E. Martínez (Eds.), *Rights of nature: The future is now* (pp. 25–29). Abya Yala.

García Salazar, J. (2007). *Ancestral territories, identity and palm: A reading from Afro-Ecuadorian communities*. Altropico. https://altropico.org.ec/en/territoriosancestrales-identidad-y-palma-una-lectura-desde-las-comunidades-afroecuatorianas-2-007/

García Salazar, J. (2010). *Territories, territoriality, and deterritorialisation: A pedagogical exercise to reflect on ancestral territories*. Altropico Foundation.

García Salazar, J., & Walsh, C. (2017). *Thinking while sowing/sowing while thinking with Grandfather Zenón*. Universidad Andina Simón Bolivar, Abya Yala.

Gilmore, R. W. (2008). Forgotten places and the seeds of grassroots planning. In R. Hale (Ed.), *Engaging contradictions: Theory, politics, and methods of activist scholarship* (pp. 31–61). University of California Press.

Gonzalez, M. (2007). Petroleum biofuel synergy, Ministry of Mines and Petroleum. http://www.ceda.org.ec/descargas/ForoBio/ 3%20PANEL/Mauro%20Gonzalez.pdf

Gran Familia Awá Binacional. (2009). *Inkal Awá Katsa Kual Wat Uzan*. Altropico.

Granda, G. (1977). *Studies on a Hispanic American dialectal area with a Black population*. Instituto Caro y Cuervo.

Green, E., & Ferndández-Flores, M. (2023). Mexican cartels are turning once-peaceful Ecuador into a narco war zone. VICE, April 17, 2023.

Grim, J., & Marglin, A. (2001). *Indigenous traditions and ecology: The interbeing of cosmology and community.* Distributed by Harvard University Press for the Centre for the Study of World Religions, Harvard Divinity School.

Grossman, Z., Parker, A., & Frank, B. (2012). *Asserting native resilience: Pacific Rim indigenous nations face the climate crisis.* Oregon State University Press.

Gudynas, E. (2009). Six key points in environment and development. In A. Acosta & E. Martínez (Eds.), *Good Living: A path to development* (pp. 39–49). Abya Yala.

Gudynas, E. (2011). Scope and content of the transitions to post-extractivism. *Ecuador Debate*, 82, 61–79.

Haran, J. (2010). Redefining hope as praxis. *Journal for Cultural Research*, 14(4), 393–408.

Harvey, D. (1989). *The limits to capital.* University of Chicago Press/Midway Reprints.

Hazlewood, J. A. (2004). *Socio-environmental consequences of market integration among the Chachis of Esmeraldas, Ecuador.* University of Florida.

Hazlewood, J. A. (2010a). Beyond the economic crisis: CO_2lonialism and geographies of hope. *Icons, Social Sciences Magazine*, (36), 81–95.

Hazlewood, J. A. (2010b). *Geographies of CO_2lonialism and hope in the Northwest Pacific Frontier territory-region of Ecuador.* [PhD Thesis]. University of Kentucky.

Hazlewood, J. A. (2012). CO_2lonialism and the "unintended consequences" of commoditising climate change: Geographies of hope amid a sea of oil palms in the northwest Ecuadorian pacific region. *Journal of Sustainable Forestry*, 31(1–2), 120–153.

Hazlewood, J. A. (2023). Be(y)on(d) the map: Collaboratively activating Geographies of (De) CO_2loniality/H_2Ope in the Ecuadorian Chocó borderlands. *Environment and Planning E: Nature and Space*, 6(3), 1463–1500.

Hazlewood, J. A., & La Chiquita and Guadualito. (2017). *Court issues ruling in world's first rights of Nature lawsuit.* Intercontinental Cry. https://intercontinentalcry.org/court-issues-ruling-worlds-first-rights-nature-lawsuit/

Hazlewood, J. A., Middleton Manning, B. R., & Casolo, J. J. (2023). Geographies of Hope-in-Praxis: Collaboratively decolonising relations and regenerating relational spaces. *Environment and Planning E: Nature and Space*, 6(3), 1417–1446. https://doi.org/10.1177/25148486231191473

Houtart, F. (2009). 21st century socialism. In A. Acosta & E. Martínez (Eds.), *Good living: A path to development* (pp. 149–168). Abya Yala.

Indigenous Environmental Network. (2007). *Carbon trading: Capitalism of the air—Conflicts with indigenous knowledge.* Distributed at the United Nations Permanent Forum for Indigenous Peoples: Sixth Session, New York.

Indigenous Environmental Network. (2015). *Support La Chiquita and Guadualito ancestral communities and Nature.* https://www.ienearth.org/support-la-chiquita-and-guadualito-ancestral-communitiesand-nature/

INEC (National Institute of Statistics and Censuses of Ecuador). (2010). *Poverty due to unsatisfied basic needs.* Population and housing census 2010. www.ecuadorencifras.gob.ec

INEC-ESPAC (National Institute of Statistics and Censuses of Ecuador-Continuous Agricultural Surface and Production Súrvey) (2016). *Surface and agricultural production survey.* https://www.ecuadorencifras.gob.ec/estadisticas-agropecuarias-2/

Interview with Awá teacher. (2009).

Interview with FEDEPAL President. (2016).

Interview with O. Nastacuaz. (2009).

Isaha Ezequiel Valencia Cuero v Palmeras de los Andes. (2017). [08100-2010-0485].

Jull, C., Redondo, P. C., Mosoti, V., & Vapnek, J. (2007). Recent trends in the law and policy of bioenergy production, promotion and use (FAO Legal Papers Online No. 68). http://www.compete-bioafrica.net/policy/Ipo68.pdf

Kauffman, C. M. (2020). *Mapping transnational rights of nature networks & laws: New global governance structures for more sustainable development.* International Studies Association Annual Conference, Toronto, Canada, March 29, 2020.

Kauffman, C. M., & Martin, P. L. (2017). Can rights of nature make development more sustainable? Why some Ecuadorian lawsuits succeed and others fail. *World Development,* 92, 130–142.

Klein, N. (2014). *This changes everything: Capitalism vs. the climate.* Simon & Schuster.

Kothari, A., Salleh, A., Escobar, A., et al. (2019). *Pluriverse: A post-development dictionary.* Tulika Books.

Lander, E. (2009). Towards another notion of wealth. In A. Acosta & E. Martínez (Eds.), *Good living: A path to development* (pp. 31–37). Abya Yala.

Lapierre Robles, M., & Macías Marín, A. (2019). *Extractivism, (neo)colonialism and organized crime in the north of Esmeraldas.* PUCE / Abya Yala / Instituto de Estudios Ecologistas del Tercer Mundo.

Lasso, G. (2012). *Factors affecting the expansion of oil-palm plantations in Ecuador: Deforestation and socio-cultural impacts.* [Master's thesis]. University of Kent.

Lasso, G. (2019). *The dispute over territories around the Ecuadorian agri-food system: Strategies of power and resistance, processes of territorialisation and deterritorialisation.* [Doctoral dissertation]. Instituto de Ciencia y Tecnología Ambientales, Universidad Autónoma de Barcelona.

Lasso, G., & Clark, P. (2016). Food sovereignty, modernisation, and neodevelopmentalism: The contradictions of agrarian policy in Ecuador's Citizen Revolution. In M. Le Quang (Ed.), *The citizen revolution: Gray scale advances, continuities, and dilemmas* (pp. 260–291). Editorial IAEN.

Leon, E., & Rosa, P. R. (2013). *Limits and boundaries: A decolonial interpretation of the territorial conflicts in the Northern Region of Esmeraldas during the late 19th century.* [PhD Thesis]. Duke University.

Marin-Burgos, V. (2014). *Access, power, and justice in commodity frontiers. The political ecology of access to land and palm oil expansion in Colombia.* University of Twente. https://doi.org/10.3990/1.9789036536851

Massey, D. (1993). Power-geometry and a progressive sense of place. In J. Bird, Ba. Curtis, T. Putnam, G. Roberston, & L. Tickner (Eds.), *Mapping the cultures: Local cultures, global change* (pp. 59–69). Routledge.

Mideros Zamora, M. A. (2010). *Palm-growing companies and the generation of local economic development in the canton of San Lorenzo, Esmeraldas (1998–2008).* [Master's thesis]. FLACSO Ecuador.

Mignolo, W. (2005). *The idea of Latin America.* Blackwell Manifestos. Wiley-Blackwell.

Mignolo, W., & Walsh, C. E. (2018). Introduction. In W. Mignolo & C. E. Walsh (Eds.), *On decoloniality: Concepts, analytics, praxis* (pp. 1–15). Duke University Press.

Minda Batallas, P. (2002). *Identity and conflict: The fight for land in the northern area of the Esmeraldas Province,* 2nd ed. Universidad Politécnica Salesiana, Ediciones Abya-Yala.

Minda Batallas, P. (2013). *Deforestation in northern Esmeraldas: The actors and their practices.* Universidad Politécnica Salesiana.

Minda Batallas, P. (2020). *Towards an environmental history of Esmeraldas: The impact of extractive economies*. Universidad Andina Simón Bolívar.

Mingorría, S., Gamboa, G., Martín-López, B., & Corbera, E. (2014). The oil palm boom: Socio-economic implications for Q'eqchi' households in the Polochic valley, Guatemala. *Environment, Development and Sustainability*, (16). https://doi.org/10.1007/s10668-014-9530-0

Misión de Verificación al Pacífico Sur de Colombia. (2009). *The palm expansion in Nariño: Report of the Verification Mission to the South Pacific*.

Moreno Parra, M. (2019). Environmental racism: Slow death and dispossession of Afro-Ecuadorian ancestral territory in Esmeraldas. *Icons—Social Sciences Magazine*, 64, 89–109.

Müller, M. (2015). Assemblages and actor-networks: Rethinking socio-material power, politics and space. *Geography Compass*, 9(1), 27–41. https://compass.onlinelibrary.wiley.com/doi/10.1111/gec3.12192

Murray Li, T. (2007). Practices of assemblage and community forest management. *Economy and Society*, 36(2), 263–293. https://doi.org/10.1080/03085140701254308

Myers, N., Mittermeier, R. A., Mittermeier, C. G., et al. (2000). Biodiversity hotspots for conservation priorities. *Nature*, 403(6772), 853–858.

Nail, T. (2017). What is an Assemblage? *SubStance*, 46(1), 21–37.

Naizot, A. L. (2011). *Nature(s), power, subject(s) in Awá territory: Bios and thanatos in the socio-environmental margin*. [Thesis]. FLACSO Sede Ecuador. http://repositorio.flacsoandes.edu.ec/handle/10469/3378

Núñez Torres, A. M. (1998). *The economic optimum of the use of agrochemicals in African palm production. The case of Santo Domingo de los Colorados*. Abya-Yala.

Núñez Torres, A. M. (2004). *Environmental monitoring of water contamination in the La Chiquita and Guadualito communities and the "La Chiquita" Wildlife Refuge due to African palm production*. Altropico Foundation-Caiman Project.

OEC World. (2023). *Palm oil in Ecuador*. The Observatory of Economic Complexity. https://oec.world/en/profile/bilateral-product/palm-oil/reporter/ecu

Ortega-Pacheco, D., & Jiang, S. (2009). Climate policy: Spatial explicit heterogeneity matters: The case of tropical deforestation at Northwestern Ecuador. *Latin American and Caribbean Association for Environmental and Natural Resource Economists Congress*. San José, Costa Rica.

Oslender, U. (2007). Violence in development: The logic of forced displacement on Colombia's pacific coast. *Development in Practice*, 17(6), 752–764.

Oslender, U. (2016). *The geographies of social movements: Afro-colombian mobilisation and the aquatic space*. Duke University Press.

Oslender, U. (2019). Geographies of the pluriverse: Decolonial thinking and ontological conflict on Colombia's pacific coast. *Annals of the American Association of Geographers*, 109(6), 1691–1705.

Pachauri, R. K., & Reisinger, A. (2007). *Climate change 2007: Synthesis report. Contribution of working groups I, II and III to the fourth assessment report of the Intergovernmental Panel on Climate Change*. Climate Change 2007. Working Groups I, II and III to the Fourth Assessment.

Pacheco, T. (2007). Inequality, environmental injustice, and racism: A fight that transcends skin colour. *Polis Latin American Magazine*, 16, 1–17. http://journals.openedition.org/polis/4754

Public Letter to President Moreno. (2020). Request for a presidential veto of the law for the strengthening and development of production, marketing, extraction, exportation, and industrialisation of oil palm and its derivatives. https://drive.google.com/file/d/1HY1eQJFrTxiFjcPY-z_LrpgubQ_Le65I/view on https://www.change.org/p/ecuadorian-president-lenin-moreno-veto-the-oil-palm-law-before-it-s-too-late?source_location=search

Pulido, L. (2017). Geographies of race and ethnicity II: Environmental racism, racial capitalism, and state-sanctioned violence. *Progress in Human Geography*, 41(4), 524–533.

Quintero, R. (2009). The conceptual innovations of the 2008 Constitution and the Sumak Kawsay. In A. Acosta & E. Martínez (Eds.), *Good living: A path to development* (pp. 75–92). Abya Yala.

Ramos, I. (2003, January). Ecuador: Oil palm and forestry companies in the Chocó Bioregion. World Rainforest Movement Bulletin No. 66. http:// www.wrm.org.uy/bulletin/66/Ecuador.html

República del Ecuador. (2004). Executive Decree No. 2332, Official Gazette No 482. Advisory Council of Biofuels of the Ecuadorian Presidency.

República del Ecuador. (2007). Executive Decree No. 28–148. Constitutional President of the Republic, Rafael Correa Delgado.

Rocheleau, D., & Roth, R. (2007). Rooted networks, relational webs and powers of connection: Rethinking human and political ecologies. *Geoforum*, 38(3), 433–437. https://doi.org/10.1016/j.geoforum.2006.10.003

Rogers, H. (2009). Slash and burn: How biofuels could destroy the planet even faster than petroleum. *Mother Jones*, March 25.

Roots & Routes IC. (2020). Defend ancestral black, Indigenous, and rural peoples in Ecuador from oil palm expansion! https://www.change.org/p/ecuadorian-president-lenin-moreno-veto-the-oil-palm-law-before-it-s-too-late?source_location=search

Roots & Routes IC and the Communities of La Chiquita and Gualdalito in collaboration with Selvas Producciones. (2024). *Together for water: Defending the Chocó Rainforest*. Teaser. www.together4water.com

Roots & Routes IC-Selvas Producciones. (2020). Documentary: Hope & struggle in the Ecuadorian Chocó Rainforest. https://www.rootsroutes.org/documentary

San Lorenzo TV. (2020). *Residents of La Chiquita denounce the pollution of the river by the Palmeras los Andes*. https://www.youtube.com/watch?v=FvDBa2vW2S4&t=300s

Santa Barbara, J. (2007). *The false promise of biofuels: A special report from the International Forum on Globalisation and the Institute for Policy Studies*. https://www.meadowviewfarmandgarden.com/resources/The%20False%20Promise%20of%20Biofuels.pdf

Silveira, M., Moreano, M., Romero, N., et al. (2017). Geographies of sacrifice and geographies of hope: Territorial tensions in plurinational Ecuador. *Journal of Latin American Geography*, 16, 69–92.

Tauli-Corpuz, V., & Lynge, A. (2008). Impact of climate change mitigation measures on indigenous peoples and their territories and lands. *United Nations Permanent Forum on Indigenous Issues: Seventh Session, New York*, April 21–May 2, 2008. Economic and Social Council Working Paper E/C.19/2008/10. http://www.un.org/esa/socdev/unpfii/en/EGM_CS08.html

Tauli-Corpuz, V., & Tamang, P. (2007). Oil palm and other commercial tree plantations, monocropping: Impacts on indigenous peoples' land tenure and resource management systems and livelihoods. United Nations Permanent Forum on Indigenous Issues: Sixth Session, New York, May 14–25, 2007. Working Paper E/C.19/2007CRP.6.

Terán, E. M. (1896). *Report to the Supreme Chief General Eloy Alfaro on the Anglo-Ecuadorian Debt*. Imprenta Nacional.

Todd, Z. (2017). Fish, kin and hope: Tending to water violations in Amiskwaciwâskahikan and treaty six territory. *Afterall: A Journal of Art, Context and Enquiry*, 43, 102–107.

Tuhiwai Smith, L. (1999). *Decolonising methodologies: Research and indigenous peoples*. Zed Books.

Tuinstra, N. A. (2008). *Sustainability analysis of African palm biodiesel in Ecuador: An environmental, socio-cultural, and artistic perspective*. Oregon State University.

United Nations Secretary-General. (2019). *Harmony with nature: Report of the Secretary-General* (pp. 1–17). United Nations. Submitted pursuant to General Assembly resolution 73/235.

US Department of Agriculture. (2023). *Production – Palm oil*. Foreign Agriculture Service. https://fas.usda.gov/data/production/commodity/4243000

Waldmueller, J. M. (2020). (In)visibilisation through decolonial delinking? Disrupting the permanently neglected disaster at the border of Colombia and Ecuador. *Disaster Prevention and Management: An International Journal*, 29(6), 1–14. https://doi.org/10.1108/DPM-01-2020-0002.

Walsh, C. (2011). Afro and Indigenous life-visions in/and politics. (De)colonial perspectives in Bolivia and Ecuador. *Bolivian Studies Journal*, 18, 49–69. https://doi.org/10.5195/BSJ.2011.43.

Walsh, C. (2013). *Decolonial pedagogies. Insurgent practices of resisting, re-existing, and re-living*. Abya-Yala Editions.

Walsh, C. (2015). Decolonial pedagogies walking and asking. Notes to Paulo Freire from Abya Yala. *International Journal of Lifelong Education*, 34(1), 9–21.

Walsh, C. (2023). *Rising up, living on: Re-existences, sowings, and decolonial cracks* (On Decoloniality). Duke University Press.

West, R. (1957). *The lowlands of the Colombian Pacific*. Louisiana State University Press.

White, H. (1998). Race, class, and environmental hazards. In D. Cuesta Camacho (Ed.), *Environmental injustices, political struggles: Race, class, and the environment* (pp. 61–81). Duke University Press.

World Bank and International Finance Corporation. (2011). *Strategic framework of the World Bank Group for its participation in the palm oils*. World Bank and International Finance Corporation.

Wray, N. (2009). The challenges of the development regime: Good Living in the constitution. In A. Acosta & E. Martínez (Eds.), *Good living: A path to development* (pp. 51–62). Abya Yala.

Wright, S. (2008). Practicing hope: Learning from social movement strategies in the Philippines. In R. Pain & S. J. Smith (Eds.), *Fear: Critical geopolitics and everyday life* (pp. 223–233). Ashgate Publishing.

Wright, S. (2010). Cultivating beyond-capitalist economies: Economic geography. *Economic Geography*, 86(3), 297–318.

Zaragocín, S. (2018). The geopolitics of the womb: Towards a decolonial feminist geopolitics in spaces of slow death. In D. Cruz & M. Bayón (Eds.), *Bodies, territories and feminisms* (pp. 81–97). Abya Yala/Third World Ecological Studies.

Zinn, H. (2006). *A power governments cannot suppress*. City Lights Publishers.

9

GOVERNANCE ARRANGEMENTS, POWER DIFFERENTIALS, AND SUSTAINABILITY IN THE HONDURAN PALM OIL INDUSTRY

Ingrid Fromm, Mélanie Feurer and Sebastian Mengel

Introduction

The palm oil industry in Honduras generates significant export revenues and employment in the agricultural sector. According to figures from the Ministry of Agriculture, in 2022, the country produced over 700,000 tonnes of palm oil, which generated USD500 million in export revenues, placing palm oil as one of the country's main exports. Palm oil is produced on over 197,000 hectares of land in Honduras, and over the last two decades, it has replaced another plantation crop—banana. The palm oil industry has created about 40,000 direct jobs, while approximately 90,000 people are indirectly employed through the industry. Although it is an industry dominated by only a few corporations, most of the production in Honduras still takes place on small-scale farms. Within areas where it has been adopted, the oil palm industry has become an important driver of land use change, which has followed a trajectory of replacing banana plantations rather than forest lands. Government interventions through targeted policies have had an important role to play in stimulating the expansion of the industry. These policies are largely justified based on the perceived value of the crop as an engine of economic development. Consequently, oil palm expansion and its related supportive policies are often seen through an economic lens only, while questions of who benefits and who controls the industry are often overlooked, and the significant social and environmental impacts of the crop's expansion are given limited attention.

The rapid expansion of palm oil production in Honduras has had an impact at the economic, social, and environmental levels. Honduras perfectly exemplifies the trajectory of rapid expansion of the palm oil industry, the world's most produced vegetable oil (Fitzherbert et al., 2008); looking at the economic trajectories of other countries and Malaysia in particular, the crop has come to be seen as an important

DOI: 10.4324/9781003459606-11

rural development mechanism (Craven, 2011). At the same time, oil palm is associated with significant environmental harm, because it grows well in tropical areas and puts pressure on forested lands that previously have not been tapped for agriculture but have high ecological value, including for biodiversity conservation, carbon sequestration, and the regulation of water cycles. Furthermore, palm oil is often associated with land grabs and social conflicts (Rist et al., 2010). Consequently, palm oil production has become a contested issue, and the companies and countries that are involved in the industry have taken steps to address these concerns.

In an analysis of the global politics of palm oil production, Dauvergne (2018) states that the palm oil industry is increasingly certifying its activities as "sustainable", "responsible", and "conflict-free". His analysis suggests that this trend does not represent a breakthrough towards better governance but primarily reflects a business strategy to channel criticism towards "unsustainable" palm oil, while promoting the value of protecting rainforests, corporate social responsibility, international trade, industrial production, and industry-guided certification. Hospes (2014) provides similar evidence for Indonesia and Brazil. Dauvergne's work draws attention to a critical challenge the industry poses to producing countries and one that has been played out with varying outcomes in the different producing states. As they do so, new players face the task of addressing governance challenges posed by the introduction of the crop and of balancing the conflicting demands of different actors and public perceptions concerning the distribution of the economic, social, and environmental costs and benefits associated with the crop.

In this chapter, we seek to analyse the dynamics of this process in detail by attempting to tackle the first, second, and fourth research questions laid out in this book: how do different governance arrangements enable and legitimise expansion; how do power differentials affect oil palm governance; and how is conflict governed and moderated in the oil palm sector. In particular, we focus on the idea that the oil palm industry constitutes a topological space in which different actors seek to negotiate and/or secure outcomes that reflect their interests. Such encounters raise the possibility of contestation and conflict. Understanding the political dimension of these negotiations and the role of power differentials in this process as they relate to the varying economic, environmental, and social impacts of the process is thus central to an adequate understanding of the industry. We illustrate this by examining the evolution of environmental governance in the palm oil industry in Honduras and analysing if sustainable palm oil production is a myth or a reality.

Central to the discussion presented in this chapter are government interventions through targeted policies to expand an industry simultaneously capable of meeting multiple objectives: Gross Domestic Product (GDP) growth, inclusion of smallholder farmers and increased incomes for them, poverty alleviation, the creation of employment opportunities mainly in rural areas, and at the same time meeting sustainability goals such as environmental protection. The main question under which this case is examined is the extent to which the industry as currently constituted, delivers on these claims. Attention is drawn to the centrality of power differentials in the way that the industry is shaped in Honduras. The dynamics of this process, including the governance of this chain, will be observed to explore how diverse elements (i.e. discourses, institutions, social groups, and physical features) have been recombined into assemblages with outcomes that were hard to predict and unintended.

For this analysis, power is understood as the capacity of certain actors to mobilise and "govern" the relationships between elements in an assemblage in ways which support their interests (Nail, 2017). Hence, power must be continually maintained through activities and coordination mechanisms through which specific relations between different objects in an assemblage are maintained. This chapter will analyse the expansion of the palm oil industry in Honduras, the governance structure and policy framework that led to this expansion, the current issues affecting this industry, in addition to the response at the local level to these power disparities.

Oil palm production in Honduras

Since the introduction of oil palm (*Elaeis guineensis*) in Honduras in 1927, the expansion of the production area has been gradual but steady. The first oil palm seeds were brought from Malaysia, initially for exhibition in the Lancetilla Botanical Garden, the first place where oil palm was planted on the entire continent. The Tela Railroad Company, a subsidiary of the United Fruit Company, which is present-day Chiquita Brands International, established the first commercial plantation. Early commercial plantations were started in 1936, and by the 1970s, 11,000 hectares had been planted by the United Fruit Company throughout the northern part of Honduras. Since then, the cultivation of oil palm across Latin America has expanded with several countries—Colombia, Ecuador, Honduras, Brazil, Costa Rica, Guatemala, Mexico, Peru, and Venezuela—adopting the crop and doubling output since 2001 (Furumo & Aide, 2017). As the banana sector in Honduras declined in terms of area and production volume, the palm oil industry has kept expanding. Since 2000, the area expansion has been five-fold (Figure 9.1) and production volumes have shown a similar trend (Figure 9.2).

For the last two decades, the expansion of the industry has been a result of targeted economic policies to promote the sector, with the idea of increasing the production of biofuels. In 2006, the Honduran government ratified an action plan to increase the competitiveness of the palm oil sector through measures such as access to credit through a special fund known as the *Fondo de Fomento Palmero*. Oil palm producers received subsidies. Although there were government subsidies given to support the industry, smallholder producers still faced difficulty in accessing credit and lacked training, especially related to the production of biodiesel (Government of the Republic of Honduras, 2018). Although these measures were taken to support oil palm producers, especially small-scale producers, the legitimacy of these

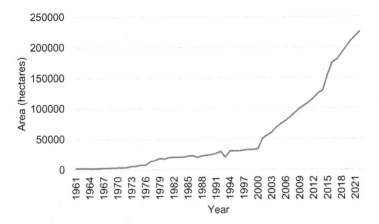

FIGURE 9.1 Oil palm production area in Honduras (FAO, 2020).

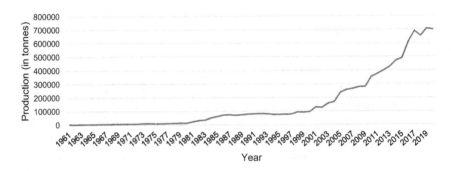

FIGURE 9.2 Oil palm production volume in Honduras (FAO, 2020).

economic policies has come into question as doubt is cast around the environmental and social sustainability of the entire industry.

Governance of the palm oil industry: Policy framework

The past 20 years have seen a growing interest in integrating more producers in the palm oil industry in Honduras, legitimised by the argument that increased biofuel production is beneficial for the national economy. According to Craven (2011), the government's legal framework was well-suited for the development of biofuels. Honduras is the only Central American country that has approved a law and a regulation that treat biodiesel and ethanol production equally. Job creation was also a key motivation underpinning the promotion of biofuel production. With a loan from the Central American Bank of Economic Integration (CABEI) of USD7 million, approved in 2005, it was possible to increase the plantation area during the first stage of the implementation of the national strategy. Additionally, there was a strong focus on integrating smallholder producers. Therefore, the apparent industry concentration, which is seen in other countries, is not as marked in Honduras, since primary production of oil palm is not dominated by large producers or transnationals but rather by small-scale producers (Picado, 2016) (see Table 9.1).

Since the early 2000s, the expansion of the palm oil industry in Honduras has been supported through targeted policies and a framework agreement promoted by the government (Furumo & Aide, 2017). The framework agreement to promote the competitiveness of the palm oil value chain was ratified in 2006. The aim of this agreement was to implement the action plan to improve the competitiveness and added value of palm oil, in particular in terms of efficiency and sustainability in the production, industrialisation, and commercialisation stages. It also aimed to consolidate and strengthen the competitive advantages through the development of a strategy and the coordinated execution of short-, medium-, and long-term actions, such as the creation of the *Fondo Palmero*, a fund that provided credits to producers and services in the sector to further its development. A programme for small and medium producers of palm oil (PRO-PALMA) was established, financed by the CABEI. Part of the capital was recovered through interest and capital quotas paid by the beneficiaries of the loans.

A year later, the Honduran government passed the Law for the Production and Consumption of Biofuels. Through this Biofuels Law, a legal basis for assigning funds and expanding the production of biofuels was put in place, which promoted the expansion of oil palm production. The legal framework also provided incentives and funds to small-scale farmers, extractors, and refineries to purchase equipment, materials, and services used for the planning, design, installation, construction, and set-up of biofuel processing pilot plants. In the initial phase of the expansion plan, which also began with the construction of a biodiesel pilot plant, 1.7 million oil palm seedlings were imported and distributed among the palm oil producers in northern Honduras. Three ministries were assigned to work on the design and implementation of policies to expand biofuels production and promote them on the market: the Ministry of Industry and Trade (SIC), the Ministry of Agriculture and Livestock (SAG), as well as the Ministry of Natural Resources (SERNA).

Despite the positive economic effects of increased palm oil production for the food processing and cosmetics industries, the expansion of plantations globally has caused significant negative environmental impacts. These include deforestation, the conversion of wetlands, habitat loss, forest fragmentation, biodiversity loss, food chain disruption, soil property changes, water and air pollution, and increased greenhouse gas emissions (Khatun et al., 2017). This has prompted a range of reactions from the scientific community and the industry (Wijedesa et al., 2017), ranging from measures to limit further industry expansion to a proposal to modify practices to mitigate the adverse effects of the industry (Sayer et al., 2012). In this context, the role of small-scale producers has been centred on debates about oil palm. There is a widespread perception that small-scale production is inherently less environmentally harmful than large-scale systems. In a report by Haupft et al. (2018), the importance of integrating small-scale producers is highlighted as a strategy to mitigate the impact of forest loss and land degradation. In relation to oil palm, Jezeer and Pasiecznik (2019) show evidence of a reduced negative environmental impact through active and inclusive smallholder participation in the certification process in Latin America, and similarly, Saadun et al. (2018) report this outcome in Malaysia. Given the positive environmental outcomes associated with smallholder farming in the oil palm sector and the limited capacity of local and national governments to enforce environmental standards, active smallholder participation and engagement have come to be regarded as a cost-effective strategy to maintain protected areas and remaining forest corridors in oil palm-dominated landscapes (Jezeer & Pasiecznik, 2019).

The Honduran government developed a range of measures to promote the growth and the integration of smallholders in the palm oil industry. All the machinery and equipment used in the production of biodiesel were exempt from tax and import duty for 15 years. Oil palm seeds from Costa Rica, Indonesia, and Malaysia were imported and given to smallholders, as well as subsidies to buy fertiliser. Other problems affecting the farmers have also been addressed, yet not fully resolved. One of the biggest constraints hindering business in Honduras is the inadequate financial market. Credit is accompanied by high interest rates, and in the case of the agricultural sector, many farmers must use their land as a guarantee before obtaining credit. The initial phase of oil palm cultivation is capital-intensive and requires large investments. As oil palms can take up to four years before they come into production, and during that time plantations require high maintenance, high interest rates on loans discourage investment in their establishment. While government subsidies to support the initial phase of production might seem beneficial, supportive financial markets are also required.

Social problems have also been linked to palm oil production in Honduras. These include unresolved land rights issues, low wages, and the uneven distribution of benefits within the industry. Examples of land rights issues are closely tied to the environmental governance of the value chain and have resulted in clashes between different interest groups (Oosterveer, 2015; Obidzinski et al., 2012). Different types of land grabbing have been observed in many places where oil palm cultivation is dominated by influential corporations associated with large-scale concession areas and smallholders with no or only customary land titles (León Araya, 2019; Kerssen, 2013; Hirsch, 2011; Sikor & Lund, 2009). Large corporations such as Dinant, which has been at the centre of land conflict in Honduras, have ongoing programmes to help smallholders acquire these land titles. However, business has been affected by the conflicts between corporations and farmer cooperatives, which have prevailed for several decades.

Global responses to concerns about palm oil

At the global level, the Roundtable on Sustainable Palm Oil (RSPO), a multi-stakeholder partnered governance approach established in 2004, has emerged as the most recognised international sustainability standard in the palm oil sector, operating a sustainability certification scheme (Dauvergne, 2018; Ivancic & Koh, 2016; von Geibler, 2013). Initiated with substantial support from Unilever and the World Wildlife Fund (WWF), the RSPO currently includes more than 4,000 members, from palm oil producers to non-governmental organisations (NGOs), consumer goods manufacturers, retailers and financial institutions (RSPO, 2023; Pacheco et al., 2017). By 2016, a Technical Committee for the Roundtable on Sustainable Palm Oil Certification was created by the Honduran Secretary of Agriculture and Livestock. This body has the task of coordinating all the efforts and resources that any public or private, national or international body or institution undertakes or provides for certification. This committee also serves as a liaison and manager with the institutions and government agencies linked to the RSPO certification, in particular the Association of Industrialists of the Oil Palm of Honduras (AIPAH) and the National Federation of African Palm Producers of Honduras (FENAPALMAH).

One of the most recent developments with an impact on the national policy framework has been the incorporation of Honduras into the Council of Palm Oil Producing Countries (CPOPC). After the 2nd Ministerial Meeting of Palm Oil Producing Countries held in November 2019 in Malaysia, the Honduran government announced a new 10-year plan to support and develop the industry. As the global price of palm oil has fallen and affected smallholder producers in Honduras, this plan will also include a normative framework for financial assistance.

Regulatory mechanisms and the Honduran palm oil industry

While Honduras participates in global regulatory frameworks and shares many of the reputational and sustainability challenges of the global palm oil industry, the industry does so within the national context with unique features and challenges. One particularity of the Honduran example is the trajectory of land use change over the last four decades. Commercial oil palm plantations have been established to some extent in former banana cultivation areas (Furumo & Aide, 2017). In this case, the agro-industrial assemblage that was organised around the production of bananas has been gradually disassembled, and components of that assemblage have now come to be reassembled in another assemblage organised around the production of oil palm (Murray Li, 2007). At the national level, it may perhaps be useful to conceptualise the regulation of the oil palm process as a means by which certain actors (especially the state) seek to "fix" relationships between objects and actors that constitute part of the Honduran oil palm assemblage in ways that reflect the interests of the state. However, besides the state, other actors seek to define the relationship between components of the oil palm assemblage in different ways. The extent to which different actors can do so reflects their ability to exercise a degree of power over the process.

Before looking at the regulatory mechanisms and the effects of the measures taken by the state to promote and regulate the palm oil industry, it is important to

TABLE 9.1 Description of the palm oil industry in Honduras

Actor	Number	Characteristics
Producers	>10,000	Small-scale producers: 1–10 hectares Medium-scale producers: 11–100 hectares Large-scale producers: >100 hectares The average price for fruit represents 12–14 percent of the international price for crude oil
Intermediaries	10	Located in all production areas Have invested in transportation and machinery Extractors have given them credits for their operation Profit margin about 30 percent of the national price for fruit They sometimes offer credit to small producers
Mills	13	Three are cooperatives and have a lower extraction capacity Others belong to the private sector and have optimal extraction capacity, as well as technical and managerial know-how
Refineries	4	Two are private sector companies High technological investments Possess highly qualified management teams
Exporters	9	Have access to national and international credit Own trucks and equipment for transportation Main markets include Mexico and Central America Sell processed products and crude oil, depending on demand

Source: Government of the Republic of Honduras (2018).

identify the different actors, their respective roles in the industry, and how they are integrated into the oil palm assemblage, in order to allow us to explore the configuration of relationships between them and how power is conferred by actors for a particular purpose. Besides the state and international bodies, the palm oil industry in Honduras includes different types of actors involved in production, trade, processing, and retailing (see Table 9.1).

These categories include varying numbers of individual entities. At the farm level, over 10,000 oil palm producers located in four different departments in Honduras produce a total of 9 million tonnes of oil palm fruit a year. Most of these producers have contracts with the extractors or are associated with producers' cooperatives. Therefore, the payment they receive for the oil palm fruit is fixed by contractual agreement, and the paid amount covers production costs and provides a level of income for the producers. Another important actor in the chain is the intermediary. About ten intermediaries are collecting fruits and providing logistics for small producers who have little access to transportation. These are autonomous traders dealing not only with producers but also with extractors. They buy from small producers and sell to extractors. These intermediaries are larger businesses that own equipment for logistics. Power differentials between smallholders, intermediaries, and extractors exist, and the regulation mechanisms include credits given by extractors to intermediaries and then to smallholders, often through informal agreements (Government of the Republic of Honduras, 2018).

The next category of actors in the industry are oil palm fruit processors. Known in Honduras in the context of the palm oil value chain as extractors, these firms are either private companies or larger farmers' cooperatives, comprised of local, smaller-scale cooperatives. Out of the 11 extractors located in the country, eight of them belong to the private sector. The other three started as farmers' cooperatives and expanded into extraction. These firms have acquired sophisticated technology throughout the years, initially relying on state subsidies, as was the case of Hondupalma and Coapalma. Another actor down the chain is the refinery facilities where palm oil is fractioned. There are four refineries nationwide, and they produce for the national and international markets. It is estimated that at least 50 percent of the national oil palm production is exported. The nine specialised exporters buy oil palm products (crude oil, concentrate, and other sub-products) and export them, mainly to Central America and Mexico (Government of the Republic of Honduras, 2018; Iscoa, 2015).

Management theorists define lead firms as the organisational integrators of dispersed economic activities in top-down processes (Lee & Gereffi, 2015). In assemblage theory, rather than organisational integrators, we can identify key firms that have employed assemblage practices to successfully organise other components of the assemblage in ways that support their goals. While the state may play an important role in the development and formal regulation of the Honduran palm oil industry, the 11 extractor companies exercise considerable control over the rest of the assemblage, stemming from their position at a key juncture between the upstream "producers" and downstream "consumers", and consequently, they play a pivotal role in shaping the relationships between other actors in the oil palm assemblage. This position is strengthened by the high concentration of these firms at this stage of the oil palm value chain and their essential function for processing a highly perishable crop (the fruit must be pressed within 24 hours of harvest), with no other use than the production of oil and resulting derivatives such as biofuel. Extractor firms play a critical, time-limited role as the sole market for the crop and the only processors, since there are only a small number of mills owned by extractors. Producers are wholly reliant on these facilities for buying their fruit, while downstream users rely on these companies for sourcing their raw materials. Most, if not all producers, irrespective of their size, are closely linked to these firms because all the production is sold to one of these extractors. Many producers belong to farmers' associations or cooperatives, so they produce under contract for the extractors.

In the two main producing areas—the Sula and Aguán valleys—the initial players and most prominent extractors are Hondupalma and Coapalma. They are examples of how extractors establish power in the assemblage. Hondupalma was officially founded in 1982 as a farmers' cooperative association. This cooperative is a result of the Honduran agrarian reform and peasant movement of the 1960s and

1970s. Presently, Hondupalma is made up of 30 smaller associations and cooperatives and has processing facilities located in Yoro, northern Honduras, an area that has traditionally had economic relevance because of the high agricultural output. Since 1985, the production facilities have been producing crude oil, refined oil, plus a variety of oils for further industrial use, and biofuels (Cartagenova, 2005). The extraction plant works at an installed capacity of 60 tonnes per hour, the refinery can produce 70,000 tonnes of crude oil a year, and the fractionation plant has an installed capacity of 150 tonnes a day (Hondupalma, 2018). The technological level of these processing facilities is high, and rigorous quality standards are implemented throughout the production processes. Hondupalma has a variety of suppliers, mostly the associated farmers, and independent producers, and it also sources from its own plantations (Cotty et al., 2002).

The Aguán Valley on the northern Atlantic coast of Honduras is another major area of agricultural production. This is where, in 1982, Coapalma Ecara was founded, the same year when Hondupalma was founded in the nearby Sula and Lean valleys in northern Honduras. It is also a farmers' cooperative, comprised of 13 smaller associated farmers' cooperatives, based in Tocoa, Colón (Coapalma, 2021). Currently, Coapalma can extract about 40 tonnes of crude oil per hour. The refinery has an installed capacity of 100 tonnes a day, and the fractionation plant likewise has a processing capacity of 100 tonnes a day. In terms of installed capacity and productivity, Coapalma lags behind Hondupalma (Muñoz, 2006).

Hondupalma and Coapalma are the biggest players in each region. Although technically both are farmers' cooperatives, they operate as companies. Both firms play a dominant role in defining the distribution of activities and monitoring the chain at the local level. Although the core of their business is extracting crude oil, both firms are engaged in defining who and what is being produced. Both firms have plantations, but they also source from associated and independent producers. Not only do these firms coordinate activities, they also monitor them. It is through the provision of extension services that they help farmers improve their production processes. In many cases, these farmers also receive assistance in the form of inputs such as fertilisers or pesticides. Both firms also engage in social development activities by providing access to education and vocational training, as well as the building of health centres and infrastructure in the local communities. This exemplifies the rendering of technical knowledge and validation to legitimise the relationship between elements of the assemblage, which appear seemingly apolitical, but nevertheless vest power in the lead firms. The extractors also conduct monitoring activities. Audits take place, and a sample is taken from each shipment of palm kernels and bunches. Producers are penalised if the bunches contain an excess of unripe or overly ripe fruit, for instance. Quality standards are already monitored at the primary production level. The extractors allocate resources to train personnel on quality and safety standards and other aspects of quality assurance.

The coordination to secure a consistent supply among several hundred farmers is complex but nevertheless efficiently conducted by the lead firms. Elements of reassembling, market efficiency, and new forms of coordination are evident here (Murray Li, 2007). Because the palm oil industry is highly concentrated and the returns are large, these firms find themselves in a position to meet the constant demands for quality, reliability, and timely diligence of the supply. The management capabilities of these firms are good. Highly skilled personnel are hired and trained. Yet the larger extractors can invest more in management systems to meet such requirements than the smallholders. In addition to coordinating activities along the value chain, these firms have taken over a stewardship role, by promoting environmental certifications. A top-down approach to the promotion and implementation of these certifications has been taken, monitored, and enforced by the lead firms.

In contrast, the profit margin for small-scale producers in the palm oil value chain is low in comparison to other actors in the chain. Those on the primary production side of the chain are the ones gaining the least, and in agriculture, this effect is magnified (Shepherd, 2016). Humphrey (2005) warned that one of the challenges small producers in developing countries face is exclusion, because of the trends in global agribusiness. The first trend is the increasing importance of large buyers in global food chains. Small producers are forced to be more competitive because of the requirements of large buyers (i.e. retailers, processors) for quality and reliability of delivery, and product differentiation has raised the level of competence required. Second, the concentration at various points in the value chain, including input suppliers (i.e. seeds, chemicals), processing, and retailing, has implications for the questions of access to agribusiness value chains and returns for small producers participating in these chains. Finally, the importance of food safety, social and environmental standards, which are becoming more stringent and closely monitored, also presents challenges for small producers.

The interactions between actors involved can determine the performance of the palm oil assemblage as they forge alignments for their interests. For producers, intermediaries, or processors to add value to farm products by adopting changes in production, handling, and processing practices, new skills and knowledge must be acquired. In the palm oil value chain, crude oil is transformed as it passes through different actors or links in the chain, where value is added until it reaches the final consumer. On the other side of the chain, there is tacit knowledge that is passed down through the different actors in the form of codified information. A key characteristic of authorising knowledge is specifying the requisite body of knowledge. The extractors and refineries have a role in the governance and coordination of activities in the chain through different mechanisms. In this interaction, trust relationships may or may not be formed, depending on the level of interaction between the actors. Arguably, farmers under contracts may form tighter trust relationships with extractors (Fromm, 2007).

Industry organisation and regulatory frameworks

The regulatory governance infrastructure in palm oil was assessed by Hamilton-Hart (2014), and she states that different facets of the industry are potentially affected by several different global and regional governmental initiatives, while a private, partnered governance institution takes the leading role in terms of setting regulatory standards for the industry. In the case of Honduras, there is evidence of a rise in compliance with environmental standards. The Ministry of Agriculture, with the support of the WWF, SNV, Solidaridad Central America, and Proforest, addressed environmental and social impacts along the palm oil supply chain in Honduras. Mining and palm oil companies formed a consortium to work towards compliance with the RSPO standards for the sustainable production of palm oil, which generated quantifiable benefits such as increased per hectare productivity, reduced threats of deforestation and environmental degradation, and the protection of the health and safety of plantation workers. Through AIPAH, nine lead firms have committed to complying with environmental standards (AIPAH, 2023). Wilmar Europe and AIPAH have established a partnership to strengthen the good agricultural and environmental practices of palm oil smallholders in Honduras. The Wilmar Smallholders Support in Honduras Programme (WISSH) was created with the aim of enhancing smallholders' knowledge and technical capacity on best agricultural management practices that incorporate principles of environmental stewardship, which will eventually lead to improved crop yields, lower input costs, and a better and more stable income for smallholders. WISSH builds upon the criteria in Wilmar's "No Deforestation, No Peat, and No Exploitation" policy (WISSH, 2016). This is an example of the range of governance models that continuously shape the development of the industry, and in this case, there is a crossover between international and national arrangements. Yet, as interest in environmental sustainability increases, social conflicts may also increase because of the conflicts of interest among different actors.

Craven (2011) affirmed that Honduras did not demonstrate the political experience or dexterity required to simultaneously balance the demands of multinational corporations (MNCs) and local stakeholders. She pointed out that Honduras did not offer international investors and MNCs a stable economic and political environment to foster investments. With reference to the events of 2009, when Honduras was submerged in a political crisis which concluded with the forceful removal of its president, Craven (2011) discussed the ramifications for the energy sector. The Ministry of Agriculture and Livestock, which was principally responsible for biofuel development, embarked on its own strategic plan for 2010–2014, discontinuing the older plan. Picado (2016) argues that after the 2009 political crisis, the ramifications for the communities in the Aguán Valley have been mostly negative, as conflict over land rights has escalated. As Nail (2017) postulates, in this assemblage, a relatively negative deterritorialisation process took place and distinct power differentials were at the root, long before this political crisis.

Power differentials: policy, lead firms, and conflict

Power differentials in the palm oil industry in Honduras are rooted in the agrarian reform and the peasant movements of the 1960s and 1970s. Even before palm oil cultivation was promoted through targeted economic policies, the seeds were planted for a fragmented agroindustry, where different actors struggled to control different discourses. The state tried to promote the agro-industrial growth of a sector that is highly regulated, while at the same time setting the stage for conflict through a poorly executed agrarian reform process. During the agrarian reform, land was distributed to peasant farmers, who often lacked the capacity to make it productive. Corporations such as Dinant in the Aguán Valley purchased idle land from farmer cooperatives who were willing to sell it (Edelman & León Araya, 2014). Yet poor regulatory systems and statutory regulations often did not fulfil the role of mediation between different interest groups, and an agrarian counter-reform process allowed for land purchases to take place in the 1990s (León Araya, 2017; Roquas, 2002). Two institutions, the National Agrarian Institute (INA), which provides technical support to farmers, and the National Development Bank (BANADESA), both created in the era of agrarian reform, were key to the transformation of agriculture, but pursued conflicting policies. While trying to boost the cultivation of subsistence crops such as maize and beans by handing out fragmented land titles to small-scale farmers, the government also indirectly promoted the large-scale cultivation of monoculture crops such as oil palm and sugar cane (Berger & Palacios, 2019).

With the establishment of commercial oil palm plantations in the 1980s, farmers' cooperatives became relevant to the success of this agroindustry but, at the same time, offered the possibility of skewed power dynamics, which led to conflict, particularly in the Aguán Valley. As more farmers started reaping economic gains through contractual arrangements with Hondupalma and Coapalma, for example, the expansion of oil palm plantations was justified, as it was perceived to be a crop that could lift smallholder farmers out of poverty and successfully integrate them into an agricultural value chain which could render high profits. Having the security of a constant income motivated farmers to abandon the traditional *milpa* production, which is a pre-Columbian method of intercropping maize, bean, and squash in small plots of land. Although the entry door to the palm oil industry was through cooperatives, issues around land tenure, particularly in the Aguán Valley, were left unresolved (Kerssen, 2013). Farmers' cooperatives actively sought to expand the industry, and by the 1990s, large corporations such as Dinant purchased land and expanded production. One key issue was the practices of assemblage—which in this case did not necessarily forge connections between the cooperatives, but rather undermined trust among actors—was the distorted definition and understanding of property rights. As Roquas (2002) points out, the state itself—as the institution that defines and protects private property through its laws and legal system—has become an actor in land conflicts because of its inability to clearly define and protect private, state, and common property.

One missing element in the entire discussion around land tenure issues in the palm oil industry, particularly in the Aguán Valley, is the market liberalisation process Honduras underwent in the 1990s. With the promotion of the structural adjustment programmes by the World Bank and the International Monetary Fund (IMF), macroeconomic and structural reforms were introduced, which included the privatisation of public enterprises. Land which was handed out to farmers or farmers' cooperatives in the agrarian reform was privatised or sold to private corporations. Regardless of the legislation, this land was considered as public land and, as such, could not be sold. The government gave ownership through private property titles to the cooperatives. Some members of these cooperatives collectively decided to sell the land, thus setting the conditions for a generation of dispossessed peasants (Diaz & Zepeda, 2012). The practices of some of the leaders of the cooperatives are questionable and reflect the anti-politics of the assemblage, especially because of the interest of this group in shifting the discussion to the actions of private enterprises only.

One critical feature of this case is that regardless of the opposition of certain groups to the expansion of commercial oil palm plantations, the opposition was not against the crop itself, but emerged as a struggle for the recognition of land rights. Rural livelihoods have benefited from the development of the palm oil sector, which has created thousands of direct and indirect jobs in historically impoverished areas of Honduras. The palm oil industry has had a redistribution effect, by also bringing in investment to these rural areas. Berger and Palacios (2019) provide evidence of farmer inclusiveness in the sector. Efforts by extractors and farmers' cooperatives, supported by policies, seem to engage smallholder farmers in sustainability efforts especially with the creation of the Technical Committee for the RSPO Certification, which groups different stakeholders in promoting sustainable agronomic and processing practices. León Araya (2019) points out that smallholders must sign contracts of exclusivity with lead firms, which provides them with financial, technical, and legal support. Smallholders had to meet certain standards to be part of the agreement, such as demonstrating land ownership, not being within a natural reserve, and adopting the technological package of the companies. Although the economic benefits and attention to environmental issues have been overall positive, the contracts with these companies and these agreements continue to eschew more power towards those in control.

Roundtables or "negotiation tables" as they are known in Honduras, which bring together private companies and farmers' cooperatives (or extractors), play a strong role in shaping the industry and distributing the benefits. Different elements such as discourses, institutions, forms of expertise, and physical features are recombined into assemblages on an ongoing basis. The inherent power differentials are central to the control of discourses by the different actors, and this dynamic will continue to shape policy and influence outcomes in future sustainability efforts.

Conclusion

Overall, this chapter has attempted to address the book's following research questions: (1) how do different governance arrangements enable and legitimise expansion; (2) how do power differentials affect oil palm governance; and (4) how is conflict governed and moderated in the oil palm sector. The Honduran palm oil industry exemplifies how governance arrangements, which seek to legitimise the expansion of an industry, must pay particular attention to power differentials to avoid mistrust among the actors. Power imbalances within the country's palm oil industry trace their origins back to the agrarian reform and peasant movements of the 1960s and 1970s. Even prior to targeted economic policies promoting palm oil cultivation, the groundwork was laid for a disjointed agroindustry, where various stakeholders vie to control different discourses. The state attempted to spur the agro-industrial expansion of a heavily regulated sector, which inadvertently fostered conflict through a flawed agrarian reform initiative. During this reform, land was distributed to peasant farmers, who often lacked the resources to render it productive. In the ensuing decades, key companies acquired unused land in the Aguán Valley from farmer cooperatives that were willing to sell. Inadequate regulatory frameworks and statutory measures frequently failed to mediate between diverse interest groups, paving the way for a counter-reform process that facilitated land acquisitions in the 1990s.

Failures can be attributed to a misguided and contradictory policy framework that actively promoted the economic growth of the sector, but was unable to resolve land rights issues that were present from the beginning of the agrarian reform. The weak regulatory frameworks and institutions, which lack the capacity to enforce and maintain the rule of law, have resulted in a heavy toll for farmers claiming their land rights. In the end, it set the stage for antagonism between different interest groups. It has tarnished the business of corporations such as Dinant, while also exerting pressure on cooperatives, caught between conflicting discourses over the quest for economic gains and the task of representing and defending the rights of the farmers they represent. For sustainability to be promoted from an economic, social, and environmental perspective in the Honduran palm oil industry, it is imperative to address and resolve these disputes.

References

AIPAH. (2023). *About us.* Asociación Industrial de Productores de Aceite de Honduras. https://aipah.org/nosotros/

Berger, V. C., & Palacios, O. (2019). *Smallholder oil palm in Honduras—A model for sustainable livelihoods and landscapes* (pp. 1–8). ETFRN News 59. https://www.solidaridadnetwork.org/wp-content/uploads/migrated-files/publications/ETFRNnews59-cohn-smallholder-oil-palm-honduras.pdf

Cartagenova, D. E. G. (2005). *Comparative analysis of the process African palm oil production: The case of Hondupalma and Coapalma Honduras* (pp. 1–31). Zamorano.

Coapalma. (2021). *Our history*. Coapalma Ecara. http://www.coapalmaecara.com/index.php/nosotros

Cotty, D. B., Estrada, I., & García, M. (2002). *Basic indicators on agricultural performance, 1971–2001*. Zamorano.

Craven, C. (2011). The Honduran palm oil industry: Employing lessons from Malaysia in the search for economically and environmentally sustainable energy solutions. *Energy Policy, 39*(11), 6943–6950. https://doi.org/10.1016/j.enpol.2010.09.028

Dauvergne, P. (2018). The global politics of the business of "sustainable" palm oil. *Global Environmental Politics, 18*(2), 34–52. https://doi.org/10.1162/glep_a_00455

Díaz, R., & Zepeda, G. (2012). *Diagnosis of the human rights problem in the Bajo Aguan*. Agencia Católica Irlandesa para el Desarrollo (Trocaire). https://issuu.com/inggallardo/docs/diagnosti-co_de_la_problematica_en_e

Edelman, M., & León Araya, A. (2014). Cycles of land grabbing in Central America: An argument for historicising and a case study of Bajo Aguán, Honduras. *Anuario de Estudios Centroamericanos, 40*(1), 195–228. https://www.redalyc.org/articulo.oa?id=15233350010

FAO. (2020). *FAOSTAT*. Food and Agriculture Organisation of the United Nations. https://www.fao.org/faostat/en/#home

Fitzherbert, E., Struebig, M., Morel, A., Danielsen, F., Bruhl, C., Donald, P., & Phalan, B. (2008). How will oil palm expansion affect biodiversity? *Trends in Ecology & Evolution, 23*(10), 538–545. https://doi.org/10.1016/j.tree.2008.06.012

Fromm, I. (2007). Integrating small-scale producers in agrifood chains: The case of the palm oil industry in Honduras. *17th Annual Food and Agribusiness Forum and Symposium*. https://ifama.org/events/conferences/2007/cmsdocs/1024_Paper.pdf

Furumo, P. R., & Aide, T. M. (2017). Characterising commercial oil palm expansion in Latin America: Land use change and trade. *Environmental Research Letters, 12*(2), 024008. https://doi.org/10.1088/1748-9326/aa5892

Government of the Republic of Honduras. (2018). *Oil palm chain*. Servicio Agrícola Y Ganadero (SAG). https://www.pronagro.sag.gob.hn/cadena-de-la-palma-aceitera/

Hamilton-Hart, N. (2014). Multilevel (mis)governance of palm oil production. *Australian Journal of International Affairs, 69*(2), 164–184. https://doi.org/10.1080/10357718.2014.978738

Hirsch, P. (2011). Titling against grabbing? Critiques and conundrums around land formalisation in Southeast Asia. *International Conference on Global Land Grabbing*. https://landportal.org/library/resources/titling-against-grabbing-critiques-and-conundrums-around-land-formalisation

Hondupalma. (2018). *A model of success*. http://www.hondupalmahn.com/video2018.html

Hospes, O. (2014). Marking the success or end of global multi-stakeholder governance? The rise of national sustainability standards in Indonesia and Brazil for palm oil and soy. *Agriculture and Human Values, 31*(3), 425–437. https://doi.org/10.1007/s10460-014-9511-9

Humphrey, J. (2005). *Shaping value chains for development: Global value chains in agribusinesses* (pp. 1–64). Deutsche Gesellschaft fur Technische Zusammenarbeit (GTZ).

Iscoa, V. (2015). *Guide to good environmental practices for oil palm cultivation in Honduras*. Secretaria de Recursos Naturales y Ambiente (SERNA) & Secretaria de agricultura y Ganaderia (SAG).

Ivancic, H., & Koh, L. P. (2016). Evolution of sustainable palm oil policy in Southeast Asia. *Cogent Environmental Science, 2*(1). https://doi.org/10.1080/23311843.2016.1195032

Jezeer, R., & Pasiecznik, N. (Eds.). (2019). *Exploring inclusive palm oil production*. ETFRN & Tropenbos International.

Kerssen, T. M. (2013). *Grabbing power: The new struggles for land, food, and democracy in northern Honduras*. Food First Books.

Khatun, R., Reza, M. I. H., Moniruzzaman, M., & Yaakob, Z. (2017). Sustainable oil palm industry: The possibilities. *Renewable and Sustainable Energy Reviews, 76*, 608–619. https://doi.org/10.1016/j.rser.2017.03.077

Lee, J., & Gereffi, G. (2015). Global value chains, rising power firms and economic and social upgrading. *Critical Perspectives on International Business, 11*(3/4), 319–339. https://doi.org/10.1108/cpoib-03-2014-0018

León Araya, A. (2017). Domesticating dispossession: African palm, land grabbing and gender in the Bajo Aguán, Honduras. *Revista Colombiana de Antropología, 53*(1), 151–185. https://doi.org/10.22380/2539472x.6

León Araya, A. (2019). The politics of dispossession in the Honduran palm oil industry: A case study of the Bajo Aguán. *Journal of Rural Studies, 71*, 134–143. https://doi.org/10.1016/j.jrurstud.2019.01.015

Muñoz, B. J. F. (2006). *Situational diagnostic productive economic African palm sector in Honduras* (pp. 1–35). Zamorano. https://bdigital.zamorano.edu/server/api/core/bitstreams/ab8aa09d-a689-4f98-bacc-cfb5f0a4b228/content

Murray Li, T. (2007). Practices of assemblage and community forest management. *Economy and Society, 36*(2), 263–293. https://doi.org/10.1080/03085140701254308

Nail, T. (2017). What is an assemblage? *SubStance, 46*(1), 21–37. https://doi.org/10.1353/sub.2017.0001

Obidzinski, K., Andriani, R., Komarudin, H., & Andrianto, A. (2012). Environmental and social impacts of oil palm plantations and their implications for biofuel production in Indonesia. *Ecology and Society, 17*(1). https://www.jstor.org/stable/26269006

Oosterveer, P. (2015). Promoting sustainable palm oil: Viewed from a global networks and flows perspective. *Journal of Cleaner Production, 107*, 146–153. https://doi.org/10.1016/j.jclepro.2014.01.019

Pacheco, P., Gnych, S., Dermawan, A., Komarudin, H., & Okarda, B. (2017). *The palm oil global value chain: Implications for economic growth and social and environmental sustainability*. Center for International Forestry Research. https://www.cifor.org/publications/pdf_files/WPapers/WP220Pacheco.pdf

Picado, H. (2016). *Expansion of oil palm plantations as state policy in Central America*. World Rainforest Movement. https://www.wrm.org.uy/es/articulos-del-boletin/expansion-de-las-plantaciones-de-palma-aceitera-como-politica-de-estado-en-centroamerica

Rist, L., Feintrenie, L., & Levang, P. (2010). The livelihood impacts of oil palm: Smallholders in Indonesia. *Biodiversity and Conservation, 19*(4), 1009–1024. https://doi.org/10.1007/s10531-010-9815-z

Roquas, E. (2002). *Stacked law: Land, property and conflict in Honduras*. Rozenberg Publishers.

RSPO. (2023). *Who we are*. Roundtable on Sustainable Palm Oil. https://rspo.org/who-we-are/

Saadun, N., Lim, E. A. L., Esa, S. M., Ngu, F., Awang, F., Gimin, A., Johari, I. H., Firdaus, M. A., Wagimin, N. I., & Azhar, B. (2018). Socio-ecological perspectives of engaging smallholders in environmental-friendly palm oil certification schemes. *Land Use Policy, 72*, 333–340. https://doi.org/10.1016/j.landusepol.2017.12.057

Sayer, J., Ghazoul, J., Nelson, P., & Klintuni Boedhihartono, A. (2012). Oil palm expansion transforms tropical landscapes and livelihoods. *Global Food Security, 1*(2), 114–119. https://doi.org/10.1016/j.gfs.2012.10.003

Shepherd, A. (2016). *Including small-scale farmers in profitable value chains: Review of case studies on factors influencing successful inclusion of small farmers in modern value chains in ACP countries*. The Technical Centre for Agricultural and Rural Cooperation.

Sikor, T., & Lund, C. (2009). Access and property: A question of power and authority. *Development and Change*, *40*(1), 1–22. https://doi.org/10.1111/j.1467-7660.2009.01503.x

von Geibler, J. (2013). Market-based governance for sustainability in value chains: Conditions for successful standard setting in the palm oil sector. *Journal of Cleaner Production*, *56*, 39–53. https://doi.org/10.1016/j.jclepro.2012.08.027

Wijedasa, L. S., Jauhiainen, J., Könönen, M., Lampela, M., Vasander, H., Leblanc, M.-C., Evers, S., Smith, T. E. L., Yule, C. M., Varkkey, H., Lupascu, M., Parish, F., Singleton, I., Clements, G. R., Aziz, S. A., Harrison, M. E., Cheyne, S., Anshari, G. Z., Meijaard, E., & Goldstein, J. E. (2017). Denial of long-term issues with agriculture on tropical peatlands will have devastating consequences. *Global Change Biology*, *23*(3), 977–982. https://doi.org/10.1111/gcb.13516

WISSH. (2016). *Wilmar Smallholders Support in Honduras Programme: First Progress Report February–May 2016* (pp. 1–4). https://www.wilmar-international.com/sustainability/wp-content/uploads/2016/06/WISSH-First-Progress-Report-final.pdf

10

PALM OIL PRODUCTION REGIMES AND RESISTANCE IN MEXICO'S OIL PALM ASSEMBLAGE

Erin C. Pischke

Introduction

Mexico produces around 230,000 metric tonnes of oil palm (*Elaeis guineensis*) per year across 115,000 hectares, in only four states—Campeche, Chiapas, Tabasco, and Veracruz—making the country the world's eighteenth largest producer of oil palm (IndexMundi, 2023; Statista, 2023). Half of Mexico's palm oil production is in the state of Chiapas (Statista, 2023). Several driving factors have led to an expansion of its production, including domestic human consumption (e.g. food and cooking oil, soap and cosmetics, as well as pharmaceutical products); a decrease in oil and gas production and a growth in renewable energy consumption globally, and its potential as a feedstock for biofuels as one method for reducing greenhouse gas emissions. This increase in demand creates pressures for expanding the Mexican oil palm sector, which has implications for the way the sector is evolving. This presents an opportunity to analyse palm oil production using assemblage theory.

As with all social processes, government processes and relationships between actors are natural, fluid processes that are less planned than they are managed. Assemblage theory sheds light on how actors, non-human entities, and intangible concepts are brought together in temporary assemblages (Murray Li, 2007). One can use the theory to study how the entities have the agency to cohere or resist coherence (Murray Li, 2007). The resulting assemblage reflects the interests of and power relations between actors and entities within it (Murray Li, 2007). In Mexico, assemblage theory can be used to study how a range of different objects are brought together in ways which are unpredictable and subject to rapid change to reveal how potential major non-state actors in palm oil production in the country can create their own rules and push back against the government within this same arrangement.

DOI: 10.4324/9781003459606-12

To understand how certain human and non-human actors managed to assemble others in the governance relationship that currently exists to reflect their strategic interests in Mexico, two research questions will be presented and answered:

1 Who or what are the key human and non-human entities involved in the governance of palm oil production in Mexico; and in what ways do various structures facilitate and legitimise the expansion of oil palm plantations?
2 How have certain actors, using different power differentials, managed to assemble others in the governance relationship that currently exists in relation to palm oil production?

This chapter is structured as follows: The first section reviews Mexico's historical background, focusing on the development of the agricultural sector and related governance structures. Next, it presents the major actors, institutions, and non-human entities involved in developing the Mexican oil palm sector. The third section looks at the use of domestic power and policies to form the current assemblage around palm oil production. The chapter will conclude by showing that while past land reforms and the historical corporatist model tended to support the interests of the national government, a range of non-state actors employed assemblage practises to redefine their relationships with other elements of the oil palm assemblage. These actors created their own rules and pushed back against the government's formal control within this same arrangement, fuelling local resistance to full governmental control of landscapes and power in the countryside.

Historical background

Land tenure and reform is an important aspect of Mexican history and land management processes. Twentieth-century Mexico was characterised by low population density in the countryside, with an agricultural economy where a large portion of the rural population practised shifting slash-and-burn subsistence agriculture (Tudela, 1989). The 1917 agrarian reform in Article 27 of the Mexican Constitution gave ownership of agricultural and forestlands to the rural poor—even in the recent era, approximately 80 percent of communal lands are owned by *ejidos* (groups of people who received land to cultivate) and *comunidades agrarias,* or agrarian communities (groups of people whose ancestral lands were returned to them) (Ponette-González & Fry, 2014; Klooster, 2003; Mexican Constitution Article 27, 1917). This constitutional provision aimed to reduce the concentration of land in the large *haciendas* or estates that characterised the pre-revolutionary landscape. It instituted a system of communal ownership, resulting in the establishment of 29,000 *ejidos* nationwide (Muñoz-Piña et al., 2003).

The 1993 national land reform, with its Programme for Certification of Ejido Rights and Titling of Plots (*Programa de Certificación de Derechos Ejidales y Titulación de Solares*, or PROCEDE), enabled *ejidos* to survey and "certify" (read: privatise) their land, granting them a title of ownership and allowing *ejidatarios* to vote whether to divide and privatise their land (Vásquez-León & Liverman, 2004). Consistent with neoliberal policies that favoured private property rights, unless *ejidatarios* certified their lands under the PROCEDE process, there was no way for them to title their land or access loans and credit from local banks (Moser et al., 2014; German et al., 2011; Vásquez-León & Liverman, 2004). The PROCEDE certification programme enabled *ejidos* to measure and certify their land; this was often impossible if the officials never visited an *ejido*. Since not every *ejido* certified their land—or their certification was never made official—communal landowners often did not have a formal proof of ownership as required by international organisations (Moser et al., 2014), and informal land rights are not usually recognised by the federal government (German et al., 2011).

While studies of the agricultural economy and agribusiness policies often focus on specific agricultural sectors or crops, such approaches are often inadequate when it comes to developing an understanding of the economic practices of resource-limited farm households. Numerous studies demonstrate that such households make a living by whatever means necessary, whether by farming (subsistence or commercial), participating in off-farm wage labour, or owning a small business. In another attempt to generate income for the state in rural areas, the national and state governments have subsidised rural smallholder farmers' participation in oil palm production by giving them access to land and, in some cases, other resources for establishing oil palm (Castellanos-Navarrete & Jansen, 2016). It does the same for international companies that have the technical knowledge and resources to establish large-scale plantations.

The people who have either diversified livelihoods (including farm and non-farm labour) or those who perform non-farm activities such as working in general stores or as mechanics, may have more stable livelihood strategies than those who depend on one industry; such income-generating activities are dynamic and evolving (Reardon et al., 2001). Over time, certain activities can fade in significance as new ones emerge and are incorporated into livelihood strategies.

Oil palm is a perennial crop that is widely regarded as being highly productive and profitable. It is also seen as being relatively labour-intensive because the fresh fruit bunches (FFBs) must be manually harvested and processed within a day of being cut (Byerlee et al., 2017). In principle, therefore, oil palm production provides an additional means by which rural dwellers may diversify incomes, gain employment, and earn stable wages—either through their own production of oil palm or via employment in the plantations of others—all of which may help secure families' resilience in the face of economic or ecological changes (Urióstegui et al., 2018; Reardon et al., 2001). Depending on the country where the oil palm is grown, people who participate in oil palm alongside other activities may do so because they perceive it as a means of safeguarding their livelihoods.

There are many experiences with oil palm in Latin American countries and people seem to have benefitted based on their socioeconomic status. In Brazil (da Silva

César & Batalha, 2013), Mexico (Castellanos-Navarrete & Jansen, 2015) and Costa Rica (Beggs & Moore, 2013), there have been increased economic benefits for poor and middle-income farmers and labourers. However, in Brazil, some small-scale farmers have earned less than expected because of money spent on inputs, labour, and equipment (Glass, 2013), while plantation workers receive low wages (Backhouse, 2013). It has been found that peasants in Guatemala also receive low wages (Alonso-Fradejas, 2015). In Honduras, smallholders have been unable to participate because of the high costs of start-ups and lack of capital (Fromm, 2007).

The oil palm experience in Southeast Asia has been slightly different. The income-generating potential of oil palm varies between households (Budidarsono et al., 2012). Many of the plantation opportunities in Indonesia and Malaysia have benefited non-local workers, foreigners, and wealthy locals who have access to financial resources (German et al., 2010) because plantations are located in sparsely populated areas (Byerlee et al., 2017). Despite this, Indonesians and legal migrants there have reported positive community benefits of oil palm, including higher wages and net employment gains (Byerlee et al., 2017; German et al., 2010). Malaysians have also reported positive employment benefits, including higher incomes, housing, and flexible work schedules (German et al., 2010). People in both countries also reported being paid regularly (German et al., 2010).

Actors with agency involved in governing oil palm

The governance of oil palm production presents opportunities to a range of actors in producing countries such as Mexico. It has implications for the way relationships between the different human and non-human elements in palm oil production are worked out. Actors involved in the governance of the oil palm system were identified primarily through the author's experience interviewing and surveying actors in the field (see Pischke et al., 2018), and a review of the literature about the development of the crop in Mexico.

Human entities

Human entities with agency in Mexican oil palm production include farmers and labourers, representatives of domestic and international companies, local and national governments, politicians, sustainability standards organisations, and consumers of palm oil products (both domestic and international). Each of these types of actors is reviewed in this section.

Farmers and labourers

There are two types of people directly involved in growing oil palm production in Mexico, namely smallholders who cultivate small plots independently or cooperatively with others, and labourers who work on the land belonging to others.

Labourers include people who do not own land and work in various roles on larger plantations that are often foreign-owned. Some labourers are also farmers who might otherwise tend to their own plots or ranches if the opportunity to work a steady job on a plantation was not there.

Farmers with small- and medium-sized acreage might have their own plots of land on which they grow the crop or work on somewhat larger, cooperatively-owned *ejidal* plots. There are smallholders in both the Chiapas (Linares-Bravo et al., 2018) and Tabasco (Abrams et al., 2019) states in Mexico. As a snapshot in time, during the mid-2010s, cooperative growers formed the following farmers' cooperatives:

- El Malayo (Abrams et al., 2019);
- Asociación de Tenosique (Femexpalma, 2017);
- Las Asociaciones de Reforma y Juárez (Femexpalma, 2017);
- Rural Association of Collective Interest (Castellanos-Navarrete & Jansen, 2017).

Representatives of domestic and international companies

Many companies support the palm oil industry in Mexico because the product is used in their manufactured goods. Some of the companies that have operated in Mexico in the past decade include PepsiCo, Oleofinos (which supplies PepsiCo), Oleopalma, and the *Federación Mexicana de Palmicultores y Extractores de Palma de Aceite AC* (Femexpalma) (Femexpalma, 2017). The Costa Rican company, Palmeras Oleaginosas del Sur (formerly doing business as Palma Tica), Grupo Propalma, and Uumbal all own oil palm plantations in Mexico (Sánchez, 2023; Soberanes, 2019; Pischke et al., 2018). As of 2017, PepsiCo was a large buyer of refined palm oil in the country (Femexpalma, 2017).

The international companies that grow and/or process oil palm are interested in gaining access to suitable land in regions with the proper climatic conditions. Likely, these companies are also interested in taking advantage of ideal growing conditions, cheap labour, and cheap land available in countries like Mexico. Since the Mexican government incentivised the establishment of processing mills—for example, in Chiapas—the companies that built them experienced a win-win situation, as they had access to land and labour (Castellanos-Navarrete & Jansen, 2016).

Other international actors that have had an interest in oil palm development in the Mexican countryside in the recent past include the High Conservation Value Resource Network (*Red de Recursos de Alto Valor de Conservación*, or HCVRN) (Femexpalma, 2017) and the non-profit Proforest, which promotes "responsible production and sourcing of palm oil" from smallholders that grow the crop (Proforest, n.d.). In Chiapas, the EU has promoted converting—through "policies of productive reconversion"—the land from cattle ranching to oil palm plantations (Linares-Bravo et al., 2018). There are also rumours that "armed narcotrafficking groups [were] in the region, and that they have witnessed incidents in which these groups have defended the interests of palm oil companies" (Soberanes, 2019).

Local and national governments and politicians

The Mexican government's interests are two-fold: (1) it wants to be somewhat independent of other countries, and (2) it wants a way to control the rural populace. Independence from other countries would come from energy independence (through oil and gas or producing bioenergy) (Radics, 2015; Rodríguez et al., 2014; Creutzig et al., 2013; Skutsch et al., 2011) or through growing crops for domestic consumption and, if possible, for export (Eastmond et al., 2014; Rodríguez et al., 2014). Control of rural land would occur by enclosing the commons and making it dependent on—and loyal to—the government (Linares-Bravo et al., 2018, Shipley, 2016).

The Mexican government has a central role in planning and managing the oil palm industry in the country (Linares-Bravo et al., 2018). The Mexican government relies on formal rules created by the executive branch of the government and gives the power to carry out policies to sectors within the federal government. Technical, financial, and infrastructural aspects of oil palm production on reconverted land in Chiapas have been supported at the state and national levels (Linares-Bravo et al., 2018).

The state has adopted several measures to support the development of a mixed palm oil industry. On the one hand, international commercial palm oil companies have gotten access to land with help from the government by incentivising international companies to plant oil palm (Soberanes, 2019). The government also subsidises smallholders' oil palm plantings through the National Institute for Forestry, Agriculture and Fisheries Research (*Instituto Nacional de Investigaciones Forestales, Agrícolas y Pecuarias*, INIFAP), which is tasked with directing "technology packages" nationwide where possible. Similarly, the ministry charged with agriculture, cattle, rural development, fisheries, and food (*Secretaría de Agricultura, Ganadería, Desarrollo Rural, Pesca y Alimentación*, SAGARPA) is charged with the strategic economic revival of certain regions in the country by establishing large oil palm plantations in the countryside (Maganda, 2008). Despite many threats to the environment and its biodiversity, there are few governmental environmental regulations (Soberanes, 2019). Where such regulations exist, the state often lacks the ability or resources to enforce compliance with them (Bryant & Bailey, 1997). On the other hand, the Federal Attorney for Environmental Protection (*Procuraduría Federal de Protección al Ambiente*, PROFEPA) took pro-environment action against Palma Tica and held the company responsible for failing to notify the local communities and stakeholders about their development plans (they fined the company USD100,000) (Soberanes, 2019).

Consumers of palm oil products

Consumers of Mexican palm oil include people residing inside and outside the country. Since the North American Free Trade Agreement (NAFTA) was signed in the early 1990s, palm oil consumption in Mexico increased at least four times over by the mid-2010s (GRAIN, 2014). While much of the oil palm that is grown to meet this demand is produced in other Latin American countries, Mexican oil palm production also contributes to this supply (GRAIN, 2014). Companies such as PepsiCo, Oleofinos, and Oleopalma buy refined palm oil for use in their products, mainly for domestic consumers (RSPO, 2018).

Unlike other countries, such as Iceland or Norway where there have been countrywide boycotts of palm oil as a food additive or as ingredients in cosmetics, there does not seem to be the same type of reaction in Mexico. One reason may be that consumers are unaware of the prevalence of the health impacts of palm oil or the potential environmental threats that oil palm plantations pose. In places where the crop is grown, employment opportunities and increased wages resulting from its production could outweigh any negative perceptions people may have (Pischke et al., 2018). International consumers may not specifically target Mexican oil palm for boycotts because the country is not one of the largest producers in the world. Since their production has largely occurred on former cattle ranches and other available lands and not on land where there had been virgin forest or highly biodiverse rainforest, as is the case in Brazil, Mexican oil palm may not seem as destructive as it could be.

However, in Mexico, communities have pushed back on the spread of oil palm plantations—if not on palm oil as an additive in food or other products—because of their environmental impacts (Sánchez, 2023). In Chiapas in 2022, community members protested pollution and other environmental degradation (Sánchez, 2023). Despite this, the country's agricultural plan through the year 2030 includes growing oil palm (Sánchez, 2023).

Sustainability standards organisations

Sustainability standards organisations have an important role in managing or mitigating the impacts of oil palm production on workers, communities, and the environment. In other oil palm-producing countries, third-party certification bodies, such as the RSPO, have been key actors concerning questions of sustainability, including environmental protection and workers' rights (Byerlee et al., 2017). Furthermore, such organisations can act as liaisons between the government and oil palm-producing companies on behalf of workers.

Recently, Mexican oil palm growers have moved towards signing agreements with third-party certification bodies. In 2018, the RSPO held a two-day meeting and granted funds to companies such as PepsiCo, Oleofinos, and Oleopalma (the latter worked towards their RSPO certification, which it received in 2020; see http://oleopalma.com.mx/certificaciones/), Femexpalma and Proforest to support 157 smallholders that grow oil palm (RSPO, 2018). There are no other active Mexican oil palm sustainability certifications at this time.

Non-human entities

Non-human entities most relevant to Mexican palm oil production include oil palm plants, animals, insects, and biodiversity, as well as water availability, all of which will be covered in this section.

Oil palm plants

Oil palm grows best in lowland humid tropics (Byerlee et al., 2017), but can grow in a wide range of soil types once irrigation and nutritional needs are met. Flooding or steep, hilly terrain prevents oil palm from being viable in locations that otherwise have ideal conditions for growing the crop (Pischke et al., 2018). For example, in Tabasco and Chiapas, Mexico, the region is hot and muggy most of the year and is semi-mountainous with large rivers running through it. Despite environmental qualities that would be ideal for the oil palm to grow, the geography—which varies from flat pasture to hilly ranch lands with high sierras (high-altitude mountains) and rocky, tree-covered outcroppings—can be a detriment to planting the crop. Inaccessible terrain that is sparsely populated makes it difficult to plant crops and find enough labour to tend to them.

Early oil palm development in most countries has concentrated on well-drained lower-lying mineral soil areas in the tropics. In many cases, this has involved the conversion of existing agricultural holdings into oil palm cultivation. However, as demand for the crop has increased, production has expanded into a range of often less suitable areas that have not been previously used for large-scale agriculture. In the case of existing agricultural areas and areas that have been newly brought into production for oil palm, the process has involved very significant landscape modification. Alongside the crop itself, this includes the construction of forest clearings, road infrastructure, drainage, and earth movement.

Hanging over the actors' relationships with each other and the power they exert is the reality that the oil palm plants themselves exert much control over where plantations are established, where government resources flow, and where human bodies must move in order to harvest, ship, and process the FFBs and the refined oil. Oil palm plants can only grow in certain locations, resulting in changed landscapes in those areas. The way the plants produce FFBs dictates the method for harvesting them (by hand, not machine), requiring labourers to live near or be able to travel to work to access plantations. There is evidence that migrant labourers from neighbouring Guatemala have moved into areas with plantations in order to take advantage of the employment opportunities (Abrams et al., 2019).

The perishable nature of the FFBs necessitates their processing within 24 hours of being harvested; this means that processing mills need to be close to plantations and accessible along paved roads so that FFBs can be reliably delivered. The small-holder farmers who want to participate in the production of oil palm by growing the crop on their own land can only do so if they are part of a cooperative that collectively sells members' FFBs to a local mill (Pischke et al., 2018) or otherwise can scale up to supply enough FFBs to a mill for processing and have an individual arrangement set up with a processor so that their produce is not wasted.

Furthermore, the lure of participating in a commodity crop market may be strong enough to shift the pattern of land sales and land use in the countryside. In Chiapas, people who want to plant and benefit from the sale of oil palm must own the land (Linares-Bravo et al., 2018). The *ejido* system limits the ownership rights to the land, while Mexican culture almost guarantees that men and their sons are typically the only people who own land (Assies, 2007). In some parts of the Chiapas state, land availability (land that had been cattle ranches before the establishment of oil palm) has allowed some farmers to shift from cattle ranching to oil palm production because they had the space and capital to do it (they could sell cattle to make money needed to cover up-front oil palm costs) (Castellanos-Navarrete & Jansen, 2016). There have been similar outcomes in Tabasco as well (Abrams et al., 2019). By pushing oil palm production in Chiapas, land that was otherwise used as farmland was converted for oil palm production, which changed the landscape from a polyculture to a monoculture; it also changed the dynamics of who used the land and how—small producers had to stop relying on subsistence agriculture or, if they could not participate in the oil palm business, had to often pay higher prices for local goods (Linares-Bravo et al., 2018).

Animals, insects, and biodiversity

In creating oil palm plantations, biodiversity is often lost or threatened and defor-estation takes place to make way for new plots, which also leads to changes in water quantity and quality, as well as a loss of carbon stocks (Sánchez, 2023, Solo-mon et al., 2015; Rodríguez et al., 2014; German et al., 2011, Fargione et al., 2010). Deforestation by biofuel companies to create monoculture plantations has been linked to the loss of wildflowers and insect populations, including important pol-linator bees (Selfa et al., 2015). Even on converted farmlands, where deforestation is not an issue, the intensive use of pesticides, herbicides and other chemical inputs linked to intensive oil palm cultivation puts people and animals at risk (Fargione et al., 2010). Wastewater and other effluents associated with oil palm production are also a concern (Alemán-Nava et al., 2014).

Water

There has been a change in the availability of local water resources in parts of Mexico due to dropping water tables attributed to drought, climate change, and oil palm plantations sucking up the available ground and river water (Soberanes, 2019). Oil palm plantations' acreage "decreased to just over 3,000 hectares by 2007 due to hurricanes and inadequate management" (Urióstegui et al., 2018, p. 26). Climatic changes can also lead to changes in the volumes available in bodies of water, movement of species across landscapes as temperatures rise and phenology (timing of seasonal variations in species), which will further impact how people cope with or adapt to their environments (Lawler, 2009).

The weather can impact the people who rely on agriculture for their livelihoods, including those who plant or work in the palm oil sector; extreme weather such as floods and drought can lead to lost crops (Urióstegui et al., 2018). Besides having an adequate climate, Mexico's oil palm-growing regions have available land that can be converted into plantations. However, only government subsidies allow growers to profit off the crop (Linares-Bravo et al., 2018). Therefore, the availability of land is not a threat to the oil palm industry; other governmentally-subsidised crops that would ideally be grown in the same area—and their related industries—would lose out when oil palm is chosen as the export crop of choice by the government. In this way, the government needs to fully plan where and how to manage the growing of all the crops and commodities that it plans to promote, making sure to coordinate with its various agencies that are tasked with planning rural projects such as oil palm plantations, cattle ranches, or reforestation efforts.

Assemblage practices and oil palm policy in Mexico

The second research question addressed in this chapter asks: How have certain actors managed to assemble others in the governance relationship that currently exists in relation to oil palm production? The government of Mexico drives the oil palm assemblage to further its key objectives of becoming more self-sustaining and developing a strategy for controlling the activities occurring in the rural countryside. Two factors ensure the viability of large-scale plantations:

1 Government control of resources (funding, equipment, technology, etc.) to strongly influence what is grown, where it is grown, as well as who harvests and processes it;
2 Past land reforms and the historical corporatist model, where the government gives rural farmers gifts in exchange for political support (a system which will self-perpetuate), work in the national government's favour.

This section will review these two driving factors and show not only how the government can steer others within the current oil palm assemblage, but also how other actors can create their own rules and push back against the government within this same arrangement.

Government control of resources

The truly powerless actors within the oil palm industry in the Mexican countryside are the labourers who do not directly participate in oil palm production on their own farms—either because they do not own land or lack the resources and equipment necessary to take advantage of government-subsidised seedlings or fertiliser—yet they may still benefit from employment and earning a regular wage (Sánchez, 2023; Pischke et al., 2018; Byerlee et al., 2017). The oil palm assemblage may be detrimental to those livelihoods if the structure of labour (growing or working long hours on oil palm plantations) is inflexible and does not allow extra time to work other jobs or grow other crops. The people who do *not* plant oil palm, even when they live in an area otherwise suitable for it—as in San Pablo Tamborel, Tabasco—often do not own their own land and only have access to communal land (Abrams et al., 2019). Furthermore, some *ejido* members did not want to plant the crop or did not have an existing relationship with the government to lure resources to their town (Abrams et al., 2019). Households that can diversify their incomes by growing their own food or having other off-farm work have more power over whether they want to work on large plantations, which dictate wages and other benefits (Byerlee et al., 2017).

In Chiapas, oil palm was first suggested, promoted, and grown by farmers when the state did not have the resources to subsidise oil palm (Castellanos-Navarrete & Jansen, 2017). Various crops, including oil palm and teak, have provided income for certain villages in Tabasco since the *La Alianza Para El Campo* subsidies were established in 1998 (Secretaría de Gobernación, 1998). The experience and familiarity with the crop on large-scale plantations that were able to receive the government's oil palm subsidies enabled farmers to use their acquired knowledge to establish their own small-scale plantations and participate in the commodity trade (Dauvergne & Neville, 2010). Structuring oil palm production on farmers' plots allows them the flexibility to grow other subsistence crops or cash crops such as coffee for supplemental income if the price of palm oil on the international market drops (Dauvergne & Neville, 2010). With small- and medium-sized farmers, domestic subsidies for crops like oil palm were not available to them, which increased the likelihood of them working as labourers on the large-scale plantations that the government, large landholders, and industry created (Secretaría de Gobernación, 1998).

Rural communities not only have come to rely on employment opportunities with large, often international, companies but are at their mercy because of the difficulty rural labourers face in finding stable employment that does not fluctuate with international market prices (Selfa et al., 2015). As has been common since free trade proliferated in North America with NAFTA, these companies offer Mexico and other countries in the Global South the chance to enter the international market while nominally improving worker and environmental conditions (Klooster, 2006). Anticipated government subsidies in the form of oil palm or other crops should be considered as one option of many for adapting to evolving institutional and social systems; and as of 2019, the federal government discontinued its support for smallholders to establish oil palm plantations (Sánchez, 2023).

Governmental support is a double-edged sword, contributing to continued dependence on the government to provide employment and resources locally, while at the same time changing the local landscape. As with other commodity crop production, farmers typically do not have the luxury of choosing what they grow—the "social sector" made up of *ejidos* and agrarian communities largely remained caught in the production of staples under increasingly adverse terms of trade: the "urban bias"' (Assies, 2007, p. 44). However, the power of the state does not mean much if it is not capable of supporting the oil palm industry or making it profitable or beneficial to the farmers growing it. As Abrams et al. (2019, p. 522) note, state governments must be competent in orchestrating the actors in the industry, including "supportive intermediary organisations, direct community ownership, or the participation of competent state authorities".

Past land reforms and the historic corporatist model

One way that Mexico's oil palm assemblage has been different from other countries is the unique characteristic that land "ownership" takes in the country, namely through its *ejidos* or commonly held land arrangements. Past land reforms, which wrested land from large landowners and some state land to create *ejidos*, spurred agricultural production in the countryside, including in the states where oil palm is now being grown. The Mexican government still gives resources to rural farmers because of the long history of "buying" political support in that way and through land redistribution (Castellanos-Navarrete & Jansen, 2016). However, since the introduction of neoliberal policies in the 1990s (in part through the privatisation of the countryside via the PROCEDE process), conditions have somewhat changed. While appealing to state interests and playing to local leaders' egos (i.e. by naming *ejidos* or projects after politicians) can help an organisation gain state resources, it also could backfire and contribute to local corruption if leaders profit personally from commonly gained resources and funding (Castellanos-Navarrete & Jansen, 2017).

Palm oil processing mill operators have the power to direct where the assemblage occurs in the establishment of the crop; plantations pop up in proximity to mills and labourers travel to work in both the plantations and mills (Abrams et al., 2019). Even in this case, however, the Mexican government has incentivised the mill operators' participation in oil palm production by awarding them land and resources to set up their mills in locations of the government's choosing (Castellanos-Navarrete & Jansen, 2016). Because of competition for smallholders' FFBs between the various mills that the government established in Chiapas, the smallholders had some power over where they would sell their produce (allowing them to earn more money this way) (Castellanos-Navarrete & Jansen, 2016). All of these actors were still dependent on the ideal growing conditions for the crop, something that is likely to change in the future due to climate change.

It seems to be the case that the Mexican government has the power to choose winners—foreign companies—to reward with land and infrastructure and practically hand them day labourers to do the work of planting and harvesting the oil palm (Castellanos-Navarrete & Jansen, 2016). Had the government not intervened in the past with land reforms, it would not have the ability today to do with land and labour as it chooses. Large landholders in Chiapas have less power than the state as they would not be able to grow oil palm if the government did not subsidise the mills that they accessed; furthermore, the government would not have subsidised the oil palm sector if there had not been peasant farmers to give subsidies to (in exchange for political support) (Castellanos-Navarrete & Jansen, 2016). While currently there may not be subsidies available to smallholders for oil palm plants, that may change if global conditions change in favour of Mexican oil palm.

Conclusion

This chapter has sought to provide insight into the first and second research questions in the book by illustrating how the different governance arrangements have enabled and legitimised the expansion of oil palm in Mexico and how power differentials within the assemblage affect oil palm governance in the country. Given the importance of agriculture in developing economies and the centrality of food security and rural development goals, it is unsurprising that national governments have a significant interest in directing the assemblage of human and non-human actors on the landscape. When the Mexican government plans new, large-scale oil palm plantations, local communities are often not consulted (Soberanes, 2019). The indigenous *ejido* communities that have been involved in oil palm in Chiapas are marginalised; so long as they want to earn money selling export crops or to companies outside of their region, they have little say in what is planted (Linares-Bravo et al., 2018). This forces certain populations to be dependent on crops that grow well in rainforest conditions and rely on the government to provide them with subsidies (e.g. one-off support, including the allotment of plants, fertilisers and in some cases, economic resources for the expenses of establishing the crop) (Linares-Bravo et al., 2018).

While it might seem that farmers and labourers are in a weak position in oil palm production unless they are growing the crop on their own land and working with a cooperative, they are not powerless agents. Farmers have taken advantage of the corporatist model of governance and benefitted from oil palm development. Castellanos-Navarrete and Jansen (2017) argue that smallholders have the agency to dictate their own relationship with nature and have the power to resist the government's attempts at governing them. The national and state governments in Mexico incentivised oil palm production. However, smallholder farmers also have used their power to take advantage of the situation for their own benefit, gaining resources and potential income from the crop (Castellanos-Navarrete & Jansen, 2017). Moreover, rural farmers were already part of the market economy before participating in oil palm production (Abrams et al., 2019). They were willing participants, both in oil palm and in other commodity crop production, because they could earn more money to provide for their families compared to other low-wage or subsistence means (Pischke et al., 2018). Smallholder farmers may be in a better position today with the growing acceptance of RSPO certification by entities involved in the oil palm industry, which promotes quality of life and respects the human rights of its smallholder growers (Proforest, 2018).

This chapter has focused on questions of resistance and the difference between formal ideas of control as expressed by the state and the capacity of other actors in the process to subvert these via assemblage practices. Entities with the most agency in Mexican palm oil production include farmers and labourers, oil palm plants, local and national governments, as well as representatives of international companies. Past land reforms and the historical corporatist model have worked in the national government's favour; however, actors can create their own rules and push back against the government within this same arrangement. Hence, this chapter has demonstrated that local resistance to full governmental control of landscapes and power has occurred within Mexico's oil palm assemblage.

References

Abrams, J., Pischke, E. C., Mesa-Jurado, M. A., Eastmond, A., Silva, C. A., & Moseley, C. (2019). Between environmental change and neoliberalism: The effects of oil palm production on livelihood resilience. *Society & Natural Resources*, 32(5), 548–565. https://doi.org/10.1080/08941920.2018.1544678

Alemán-Nava, G. S., Meneses-Jácome, A., Cárdenas-Chávez, D. L., Díaz-Chavez, R., Scarlat, N., Dallemand, J., Ornelas-Soto, N., García-Arrazola, R., & Parra, R. (2014). Bioenergy in Mexico: Status and perspective. *Biofuels, Bioproducts and Biorefining*, 9(1), 8–20. https://doi.org/10.1002/bbb.1523

Alonso-Fradejas, A. (2015). Anything but a story foretold: Multiple politics of resistance to the agrarian extractivist project in Guatemala. *The Journal of Peasant Studies*, 42(3–4), 489–515. https://doi.org/10.1080/03066150.2015.1013468

Assies, W. (2007). Land tenure and tenure regimes in Mexico: An overview. *Journal of Agrarian Change*, 8(1), 33–63. https://doi.org/10.1111/j.1471-0366.2007.00162.x

Backhouse, M. (2013). The sustainable dispossession of the Amazon. The case for investments in oil palm in Pará. *Fair Fuels?* Working Paper 6. https://d-nb.info/1276600720/34

Beggs, E., & Moore, E. (2013). *The social landscape of African oil palm production in the Osa and Golfito region, Costa Rica* (pp. 1–31). Iniciativa de Osa y Golfito & Stanford Woods Institute for the Environment, Stanford University. https://inogo.stanford.edu/sites/default/files/African%20palm%20social%20landscape%20INOGO%20June%202013.pdf

Bryant, R. L., & Bailey, S. (1997). *Third world political ecology*. Routledge.

Budidarsono, S., Dewi, S., Sofiyuddin, M., & Rahmanulloh, A. (2012). *Socioeconomic impact assessment of palm oil production* (pp. 1–4). World Agroforestry Centre—(ICRAF). https://apps.worldagroforestry.org/downloads/Publications/PDFS/TB12053.PDF. Technical Brief No. 27: Palm oil series.

Byerlee, D., Falcon, W. P., & Naylor, R. (2017). *The tropical oil crop revolution: Food, feed, fuel, and forests*. Oxford University Press.

Castellanos-Navarrete, A., & Jansen, K. (2015). Oil palm expansion without enclosure: Smallholders and environmental narratives. *The Journal of Peasant Studies*, 42(3–4), 791–816. https://doi.org/10.1080/03066150.2015.1016920

Castellanos-Navarrete, A., & Jansen, K. (2016). Is oil palm expansion a challenge to agro-ecology? Smallholders practising industrial farming in Mexico. *Journal of Agrarian Change, 18*(1), 132–155. https://doi.org/10.1111/joac.12195

Castellanos-Navarrete, A., & Jansen, K. (2017). Why do smallholders plant biofuel crops? The "politics of consent" in Mexico. *Geoforum, 87*, 15–27. https://doi.org/10.1016/j.geoforum.2017.09.019

Creutzig, F., von Stechow, C., Klein, D., Hunsberger, C., Bauer, N., Popp, A., & Edenhofer, O. (2012). Can bioenergy assessments deliver? *Economics of Energy & Environmental Policy, 1*(2). https://doi.org/10.5547/2160-5890.1.2.5

da Silva César, A., & Batalha, M. O. (2013). Brazilian biodiesel: The case of the palm's social projects. *Energy Policy, 56*, 165–174. https://doi.org/10.1016/j.enpol.2012.12.014

Dauvergne, P., & Neville, K. J. (2010). Forests, food, and fuel in the tropics: The uneven social and ecological consequences of the emerging political economy of biofuels. *The Journal of Peasant Studies, 37*(4), 631–660. https://doi.org/10.1080/03066150.2010.512451

Eastmond, A., García, C., Fuentes, A., & Becerril-García, J. (2014). Mexico. In B. D. Solomon & R. Bailis (Eds.), *Sustainable Development of Biofuels in Latin America and the Caribbean* (pp. 203–222). Springer.

Fargione, J. E., Plevin, R. J., & Hill, J. D. (2010). The ecological impact of biofuels. *Annual Review of Ecology, Evolution, and Systematics, 41*(1), 351–377. https://doi.org/10.1146/annurev-ecolsys-102209-144720

Femexpalma. (2017). *Federación Mexicana de Palmicultores y Extractores de Palma de Aceite A.C.* https://www.rspo.org/acop/2017/federacin-mexicana-de-palmicultores-y-extractores-de-palma-de-aceite-a.c./CV%20FEMEXPALMA%202018.pdf

Fromm, I. (2007). Upgrading in agricultural value chains: The case of small producers in Honduras. *SSRN Electronic Journal, 64*, 1–32. https://doi.org/10.2139/ssrn.1071669

German, L., Schoneveld, G., Skutch, M., Andriani, R., Obidzinski, K., Pacheco, P., Komarudin, H., Andrianto, A., Lima, M., & Dayang Norwana, A. A. B. (2010). *The local social and environmental impacts of biofuel feedstock expansion: A synthesis of case studies from Asia, Africa and Latin America* (pp. 1–12). Center for International Forestry Research. https://www.jstor.org/stable/resrep01888

German, L., Schoneveld, G. C., & Pacheco, P. (2011). The social and environmental impacts of biofuel feedstock cultivation: Evidence from multi-site research in the forest frontier. *Ecology and Society, 16*(3). https://www.jstor.org/stable/26268914

Glass, V. (2013). *Expansion of oil palm in the Brazilian Amazon: Elements for an analysis of the impacts on family farming in the northeast of Pará* (pp. 1–15). Reporter Brasil. https://reporterbrasil.org.br/documentos/Dende2013.pdf

GRAIN. (2014, September 22). *New frontiers for oil palm.* https://grain.org/article/entries/5036-new-frontiers-for-oil-palm

Government of Mexico, (1917). Constitution of Mexico - Chapter 1: Individual guarantees, Article 27.

IndexMundi. (2023). *Palm oil production by country in 1000 MT.* https://www.indexmundi.com/agriculture/?commodity=palm-oil&graph=production

Klooster, D. (2003). Campesinos and Mexican forest policy during the twentieth century. *Latin American Research Review, 38*(2), 94–126. https://www.jstor.org/stable/1555421

Klooster, D. (2006). Environmental certification of forests in Mexico: The political ecology of a nongovernmental market intervention. *Annals of the Association of American Geographers, 96*(3), 541–565. https://doi.org/10.1111/j.1467-8306.2006.00705.x

Lawler, J. J. (2009). Climate change adaptation strategies for resource management and conservation planning. *Annals of the New York Academy of Sciences, 1162*(1), 79–98. https://doi.org/10.1111/j.1749-6632.2009.04147.x

Linares-Bravo, B. C., Zapata-Martelo, e, Nazar-Beutelspacher, A., & Suárez-San Román, B. (2018). Productive reconversion to oil palm in the Tulijá Valley, Chiapas, Mexico: Impact differentiated by gender. *Agricultura, Sociedad Y Sesarrollo, 15*(4), 487–506.

Maganda, C. (2008). The Latin American water tribunal and the need for public spaces for social participation in water governance. In J. Feyen, K. Shannon, & M. Neville (Eds.), *Water and Urban Development Paradigms* (pp. 705–710). CRC Press.

Moser, C., Hildebrandt, T., & Bailis, R. (2014). International sustainability standards and certification. In B. D. Solomon & R. Bailis (Eds.), *Sustainable Development of Biofuels in Latin America and the Caribbean* (pp. 27–69). Springer.

Muñoz-Piña, C., de Janvry, A., & Sadoulet, E. (2003). Recrafting rights over common property resources in Mexico. *Economic Development and Cultural Change, 52*(1), 129–158. https://doi.org/10.1086/380104

Murray Li, T. (2007). Practices of assemblage and community forest management. *Economy and Society, 36*(2), 263–293. https://doi.org/10.1080/03085140701254308

Pischke, E. C., Rouleau, M. D., & Halvorsen, K. E. (2018). Public perceptions towards oil palm cultivation in Tabasco, Mexico. *Biomass and Bioenergy, 112*, 1–10. https://doi.org/10.1016/j.biombioe.2018.02.010

Ponette-González, A. G., & Fry, M. (2014). Enduring footprint of historical land tenure on modern land cover in eastern Mexico: Implications for environmental services programmes. *Area, 46*(4), 398–409. https://doi.org/10.1111/area.12125

Proforest. (n.d.). *Responsible sourcing for sustainable livelihoods.* https://www.proforest.net/

Radics, R. I., Dasmohapatra, S., & Kelley, S. (2015). Systematic review of bioenergy perception studies. *BioResources, 10*(4). https://doi.org/10.15376/biores.10.4.radics

Reardon, T., Berdegué, J., & Escobar, G. (2001). Rural nonfarm employment and incomes in Latin America: Overview and policy implications. *World Development, 29*(3), 395–409. https://doi.org/10.1016/s0305-750x(00)00112-1

Rodríguez, O. A. V., Vázquez, A. P., & Muñoz Gamboa, C. (2014). Drivers and consequences of the first Jatropha curcas plantations in Mexico. *Sustainability, 6*(6), 3732–3746. https://doi.org/10.3390/su6063732

RSPO. (2018, February 26). *Femexpalma conference and smallholders in Mexico receive grant from RSPO Smallholder Support Fund (RSSF).* https://rspo.org/femexpalma-conference-and-smallholders-in-mexico-receive-grant-from-rspo-smallholder-support-fund-rssf/

Sánchez, A. (2023, June 20). *Palm oil: The crop that cuts into southeastern Mexico's jungles and mangroves.* Mongabay. https://news.mongabay.com/2023/06/palm-oil-the-crop-that-cuts-into-southeastern-mexicos-jungles-and-mangroves/

Secretaría de Gobernación. (1998). *Rules of operation of the Alliance for the Countryside 1998, for the Agricultural Development, Livestock, Rural Development and Agricultural Health Programme.* Diario Oficial de La Federación. https://dof.gob.mx/nota_detalle.php?codigo=4881535&fecha=03/06/1998#gsc.tab=0

Selfa, T., Bain, C., Moreno, R., Eastmond, A., Sweitz, S., Bailey, C., Pereira, G. S., Souza, T., & Medeiros, R. (2015). Interrogating social sustainability in the biofuels sector in Latin America: Tensions between global standards and local experiences in Mexico, Brazil, and Colombia. *Environmental Management, 56*(6), 1315–1329. https://doi.org/10.1007/s00267-015-0535-8

Shipley, T. (2016). Enclosing the commons in Honduras. *American Journal of Economics and Sociology*, *75*(2), 456–487. https://doi.org/10.1111/ajes.12146

Skutsch, M., de los Rios, E., Solis, S., Riegelhaupt, E., Hinojosa, D., Gerfert, S., Gao, Y., & Masera, O. (2011). Jatropha in Mexico: Environmental and social impacts of an incipient biofuel programme. *Ecology and Society*, *16*(4). https://www.jstor.org/stable/26268964

Soberanes, R. (2019, November 14). *Mexico plans huge increase in palm oil production in sensitive ecosystems*. Mongabay. https://news.mongabay.com/2019/11/mexico-plans-huge-increase-in-palm-oil-production-in-sensitive-ecosystems/

Solomon, B. D., Banerjee, A., Acevedo, A., Halvorsen, K. E., & Eastmond, A. (2015). Policies for the sustainable development of biofuels in the Pan American region: A review and synthesis of five cuntries. *Environmental Management*, *56*(6), 1276–1294. https://doi.org/10.1007/s00267-014-0424-6

Statista. (2023). *Leading states in planted area of oil palm in Mexico in 2022 (in 1,000 hectares)*. https://www.statista.com/statistics/898284/mexico-oil-palm-planted-area-state/

Tudela, F. (1989). *The forced modernisation of the tropics: The case of Tabasco*. El Colegio de Mexico.

Uriustegui, I. F. R., Pat Fernández, J. M., Pat Fernández, L. A., & van der Wal, J. C. (2018). The effect of oil palm on income strategies and food security of households in rural communities in Campeche, Mexico. *Acta Universitaria*, *28*(2), 25–32. https://doi.org/10.15174/au.2018.1553

Vásquez-León, M., & Liverman, D. (2004). The political ecology of land-use change: Affluent ranchers and destitute farmers in the Mexican municipio of Alamos. *Human Organisation*, *63*(1), 21–33. https://www.jstor.org/stable/44126988

SECTION 3
Outlook

11

LEVERAGING PALM OIL FOR THE SOCIO-ECOLOGICAL TRANSFORMATION OF AFRICAN EMERGING ECONOMIES

Olawale Emmanuel Olayide and Patrick O'Reilly

Introduction

The oil palm tree (*Elaeis guineensis*) is indigenous to West and Central Africa. Historical records indicate that the plants' fruits and kernels have long been used in Africa as a vegetable-based oil for cooking and lighting and as an ingredient in the production of a number of products, including soaps and medicines in traditions that have dated back to ancient Egypt. However, while the plant is of African origin, the emergence of the crop as a commercial opportunity is largely linked to events before and during the European colonial period, during which palm oil drew the attention of European botanists. Subsequently, this gave rise to efforts to commercialise the crop in Africa via large-scale plantations in the early twentieth century (Ordway et al., 2019; Byerlee et al., 2017; Aghalino, 2000).

As recently as the 1960s, Africa dominated world palm oil production, accounting for more than 95 percent of total global output. However, the heterogeneous impacts of diffuse decolonisation events and processes in Africa and Southeast Asia had a profound impact on oil palm cultivation, palm oil production, and distribution regimes. Among the impacts of these was post-war stagnation in plantation-based colonial commodity industries. This saw African palm oil production enter into a decline at the same time as newly-independent Southeast Asian countries were beginning to implement policies to replace their own ailing plantation sectors (in particular, rubber).

Throughout the 1960s, Malaysia invested significant economic and research resources in programmes to acquire promising African palm oil germplasm, using this in the implementation of breeding programmes (Basiron, 2007). These were successful in creating hybrid types with enhanced beneficial agronomic traits (e.g. improved yield volumes, dwarfing, reducing time to first harvest). Midway through

DOI: 10.4324/9781003459606-14

the 1960s, Malaysian factories also began to adopt enhanced technologies, such as the new screw presses first developed in the Belgian Congo. These developments were informed by and advanced postcolonial political agendas in Malaysia and later, Indonesia, whereby the growth of nationally owned and controlled commodity industries assumed huge ideological significance (Haiven, 2022; Varkkey & O'Reilly, 2020). Through a range of initiatives, including the establishment of the Federal Land Development Agency (FELDA), the Malaysian government started the process of converting a commodity agriculture sector which was formerly dominated by a declining rubber industry into one which was primarily organised around the far more lucrative oil palm sector.

In both Malaysia and subsequently Indonesia, palm oil was, and to a very large degree continues to be, treated as something of a developmental panacea, supposedly capable of supporting the development of a globally influential corporate sector and growing Gross Domestic Product (GDP), while simultaneously advancing rural development objectives such as enhancing village infrastructure and boosting the incomes and wellbeing of rural households. While these assertions are often contested, the scale of the impact of palm oil on the physical, social, political, and economic landscape of Southeast Asia is, for good or bad, undeniable. Southeast Asia first surpassed Africa in terms of palm oil production in 1966, firmly establishing Southeast Asia as the industry's centre, a position it still holds today (Rival & Levang, 2014).

It is now widely and unambiguously recognised that the oil palm is the most productive of all current oil crops (Cramb & Curry, 2012). As its advocates regularly point out, this means that the crop produces substantially more vegetable oil per hectare than any other crop; a fact that has long featured in the industry's refutations of those who have attacked it on environmental grounds. Furthermore, the oil produced (or rather oils, as the crop produces different types of oils) represents a valuable commodity in the global market due to high demand driven by its variety of uses in industries such as food, cosmetics, and pharmaceuticals (Corley & Tinker, 2016). Again, its advocates and industry-linked actors have been quick to claim that it has the potential to serve as an alternative to mineral oil in fuels, lubricants, and other sectors. These facts are reflected in the dramatic expansion of the industry. The area of the world where oil palm is grown has increased from less than 5 million hectares in 1980 to more than 25 million hectares in 2018 (Figure 11.1), an area which continues to expand as new countries enter the "boom" (O'Reilly & Varkkey, 2020; Euler et al., 2017; Krishna et al., 2017).

As is the case elsewhere, the nature of the oil palm itself means that, given adequate water and nutrients, the tree can be cultivated in large areas of the tropical belt. The structure of the global industry reflects this; perhaps more than is the case with any other major commodity crop, the palm oil industry is headquartered in the Global South. Equally, however, the relatively late emergence of the crop also appears to be a factor here. As we have already seen, the intensive commercialisation of oil palm began in earnest due to initiatives implemented by

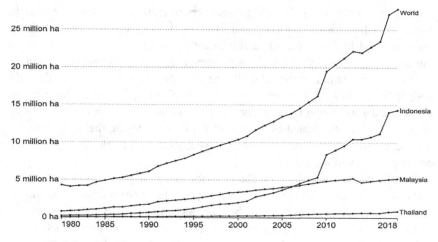

FIGURE 11.1   Land use for palm oil production between 1980 and 2018.

Source: FAO.

newly-independent Malaysia, which, besides breeding programmes, played a significant role in the development of the palm oil supply chain and market. Oil palm cultivation and its related activities have provided a source of income and livelihood for smallholder farmers and labourers, which has helped bring millions of people out of poverty—in large part, a claim that is often attributed to the pioneering work of Malaysian researchers, plantation companies, and policymakers. This is not something that is lost on Malaysians themselves, who are quick to challenge negative Western portrayals of the industry. These efforts are not only seen as denigrating but also perceived as critical to an outstanding Malaysian success story. It is perhaps deeply ironic that while the oil palm has been central to transitions away from a colonial rural landscape in Southeast Asia, researchers and governments based in former colonial powers that previously presided over plantation monocultures in that region now contrive to impose a new set of values and interests on the Global South that reflect their newfound concerns linked to conservation and environmental protection.

The relationship between African nations, including those in which the crop originated, and the industry is far more contested and its future is far less certain. Currently, the continent accounts for around only 10 percent of the world's palm oil, according to the Food and Agriculture Organisation (FAO), a far cry from the 1960s when it accounted for 95 percent of all global production (FAO, 2020). Nigeria produces the most palm oil in Africa, which is an estimated 38 percent of the continent's total production. Other producing states include Côte d'Ivoire, Cameroon, Ghana, and the Democratic Republic of the Congo.

However, the continent of Africa remains a net importer of palm oil. In 2019, Africa imported about 6.3 million tonnes of palm oil, exceeding its domestic production (Figure 11.2). The main consumers of palm oil in Africa are Nigeria, Egypt, South Africa, Morocco, and Algeria, among others. Unsurprisingly, Indonesia and Malaysia provide a significant portion of the palm oil that Africa imports. Thus, the story of the oil palm assemblage in Africa generally stands in marked contrast to that of the palm oil assemblage in Southeast Asia. While in the case of Southeast Asia, the process of "becoming" a palm oil region has resulted in spectacular industry growth that has supported a transition away from a colonial commodity base dominated by rubber, in the case of Africa, the story is one of decline, in which the region from which the palm oil itself originated has been left on the periphery of the industry.

Despite its current position in relation to the global oil palm assemblage, Africa still possesses a considerable bank of "untapped" land which fulfils the agronomic conditions necessary for cultivating the crop. As demand for oil palm and its products continues to rise, it is reasonable to observe that the future prospects for palm oil in Africa are, at the very least, a tempting option for agricultural expansion for numerous actors including nation states, non-governmental organisations (NGOs), large and small corporate entities, small farmers, as well as sundry suppliers and potential beneficiaries. Simultaneously, however, the possibility of expanding the sector faces multiple challenges, many of which simply did not exist or were of much less significance during the initial period of palm oil development in Southeast Asia. Of particular note in this respect is the growing import of environmental concerns, and with these the emergence of a vocal "anti-palm oil" movement. An interesting question thus presents itself concerning the future relationship between Africa and the palm oil industry; how should African nations position themselves to leverage benefits from the palm oil assemblage in a rapidly changing global environmental and political economy context?

While Africa may not hold a prominent position within the palm oil industry compared to regions such as Southeast Asia and Latin America, it grapples with analogous challenges stemming from oil palm expansion. Similar to its counterparts, Africa faces concerns related to environmental degradation, social injustices, and other negative repercussions associated with the expansion of this industry. Issues related to power differentials, the different governance arrangements which legitimise oil palm expansion, and conflict moderation characterise the African oil palm landscape. Smallholders are still struggling to prosper through palm oil production and trade due to pockets of large-scale investments that have often come in the form of land grabbing, land conversion, and deforestation. Besides that, the growing level of importation is only symptomatic of the root causes of the critical challenges: rudimentary production methods, poor development of the value chains, and low investments. These issues result in the continued impoverishment of smallholders and low-capacity utilisation, albeit the impact of palm oil production on smallholders and livelihoods has been described as positive (Ahmed et al., 2019; Balde et al., 2019). The future growth and development in Africa of the palm oil sector will need to overcome these challenges related to the political economy

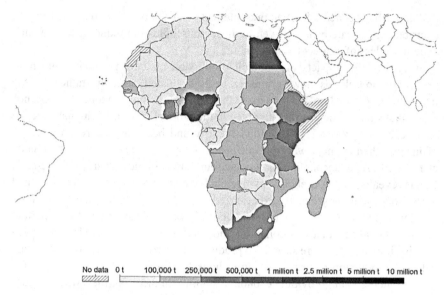

FIGURE 11.2 Palm oil imports in Africa, 2019.

Source: FAO.

of the crop, as well as practical issues such as low investments, divestment in green business models, and other issues presented through technological advancements and emerging bio-circular economic systems of production.

Palm oil and socio-ecological transformation in Africa: Opportunities and challenges

Globally, the expansion of oil palm production is associated with transformative impacts on land productivity and substantial increases in revenues generated from agricultural activity in converted areas. A particular feature of the increase—and indeed as promoted by the industry itself—is that it is reported to deliver income benefits for multiple supply chain participants. While evidence is relatively limited, it is generally suggested that oil palm expansion in Africa has delivered large revenue increases for farmers, workers, and others such as traders, middlemen, and small-scale processors. It is broadly believed that increases in farm income linked to palm oil are particularly beneficial in delivering broader development benefits in producing localities, including increased job opportunities and investment. In addition, palm oil expansion also facilitates the delivery of public goods in the form of improved infrastructure, which helps rural households and communities (Edwards, 2019a, 2019b; Naylor et al., 2019; Gatto et al., 2017; Obidzinski et al., 2012; Feintrenie et al., 2010; Rist et al., 2010; Obado et al., 2009). While it may be the case that not all households and communities enjoy the same level of advantage (Santika et al., 2019; Obidzinski et al., 2012; McCarthy, 2010), its capacity to deliver these benefits is often presented as evidence of the industry's ability to contribute aggregate benefits, even to those households in producing areas which experience less of a direct financial gain.

In Africa, the impacts of palm oil production on both producing and non-producing households have primarily been evaluated through studies on the involvement of smallholder farmers in the industry. A study conducted in Guinea found that small-scale growers involved in the production of oil palm have indeed experienced improved levels of food security and better and more stable flows of income than farmers who did not grow oil palm or other cash crops (Balde et al., 2019). After accounting for any confounding variables, a study conducted in Ghana revealed that households involved in the cultivation of oil palm also earned more money and consequently experienced lower levels of multidimensional poverty relative to non-producers (Ahmed et al., 2019). This latter point is important because it confirms that palm oil cultivation and the income derived from it have positive impacts which are specific to producing households. Hence, it appears that palm oil production contributes to farm household incomes favourably by causing it to increase and providing economic stability to these families, which in turn helps with annual budgets and consumption smoothing.

In addition, while comparatively less labour intensive than some cash crops (in particular, short cycle food crops), it remains the case that the majority of tasks linked to palm oil cultivation are done by hand (van Noordwijk et al., 2001). In most cases, this work is not considered to be highly skilled. Consequently, as regions have become involved in the industry, rapid increases in the area under palm oil have resulted in large increases in the availability of employment, which can be undertaken by people who have received minimal training in areas where alternative forms of wage labour may be scarce or non-existent. This provides job opportunities for landless people as well as an additional source of income for small farmers. Ahmed et al. (2019) found that many rural households in Ghana rely heavily on income from the oil palm industry. Those who worked in the oil palm industry generally experienced better financial circumstances than those employed in other agricultural subsectors (Ahmed et al., 2019). Unsurprisingly, given the relatively sparse populations in many palm oil-producing areas, the availability of lucrative new work prospects results in the movement of people into these areas. A study of young adult workers in Uganda, for example, confirmed that significant numbers moved to oil palm districts for employment (Ssemmanda & Opige, 2019).

Such findings are consistent with work in other parts of the world in which oil palm is cultivated, indicating that in general, those who cultivate the crop—either as smallholder farmers themselves or as employees of those who do—receive some direct income benefit, which is generally in excess of what can be accrued from alternative activities currently available in these areas. This reflects research elsewhere, which has found that the development of oil palm tends to rapidly exhaust the local labour pool. However, at the same time, there is also a growing recognition that such benefits are neither uniform nor universal. Elsewhere in this volume, it has been suggested that the scale of such benefits is variable and in some cases non-existent, and it is also closely linked to the way the industry is structured in different locations. There is evidence that specific features of the oil palm industry supply chain in Africa vary significantly from the model that emerged in the core producing countries in Southeast Asia. In the case of both Indonesia and Malaysia in particular, the model for the palm oil industry has strongly favoured the large-scale production of crude palm oil (CPO), kernel oil, and products via sizeable mills. It is arguably the case that the size and investment involved in the development of these industrial facilities have a very significant—if not the most significant—bearing on the spatial, economic, and organisational arrangement of the oil palm assemblage in both countries. The size of these plants dictates the scale of producing areas and concessions (which must be sufficient to ensure that the mills operate profitably). Likewise, the scale of financing involved also has a strong bearing on the entities that can realistically mobilise the know-how and financial resources to develop these plants, which in turn, dictates the relationships these owners are likely to have with their suppliers, the growers (who must be subordinated to the requirements of the milling process). By contrast, studies conducted in Ghana and Cameroon (Nkongho et al., 2014; Awusabo-Asare & Tanle, 2008), have identified the key role played by small-scale businesses, particularly those developed, owned, and operated by impoverished rural women in the development of the African palm oil industry. In these studies, it was found that increased demand for palm oil prompted women, in particular, to invest in the construction of smaller-scale artisanal pro-cessing mills. According to a study conducted in Nigeria (Ohimain & Izah, 2014), small-scale processing facilities also provide employment for rural households.

However, despite higher wages, employment in the palm oil sector does not always result in an improvement in welfare in terms of food security and other non-income dimensions. This is because these factors also depend on the acces-sibility of sufficient food of adequate quality, the effectiveness of the food market, intra-household gender roles, as well as wider issues linked to infrastruc-ture development in what are often areas in which prior access has been poor (Castellanos-Navarrete et al., 2019; Hamann, 2017).

Moreover, despite the observed revenue benefits that accrue to households from their involvement in oil palm development, aggregate improvements in income may, to some extent, mask enduring and increasing income inequality related to the industry. This situation may be particularly stark in the case of landless migrants who move into oil palm regions for work (Ssemmanda & Opige, 2019). Indeed, such groups suffer multiple disadvantages throughout the palm oil industry. With low rates of pay and often poor working conditions allied to a lack of local social infrastructure and limited political representation (Pye, 2021), this has contributed to the emergence of a poorly paid "class" of landless labour.

The inequality engendered by the industry extends to smallholders themselves, who may lack access to the financial and technical know-how required to construct oil palm farms or—where they do—succeed in establishing an oil palm venture which is able to achieve the efficiencies needed to maximise income. Because of this, farm households with better access to cash tend to have the ability and willingness to embrace oil palm earlier and with more success (Obidzinski et al., 2012; Colchester, 2010; McCarthy, 2010). Addressing the challenges that smallholders face in becoming involved in the palm oil industry is not an exclusively African problem. Indeed, it has been a key concern of Southeast Asian producing countries, which have striven to simultaneously grow a global commodity industry while also demonstrating to both domestic and global audiences that this business delivers for small farmers. Initially, both Indonesia and Malaysia sought to resolve the financial and technical challenges facing smallholders who wished to enter the industry via arrangements that linked large-scale commercial core operations in the corporate, public, or often hybrid state-corporate sector, which exchanged financial and technical advice for supply commitments. More recently, however, the emergence of a more independent smallholder sector has been observed, although the question of what that sector is and who is involved in it is a matter of some debate. In the African context, there is very limited evidence that similar efforts to foster small-scale palm oil have been pursued. Rather, the industry is developing in a less coordinated manner and in a more piecemeal fashion (Oxford Analytica, 2016).

Oil palm expansion in Africa: Land management and the environmental question

While the challenges of managing the industry to ensure effective participation and reasonable rewards are achieved by all participants still remain, additional issues prevent the industry from expanding in Africa. Apart from exacerbating the challenge of delivering benefits to multiple industry participants, variations in agro-ecological and socio-economic situations inevitably benefit some more than others (Edwards, 2019a). This may fuel resentment towards the industry, besides being the basis for the conflicts over land use that have characterised oil palm production globally. Such conflicts may arise when palm oil corporations receive government land concessions that cross community land (Fitzpatrick, 1997). They may be exacerbated in circumstances where conflicting local and national legal systems and other norms related to land use come into conflict. Discussions involving compensation measures and/or the participation of neighbourhood communities through outgrower programmes can occasionally assist in resolving disputes, but this necessitates the willingness of both sides to engage in mediation, which is not always the case (Rist et al., 2010). A particularly challenging issue in Africa in this respect is conflicts between state-based property claims and those of indigenous communities.

In more recent years, a greater level of public interest and global opposition to land conversion means that increasingly, these local land disputes are fought out in the context of global concerns over the threat that the oil palm industry poses to lands considered to be of "high ecological value". In the context of Africa, particular concerns often focus on the impact of palm oil expansion on the habitats of iconic species, in particular, that of the great apes (Meijaard et al., 2020; Wich et al., 2014; Humle & Matsuzawa, 2004). Indeed, it is the question of the environmental impacts of palm oil expansion that is likely to play a key role in shaping the future of the industry in Africa.

It is undoubtedly the case that the scale of damage linked to palm oil has come to be a key feature of the industry, alongside its global expansion. This has prompted a significant backlash to the industry, particularly in countries located in the Northern Hemisphere. This has led to measures such as a European Union (EU) moratorium on biofuel targets and palm oil import restrictions. For its part, the industry may point to oil palm productivity, meaning that less land is required to produce palm oil than is the case for any of the alternative oil crops. They also point to the industry's efforts to improve its sustainability via the Roundtable on Sustainable Palm Oil (RSPO) (Parish et al., 2021). Albeit that, claims regarding land use seldom take into account the environmental value of the land used for different oils or of the effectiveness of initiatives such as the RSPO (Afrizal et al., 2022; Kusumaningtyas, 2017; Ruysschaert & Salles, 2016). The environmental issues associated with oil palm production include (1) deforestation and biodiversity loss, (2) greenhouse gas emissions, in addition to (3) soil erosion and water pollution.

While by no means as dramatic as the level of forest loss due to oil palm conversion in Southeast Asia (Meijaard et al., 2020), Africa lost about 2.8 million hectares of forest cover between 2000 and 2012 due to oil palm expansion (Hansen et al., 2013). This represents a loss of about 7 percent of the total forest area in the continent. Deforestation not only reduces carbon storage capacity but also affects the habitat and survival of many plant and animal species. For instance, oil palm plantations have been linked to the decline of endangered species such as chimpanzees, gorillas, elephants, and forest antelopes in Central and West Africa (Meijaard et al., 2020). Moreover, deforestation can also alter the hydrological cycle and increase the risk of flooding and drought.

Another environmental impact of palm oil production in Africa is greenhouse gas emissions. Oil palm plantations emit carbon dioxide (CO_2) through land use change, biomass burning, fertiliser use, and processing activities. According to a report by the International Union for Conservation of Nature (IUCN), oil palm plantations in Africa emit about 0.4 gigatons of CO_2 per year, which is equivalent to about 1 percent of the global total. Oil palm plantations can also release methane (CH4) and nitrous oxide (N_2O), which are more potent greenhouse gases than CO_2. For example, a study by Kritee et al. (2015) found that oil palm plantations in Nigeria emitted about 0.8 megatons of methane per year, which is comparable to the emissions from rice paddies. The implications of deforestation and the related changes in hydrology are particularly severe in the case of tropical peatlands, of which Africa's Congo Basin possesses some of the world's largest intact areas. If converted for agricultural use, changes in the hydrology of such locations have been found to lead to significant releases of greenhouse gases previously stored in the soil. This would effectively lead to the possible transformation of globally important carbon sinks such as the Congo into carbon-emitting regions (Page et al., 2022).

Oil palm plantations can degrade soil quality and fertility by removing organic matter and nutrients from the land. This can lead to soil erosion and nutrient leaching, which can affect the productivity and sustainability of the crop. Moreover, oil palm plantations can also pollute water resources by discharging wastewater, pesticides, and fertilisers into rivers and streams. This can contaminate the water quality and affect aquatic life and human health. For instance, a study by Obidzinski et al. (2012) discovered that oil palm farms in Cameroon contributed to high levels of nitrogen, total suspended solids, and biochemical oxygen demand in water bodies. Overall, social change, climate change impacts, and land resource constraints constitute both direct and indirect effects of the socio-ecological transformation of the oil palm industry. Socio-ecological transformation is both a necessary and sufficient condition for sustainable development in the era of economic and environmental changes (Olayide, 2018).

The abuse of agrochemicals is one example of how oil palm cultivation has wider negative environmental externalities that can lead to socioeconomic issues beyond the industry itself and the locations in which it is based. Studies conducted in Indonesia and Uganda discovered negative effects on the local fishing sector and water quality (Ssemmanda & Opige, 2019; Fearnside, 1997).

Towards a sustainable business model for socio-ecological transformation of the palm oil industry in Africa

Despite the widespread evidence that palm oil does indeed improve income and that this income benefits a very wide range of actors involved in oil palm assemblages, the successful development of the oil palm industry in Africa's suitable growing regions faces multiple challenges and difficulties. Of particular note is the question of how to develop the industry in the face of a growing backlash and demands that the industry become ever more sustainable. In some cases, these challenges also concern the practicalities of developing a new or expanded palm oil industry from what remains an exceptionally low base. The critical issues that the industry faces in Africa are numerous and will be examined below.

Mobilising finance

The oil palm is a long-term crop. While the ongoing maintenance costs associated with the crop are regarded as manageable, establishment costs are significant. Investment is required to clear, drain, and/or irrigate sites of a realistic size, acquire planting material, as well as to cover current costs such as labour, agrochemicals, and fertiliser. These costs are particularly onerous in the time between planting and first harvest, which typically ranges from 3 to 5 years. Besides the costs of these investments, the perceived opportunity costs of opting to forego other income opportunities on these lands must also be taken into account (although this can be mitigated through timber sales and the cultivation of secondary crops where the expertise and opportunity to produce these exist).

Naturally, there is a perception of risk that may accompany a decision to embark on such a long-term venture. In Southeast Asia in particular, these issues were to some extent addressed through the development of parastatals, such as FELDA in Malaysia and outgrower schemes. These arrangements provided investment funds and oversee much of the governance and day-to-day management of small farm palm oil production via public–private partnerships, whereby large companies' access to land and permissions to cultivate palm oil were linked to their willingness to participate in the outgrower schemes, most famously in the form of the Indonesian "nucleus and plasma" model (Asian Agri, 2018).

While there is no doubt that such programmes were very influential in the early development of the industry in that region, the extent and distribution of benefits within this system have often been called into question, with concerns raised about the extent of the fees charged and prices offered to smallholders involved in these schemes. More recently, Indonesia has seen growth in the so-called independent smallholder sector, particularly in the form of a "small plantation" sector made up of small, locally-owned commercial plantations (Nashr et al., 2021; Jelsma et al., 2017). In other countries, smallholders have always had more control and greater power in the oil palm assemblage. There can be little doubt that the particular Malaysian and Indonesian "tied" smallholder system evolved in large part because of the degree of control exercised by the state in both countries.

In Africa, the challenges may be twofold: to create a viable financing model that generates adequate investment, while at the same time facilitating local participation and local entrepreneurship in a context where governance and the power of the state to mobilise such resources are limited.

Prioritising smallholders

Above and beyond the availability of finance, nations with an interest in expanding palm oil businesses require appropriate procedures to engage smallholders. It has generally been observed that smallholder oil palm cultivation in diverse landscapes can be more environmentally benign than large-scale plantations, in addition to ensuring a better distribution of benefits and socioeconomic advantages (Potter, 2015; Feintrenie et al., 2010; McCarthy, 2010; Rist et al., 2010). From a global point of view, smallholder production is often regarded as being more palatable than large-scale producers as it is perceived as having a more genuine development merit in producing nations. It is perhaps for these reasons that, overall, palm oil production by smallholders appears to currently be in the ascendancy.

At the same time, however, smallholder production is strongly implicated in unsustainable industry practices, lower productivity, and quality issues. Without the right policies and technical support in place, it is not necessarily more rainforest-preserving or economically and socially sustainable (Kubitza et al., 2018a, 2018b). Furthermore, being a smallholder does not necessarily protect rural households from exploitation. Smallholder palm oil producers are particularly vulnerable to asymmetrical market relations, problems with accessing transport, price instability, as well as reliable access to good quality inputs. A healthy, vibrant, and sustainable smallholder sector is not likely to emerge spontaneously. It requires significant regulatory, financial, and policy intervention. Policies on secure land titles, access to credit, and market data, in addition to technical support to enhance yields and improve environmental performance, are all critical to the expansion of smallholder palm oil development (Bennett et al., 2019).

To address these challenges, more effort is required to promote fair trade and ensure shared prosperity in the palm oil industry. For instance, the development of sustainable finance mechanisms to support smallholder farmers and the promotion of community-based land tenure systems are needed to ensure that local communities benefit from palm oil production. Much will depend on developing a workable model for the industry in Africa and addressing the problems of smallholders without overly reducing their bargaining power. A recent study conducted in Ghana suggests that the contract model of farming, which has been successful in supporting smallholders in other agricultural activities in Africa, may offer a possible option here. Smallholder oil palm farmers who have contracts to provide palm fruits may fare significantly better than those who do not (Ruml & Qaim, 2020). This study supports the finding that significant intervention by the state—which acts as a guarantor and promoter of such programmes, alongside the implementation of effective support policies to ease the availability of credit and technical assistance—can address some of the concerns linked to a sustainable business model for smallholder farmers in Africa.

Increasing oil palm productivity

There exists a yield gap in palm oil production, especially in Africa. Less than half of the potential 8 tonnes per hectare per year is currently obtained (Corley & Tinker, 2016). However, by raising oil palm yields or by extending the oil palm area, the rising worldwide demand can be satisfied. More agronomic practices (seeding rate, fertiliser use, and water management) and enhanced extension services would be needed to close the yield gap and aid expansion on a broader scale (Darras et al., 2019). Similarly, oil palm variants that are highly prolific and more resilient to altitude and climate challenges may be developed with the use of contemporary breeding techniques (Corley et al., 2018).

Forest protection and land property rights

It is imperative to protect the forest and ensure property rights. Clear property rights for agricultural land are just as crucial to reducing deforestation as clearly defined property rights for forest land. With the aid of data from Indonesia, Kubitza et al. (2018b) demonstrated how formal land titles can boost crop output while lowering farmers' propensity to clear more forest land for cultivation. Farmers close to the forest boundaries, however, rarely have access to land titles because of policy limitations, and thus they are more likely to advance into the forest to increase crop production (Kubitza et al., 2018b). To promote more sustainable land resource development in Africa, it could be necessary to reevaluate current land rules and compensation rates.

Mosaic scenery

Land-sharing strategies can frequently be useful complementing tools, even though land-sparing methods with clearly delineated intensive agriculture and virgin forests have an important role to play in protecting biodiversity and other ecosystem services (Mertz & Mertens, 2017). The creation of mosaic landscapes made up of a variety of agricultural and agroforestry plots, forest patches, and other natural landscape aspects is required to combine land-saving and land-sharing strategies (Grass et al., 2020). Adding trees and other natural components to oil palm plantations may increase biodiversity significantly while just slightly reducing oil production per unit of land. It is expedient to create mosaic landscapes that can balance economic, social, and environmental goals and create policies to put them into practice on a larger scale (Mertz & Mertens, 2017).

Fair trade, shared prosperity and just transition

To ensure fair trade, shared prosperity and just transition policies are required to be implemented to promote sustainable palm oil production. This will involve the creation of a sustainability standard and a certification system for palm oil. These policies in place will ensure that palm oil production is environmentally sustainable and socially responsible. A just transition in the sector is needed so as to ensure the participation of smallholder farmers in the value chain guarantees access to opportunities on fair and equitable terms. For many African economies, fair trade, shared prosperity, and a just transition are important considerations; however, implementation is still being debated for many of these nations. While they are being made, there are also challenges, such as the lack of transparency and accountability in the palm oil supply chain, that need to be addressed to ensure that workers are treated fairly and that environmental and social standards are being met.

Research, infrastructure, and technological advancement

Inadequate research into production and processing is a major bane to the advancement of the oil palm industry in Africa. Despite the fact that the palm tree is indigenous to Africa, other nations and regions (such as Malaysia, Indonesia, and Latin

America) that imported germplasm from Africa have advanced in terms of production and value addition through adaptable research, infrastructure, and technological advancement, while Africa still lags behind. Africa can re-emerge by taking advantage of the convergence of technology and through enhancing its infrastructure.

Conclusion: Some reflections on the emerging palm oil bio-circular economy in Africa

The palm oil sector is a potentially valuable means for transforming economies in Africa. A very large proportion of Africa's economy is agriculture-based. The agriculture sector provides food, fibre, feed, employment, opportunities for women's empowerment, agri-food systems, climate resilience, and sustainability. Moreover, by its very nature and mode in Africa, changes in agricultural income and practice for better or worse have a disproportionate impact on those with the lowest income. For these reasons, the agrarian sector in Africa is recognised as a critical sector for transforming livelihoods for shared prosperity as envisioned by the Comprehensive Africa Agriculture Development Programme (CAADP) which has a focus on economic opportunities, poverty eradication, and shared prosperity (African Union Commission, 2015).

Other sectors identified as having high potential for generating economic growth, such as petroleum and mineral exploration, have historical and significant negative economic, social, and environmental externalities (an aspect of Africa's "resource curse"). For instance, the mono-product of the crude oil economy has impacted the environment negatively, resulting in militancy and impoverishment in Nigeria, which is one of the largest producers and exporters of crude oil in Africa. The environmental pollution of crude oil production has taken a significant toll on livelihoods and the economy there. Arguably, these repercussions are unavoidable, although they could be reduced through better environmental regulation. Overall, however, it is widely understood that while agriculture has climate change impacts and land resource implications, the potential for environmental harm reduction and elimination is much greater, including in the oil palm industry.

The emerging oil palm industry is both lucrative and is one in which Africa, along with other countries in the tropical belt, has a range of comparative advantages. If properly managed, the palm oil industry has the potential to contribute to the development of a bio-circular economy, based on the long-recognised and widely appreciated versatility and utility of the oil palm tree and its products. Besides the oil itself, there are multiple uses for the plant (trunk, leaves, fruits, kernel, etc.), with significant progress also being made in the use of the by-products of the industry, creating opportunities to reduce waste. The utility of these various new and existing products is recognised in multiple sectors including pharmaceuticals, tourism, road transport and aviation, housing, and the culinary world. Hence, the emerging palm oil industry in Africa offers the potential to form the basis of a genuine bio-circular economy, offering prospects for livelihood enhancement and simultaneous environmental bioremediation.

While the industry in core producing countries boasts some very large corporate and government-linked commercial entities, the global palm oil bio-economy involves a large range of actors, many of whom operate at a significantly smaller scale. This is particularly promising in the context of discussions concerning socio-ecological transformation and sustainability. There is currently no cartel in the African oil palm industry, which is equivalent to what exists in the petroleum industry. Indeed, efforts to centralise control of the industry have struggled to gain traction. While in some instances this may result in what may be problematic outcomes from a sustainability outlook—take, for example, issues with the implementation of the RSPO and competing sustainability standards—at the same time, this provides scope and opportunity for new players, innovation, and even products to thrive. The emerging palm oil bio-economy provides a potentially vibrant platform for competitiveness, entrepreneurship, and technological innovation.

However, while palm oil can potentially be leveraged for socio-ecological transformation, modifying the industry into a more "bio-circular" model for production, consumption, and trade requires significant intervention. As we have already seen, two critical factors (finance and know-how) are integral to the way that global and national oil palm assemblages have been shaped. It was the research and development investment of Malaysia that paved the way for the country's early leadership in the palm oil boom. Equally, it was the central role envisioned for large industrial-scale mills that presaged the pivotal position of larger corporations in the Southeast Asian palm oil industry. The development of the industry has also required significant state intervention, involving government policies, regulations, and practices at all levels—local, national, and international. However, this book suggests the specific shape of resultant national palm oil assemblages exhibit significant variation.

The industry's history suggests that the government and private sector have played symbiotic roles in the expansion and transformation of the industry, as well as in the adoption of new technologies and business models. As palm oil has been pegged as the renewable oil for the emerging bio-circular economy in Africa, we would suggest that, regardless of the possible ideological squeamishness of market economists, the successful development of a bio-circular palm oil economy would require a similar strong state-private intervention, including the implementation of effective support policies (secure land titles, access to credit, and technical support) through which smallholder yield gaps might be closed, saving the rainforests, and promoting the sustainability of the economy, society, and environment.

As the amount of available suitable land for oil palm development is further reduced, pressure on African states to support the conversion of large land areas for this end is almost inevitable and is likely to increase (Pirker et al., 2016). Therefore, it is likely that the threat the oil palm industry poses to lands considered to be of "high ecological value" will increase disproportionately. There is an irony here in that such areas are likely to be seen as suitable for the industry and targeted for use by industry actors specifically because of the features that render them of high ecological value—they are sparsely populated and offer relatively limited local opposition or problems with prior landowners.

As the discussion above illustrates, the undoubted potential that the industry offers to deliver on core objectives in relation to agrarian and rural development is something that no African government can be reasonably expected to ignore. The critical question concerns the manner in which African palm oil is developed and the resources that can be mobilised to ensure that this development is undertaken in ways that are consistent with the global desire to protect what many regard as global public goods. Whether such a characterisation is fair or not is debatable; however, what is absolutely clear is that such goods would be significantly impacted by unregulated palm oil expansion. The question for African countries is not so much if, but how palm oil expansion can be managed. Similarly, the question for those in the international community should be how, and not if they should support those efforts. In this regard, the emerging African economies would be based on the bio-circular palm oil industry, and not the fossil fuel industry.

Reference list

African Union Commission. (2015). *Agenda 2063: The Africa we want* (pp. 1–172).

Afrizal, A., Hospes, O., Berenschot, W., Dhiaulhaq, A., Adriana, R., & Poetry, E. (2022). Unequal access to justice: An evaluation of RSPO's capacity to resolve palm oil conflicts in Indonesia. *Agriculture and Human Values*, *40*, 291–304. https://doi.org/10.1007/s10460-022-10360-z.

Aghalino, S. O. (2000). British colonial policies and the oil palm industry in the Niger Delta region of Nigeria, 1900–1960. *African Study Monographs*, *21*(1), 19–33. https://doi.org/10.14989/68190.

Ahmed, A., Dompreh, E., & Gasparatos, A. (2019). Human wellbeing outcomes of involvement in industrial crop production: Evidence from sugarcane, oil palm and jatropha sites in Ghana. *PLOS One*, *14*(4), e0215433. https://doi.org/10.1371/journal.pone.0215433.

Asian Agri. (2018, May 25). *Indonesia's plasma farmer scheme explained*. https://www.asianagri.com/en/media-publications/articles/indonesia-s-plasma-farmer-scheme-explained.

Awusabo-Asare, K., & Tanle, A. (2008). Eking a living: Women entrepreneurship and poverty reduction strategies: The case of palm kernel oil processing in the Central Region of Ghana. *Norwegian Journal of Geography*, *62*(3), 149–160. https://doi.org/10.1080/00291950802335525.

Balde, B., Diawara, M., Rossignoli, C., & Gasparatos, A. (2019). Smallholder-based oil palm and rubber production in the forest region of Guinea: An exploratory analysis of household food security outcomes. *Agriculture*, *9*(2), 41. https://doi.org/10.3390/agriculture9020041.

Basiron, Y. (2007). Palm oil production through sustainable plantations. *European Journal of Lipid Science and Technology*, *109*(4), 289–295. https://doi.org/10.1002/ejlt.200600223.

Bennett, A., Ravikumar, A., McDermott, C., & Malhi, Y. (2019). Smallholder oil palm production in the Peruvian Amazon: Rethinking the promise of associations and partnerships for economically sustainable livelihoods. *Frontiers in Forests and Global Change*, 2. https://doi.org/10.3389/ffgc.2019.00014.

Byerlee, D., Falcon, W. P., & Naylor, R. (2017). *The tropical oil crop revolution: Food, feed, fuel, and forests*. Oxford University Press.

Castellanos-Navarrete, A., Tobar-Tomás, W. V., & López-Monzón, C. E. (2019). Development without change: Oil palm labour regimes, development narratives, and disputed moral economies in Mesoamerica. *Journal of Rural Studies, 71*, 169–180. https://doi.org/10.1016/j.jrurstud.2018.08.011.

Colchester, M. (2010). *Palm oil and indigenous peoples of South East Asia: Land acquisition, human rights violations and indigenous peoples on the palm oil frontier* (pp. 1–22). Forest Peoples Programme.

Corley, R. H. V., Rao, V., Palat, T., & Praiwan, T. (2018). Breeding for drought tolerance in oil palm. *Journal of Oil Palm Research, 30*(1), 26–35. https://doi.org/10.21894/jopr.2017.0011

Corley, R. H. V., & Tinker, P. B. H. (2016). *The oil palm*. Wiley-Blackwell.

Cramb, R. A., & Curry, G. N. (2012). Oil palm and rural livelihoods in the Asia-Pacific region: An overview. *Asia Pacific Viewpoint, 53*(3), 223–239. https://doi.org/10.1111/j.1467-8373.2012.01495.x.

Darras, K. F. A., Corre, M. D., Formaglio, G., Tjoa, A., Potapov, A., Brambach, F., Sibhatu, K. T., Grass, I., Rubiano, A. A., Buchori, D., Drescher, J., Fardiansah, R., Hölscher, D., Irawan, B., Kneib, T., Krashevska, V., Krause, A., Kreft, H., Li, K., … Veldkamp, E. (2019). Reducing Fertilizer and Avoiding Herbicides in Oil Palm Plantations—Ecological and Economic Valuations. *Frontiers in Forests and Global Change, 2*. https://doi.org/10.3389/ffgc.2019.00065

Edwards, R. B. (2019a). *Export agriculture and rural poverty: Evidence from Indonesian palm oil*. Dartmouth College.

Edwards, R. B. (2019b). *Spillovers from agricultural processing*. Darthmouth College.

Euler, M., Krishna, V., Schwarze, S., Siregar, H., & Qaim, M. (2017). Oil palm adoption, household welfare, and nutrition among smallholder farmers in Indonesia. *World Development, 93*, 219–235. https://doi.org/10.1016/j.worlddev.2016.12.019.

FAO. (2020). *FAOSTAT*. Food and Agriculture Organisation of the United Nations. https://www.fao.org/faostat/en/#home.

Fearnside, P. M. (1997). Transmigration in Indonesia: Lessons from its environmental and social impacts. *Environmental Management, 21*(4), 553–570. https://doi.org/10.1007/s002679900049.

Feintrenie, L., Chong, W. K., & Levang, P. (2010). Why do farmers prefer oil palm? Lessons learnt from Bungo district, Indonesia. *Small-Scale Forestry, 9*(3), 379–396. https://doi.org/10.1007/s11842-010-9122-2.

Fitzpatrick, D. (1997). Disputes and pluralism in modern Indonesian land law. *Social Science Research Network, 22*, 171–212.

Gatto, M., Wollni, M., Asnawi, R., & Qaim, M. (2017). Oil palm boom, contract farming, and rural economic development: Village-level evidence from Indonesia. *World Development, 95*, 127–140. https://doi.org/10.1016/j.worlddev.2017.02.013.

Grass, I., Kubitza, C., Krishna, V. V., Corre, M. D., Mußhoff, O., Pütz, P., Drescher, J., Rembold, K., Ariyanti, E. S., Barnes, A. D., Brinkmann, N., Brose, U., Brümmer, B., Buchori, D., Daniel, R., Darras, K. F. A., Faust, H., Fehrmann, L., Hein, J., & Hennings, N. (2020).

Trade-offs between multifunctionality and profit in tropical smallholder landscapes. *Nature Communications, 11*(1), 1186. https://doi.org/10.1038/s41467-020-15013-5.

Haiven, M. (2022). *Palm oil: The grease of empire*. Pluto Press.

Hamann, S. (2017). Agro-industrialisation and food security: Dietary diversity and food access of workers in Cameroon's palm oil sector. *Canadian Journal of Development Studies, 39*(1), 72–88. https://doi.org/10.1080/02255189.2017.1336079.

Hansen, M. C., Potapov, P. V., Moore, R., Hancher, M., Turubanova, S. A., Tyukavina, A., Thau, D., Stehman, S. V., Goetz, S. J., Loveland, T. R., Kommareddy, A., Egorov, A., Chini, L., Justice, C. O., & Townshend, J. R. G. (2013). High-Resolution global maps of 21st-century forest cover change. *Science, 342*(6160), 850–853. https://doi.org/10.1126/science.1244693.

Humle, T., & Matsuzawa, T. (2004). Oil palm use by adjacent communities of chimpanzees at Bossou and Nimba Mountains, West Africa. *International Journal of Primatology, 25*(3), 551–581. https://doi.org/10.1023/b:ijop.0000023575.93644.f4.

Jelsma, I., Schoneveld, G. C., Zoomers, A., & van Westen, A. C. M. (2017). Unpacking Indonesia's independent oil palm smallholders: An actor-disaggregated approach to identifying environmental and social performance challenges. *Land Use Policy, 69*, 281–297. https://doi.org/10.1016/j.landusepol.2017.08.012.

Kritee, K., Nair, D., Tiwari, R., Rudek, J., Ahuja, R., Adhya, T., Loecke, T., Hamburg, S., Tetaert, F., Reddy, S., & Dava, O. (2015). Groundnut cultivation in semi-arid peninsular India for yield scaled nitrous oxide emission reduction. *Nutrient Cycling in Agroecosystems, 103*, 115–129.

Krishna, V., Euler, M., Siregar, H., & Qaim, M. (2017). Differential livelihood impacts of oil palm expansion in Indonesia. *Agricultural Economics, 48*(5), 639–653. https://doi.org/10.1111/agec.12363

Kubitza, C., Krishna, V. V., Alamsyah, Z., & Qaim, M. (2018a). The economics behind an ecological crisis: Livelihood effects of oil palm expansion in Sumatra, Indonesia. *Human Ecology, 46*(1), 107–116. https://doi.org/10.1007/s10745-017-9965-7.

Kubitza, C., Krishna, V. V., Urban, K., Alamsyah, Z., & Qaim, M. (2018b). Land property rights, agricultural intensification, and deforestation in Indonesia. *Ecological Economics, 147*, 312–321. https://doi.org/10.1016/j.ecolecon.2018.01.021.

Kusumaningtyas, R. (2017). *External concern on the ISPO and RSPO certification schemes*. Profundo.

McCarthy, J. F. (2010). Processes of inclusion and adverse incorporation: Oil palm and agrarian change in Sumatra, Indonesia. *The Journal of Peasant Studies, 37*(4), 821–850. https://doi.org/10.1080/03066150.2010.512460.

Meijaard, E., Brooks, T. M., Carlson, K. M., Slade, E. M., Garcia-Ulloa, J., Gaveau, D. L. A., Lee, J. S. H., Santika, T., Juffe-Bignoli, D., Struebig, M. J., Wich, S. A., Ancrenaz, M., Koh, L. P., Zamira, N., Abrams, J. F., Prins, H. H. T., Sendashonga, C. N., Murdiyarso, D., Furumo, P. R., ... Sheil, D. (2020). The environmental impacts of palm oil in context. *Nature Plants, 6*(12), 1418–1426. https://doi.org/10.1038/s41477-020-00813-w.

Mertz, O., & Mertens, C. F. (2017). Land sparing and land sharing policies in developing countries – drivers and linkages to scientific debates. *World Development, 98*, 523–535. https://doi.org/10.1016/j.worlddev.2017.05.002

Nashr, F., Putri, E. I. K., Dharmawan, A. H., & Fauzi, A. (2021). The sustainability of independent palm oil smallholders in multi-tier supply chains in East Kalimantan, Indonesia. *International Journal of Sustainable Development and Planning, 16*(4), 771–781. https://doi.org/10.18280/ijsdp.160418.

Naylor, R. L., Higgins, M. M., Edwards, R. B., & Falcon, W. P. (2019). Decentralisation and the environment: Assessing smallholder oil palm development in Indonesia. *Ambio, 48*(10), 1195–1208. https://doi.org/10.1007/s13280-018-1135-7.

Nkongho, R. N., Feintrenie, L., & Levang, P. (2014). *The non-industrial palm oil sector in Cameroon* (pp. 1–22). Center for International Forestry Research. https://www.cifor.org/publications/pdf_files/WPapers/WPaper139Nkongho.pdf.

O'Reilley, P., & Varkkey, H. (2020). Palm oil governance in different locations: Using the assemblage approach to understand a "complex" sector. *International Review of Modern Sociology, 46*(1–2), 1–17.

Obado, J., Syaukat, Y., & Siregar, H. (2009). The impacts of export tax policy on the Indonesian crude palm oil industry. *International Society for Southeast Asian Agricultural Sciences Journal, 15*(2), 107–119.

Obidzinski, K., Andriani, R., Komarudin, H., & Andrianto, A. (2012). Environmental and social impacts of oil palm plantations and their implications for biofuel production in Indonesia. *Ecology and Society, 17*(1). https://doi.org/10.5751/es-04775-170125.

Ohimain, E. I., & Izah, S. C. (2014). Energy self-sufficiency of smallholder oil palm processing in Nigeria. *Renewable Energy, 63*, 426–431. https://doi.org/10.1016/j.renene.2013.10.007.

Olayide, O. E. (2018). Understanding the context of climate change and socio-ecological transformation in Nigeria. In *Climate Change and Socio-ecological Transformation in Nigeria: Challenges and Opportunities* (pp. 1–16). Friedrich-Ebert-Stiftung. ISBN 978-987-971-018-8.

Ordway, E. M., Naylor, R. L., Nkongho, R. N., & Lambin, E. F. (2019). Oil palm expansion and deforestation in Southwest Cameroon associated with proliferation of informal mills. *Nature Communications, 10*(1). https://doi.org/10.1038/s41467-018-07915-2.

Oxford Analytica. (2016). Investment in Africa's palm oil sector will grow. *Expert Briefings.* https://doi.org/10.1108/OXAN-DB212084.

Page, S., Mishra, S., Agus, F., Anshari, G., Dargie, G., Evers, S., Jauhiainen, J., Jaya, A., Jovani-Sancho, A. J., Laurén, A., Sjögersten, S., Suspense, I. A., Wijedasa, L. S., & Evans, C. D. (2022). Anthropogenic impacts on lowland tropical peatland biogeochemistry. *Nature Reviews Earth & Environment, 3*(7), 426–443. https://doi.org/10.1038/s43017-022-00289-6.

Parish, F., Afham, A., & Yew, S. Y. (2021). Role of the Roundtable on Sustainable Palm Oil (RSPO) in tropical peatland management. In M. Osaki, N. Tsuji, N. Foead, & J. Rieley (Eds.), *Tropical Peatland Eco-management* (pp. 509–533). Springer.

Pirker, J., Mosnier, A., Kraxner, F., Havlík, P., & Obersteiner, M. (2016). What are the limits to oil palm expansion? *Global Environmental Change, 40*, 73–81. https://doi.org/10.1016/j.gloenvcha.2016.06.007.

Potter, L. (2015). *Managing oil palm landscapes: A seven-country survey of the modern palm oil industry in Southeast Asia, Latin America and West Africa* (pp. 1–144). Centre for International Forestry Research.

Pye, O. (2019). Agrarian Marxism and the proletariat: A palm oil manifesto. *The Journal of Peasant Studies, 48*(4), 1–20. https://doi.org/10.1080/03066150.2019.1667772.

Rist, L., Feintrenie, L., & Levang, P. (2010). The livelihood impacts of oil palm: Smallholders in Indonesia. *Biodiversity and Conservation, 19*(4), 1009–1024. https://doi.org/10.1007/s10531-010-9815-z.

Rival, A., & Levang, P. (2014). *Palms of controversies.* Centre for International Forestry Research.

Ruml, A., & Qaim, M. (2020). Effects of marketing contracts and resource-providing contracts in the African small farm sector: Insights from oil palm production in Ghana. *World Development, 136*, 105110. https://doi.org/10.1016/j.worlddev.2020.105110.

Ruysschaert, D., & Salles, D. (2016). The strategies and effectiveness of conservation NGOs in the global voluntary standards: The case of the Roundtable on Sustainable Palm Oil. *Conservation and Society, 14*(2), 73–85. https://doi.org/10.4103/0972-4923.186332.

Santika, T., Wilson, K. A., Budiharta, S., Law, E. A., Poh, T. M., Ancrenaz, M., Struebig, M. J., & Meijaard, E. (2019). Does oil palm agriculture help alleviate poverty? A multidimensional counterfactual assessment of oil palm development in Indonesia. *World Development, 120*, 105–117. https://doi.org/10.1016/j.worlddev.2019.04.012.

Ssemmanda, R., & Opige, M. (2019). *Impacts and implications of oil palm on landscapes and livelihoods in Uganda's Lake Victoria islands—An overview of recent research* (pp. 1–6). Tropenbos International.

van Noordwijk, M., Hairiah, K., & Weise, S. (2001). *Sustainability of tropical land use systems following forest conversion* (pp. 1–27). International Centre for Research in Agroforestry.

Varkkey, H., & O'Reilly, P. (2020). Sociopolitical responses toward transboundary haze: The oil palm in Malaysia's discourse. In S. Kukreja (Ed.), *Southeast Asia and Environmental Sustainability in Context* (pp. 65–88). Lexington Books.

Wich, S. A., Garcia-Ulloa, J., Kühl, H. S., Humle, T., Lee, J. S. H., & Koh, L. (2014). Will oil palm's homecoming spell doom for Africa's great apes? *Current Biology, 24*(14), 1659–1663. https://doi.org/10.1016/j.cub.2014.05.077.

12

THE GOLDEN CROP THROUGH ITS ASSEMBLAGES

Understanding and reconciling national variations of palm oil governance across the tropical belt

Patrick O'Reilly, Helena Varkkey and Sarah Ali

Ideational framings of the palm oil debate

Across the entire field of discussion and analysis of the palm oil industry is the sense that palm oil generates significant and unique "questions" that producers and potential producer countries must address in their encounters with the industry. It is one that permeates across all of the studies in this volume. As we stated at the outset of this book, the framing of this question as a global one is well understood and constitutes the focus of much of the academic and policy discourse around oil palm. We suggested that there is a tendency for such debates to be dominated by a specific series of questions, one set of which addresses the theme of oil palm productivity/development. Such questions speak of the crop's ability to meet what is often described as the world's ever-growing thirst for vegetable oils (see for example Fry & Fitton, 2010; Corley et al., 2009), the observation that oil palm represents the most efficient means of producing these oils (Fry & Fitton, 2010), and the fact that the crop's biophysical characteristics mean that it is best cultivated in tropical belt countries which in many cases face significant challenges in achieving economic growth and associated socio-economic development goals (see for example Syahza et al., 2013). So axiomatic have these ideas become that it is scarcely possible to discuss the topic of palm oil expansion without tipping a metaphorical hat towards them. These ideas underpin and are reflexively supported by an ideational bias in which the world's "unquenchable thirst" for oils drives the industry via a necessity for ever-increasing expansion. In doing so, the crop is portrayed as offering substantial development benefits to countries with large agrarian populations which badly need them (see for example Saleh et al., 2018).

Alongside these, there is a second set of questions associated with issues of degradation/conservation. These themes are also informed by ideas about the crop's

DOI: 10.4324/9781003459606-15

productivity and earning potential. However, as opposed to the benefits associated with these characteristics, the focus here has been on their negative implications (see for example Meijaard et al., 2020; Syahza et al., 2020; Naylor, 2019). These form a separate and contradictory ideational bias, in which the crop's productivity and profitability create enormous incentives to drive increased production. This, in turn, requires a significant intensification of human activity in formerly "untouched" areas. In particular, it involves changes in land use that result in adverse environmental impacts. The fact that oil palm grows in areas in the tropics exacerbates this situation, creating circumstances in which weak, development-hungry states with limited capacity to exert effective control over land and environmental governance are perceived to be vulnerable to being either compelled or induced into decisions that result in the widespread conversion of large areas of land (see for example Dauvergne, 2018). Of particular concern, this tends to impact areas considered to be of high ecological value, especially hard. It is precisely those characteristics that contribute to this high value that make them particularly appealing for oil palm cultivation. The growing conditions that support their biodiversity are ideal for oil palm, and at the same time, perceptions of these areas as having low development potential have resulted in limited human activity. As a consequence, these areas are easily portrayed as being "available" or "wasted", or simply "unused" (Padfield et al., 2023; Manzo et al., 2019; McCarthy, 2010). As the chapters by Hazlewood et al., Pido et al., and Fromm et al. in this volume demonstrate, the consequences of their incorporation into the oil palm industry can result in severe, even catastrophic impacts on these areas, their biodiversity, ecosystem service provision, their role in climate regulation, and on the social and human rights of the communities who live in these places.

These biases have a significant bearing in debates and research relating to the palm oil industry, underpinning the dominating problematic of development versus the environment, a situation exacerbated by the historical evolution both of the industry itself and of research about the industry. There is in the literature a large portion of work aimed at reconciling the elements of this problematic (see, for example, Purnomo et al., 2020). We suggested in the introduction that these efforts at reconciliation enact a further bias, one that presumes that the reconciliation of the two dimensions of this binary represents the overarching theme of oil palm research and policy, and assumes that such a reconciliation exists "out there," awaiting discovery.

Chapters by Olayide and O'Reilly, Varkkey and Ali, as well as O'Reilly et al. in this volume, illustrate the trajectory of the development of the industry around the world. Under the impetus of post-colonial political imperatives, Malaysian researchers oversaw transfers of germplasm and subsequent breeding and agronomic programmes. As a result, the oil palm—a relatively minor component of colonial economic assemblages in parts of West Africa as well as of plantation agriculture elsewhere (Haiven, 2022)—was transformed into the basis of the new postcolonial oil palm assemblage, one which has—with some justification—been credited with a major role in the transformation of national economies, agrarian livelihoods, and rural landscapes in these core Southeast Asian producing states.

Given the dominance of these events in the palm oil story, it is not surprising that the ills and benefits of the industry in these countries have come to be more widely applied to the industry as a whole. In simple numerical terms, whatever happens in these nations continues to be largely responsible for the global impact of palm oil, for good and bad. For those interested in both promoting the industry and those concerned with its adverse global impacts, there are, therefore, very understandable reasons for presenting the industry in those countries as a model and an industry archetype. Or, at the very least, for justifying a focus in debates about the value and problems associated with the Southeast Asian "palm oil complex" simply because in global terms, this is the most significant form that the industry has taken. This view is not untypical; the briefest surveys of the literature on sustainable palm oil rapidly reveal a bias towards Southeast Asian circumstances and stakeholders, and within this, a penchant on the part of policymakers and researchers alike for time spent with "key decision makers" in hotel breakout rooms in Indonesia and Malaysia. The industry in Southeast Asia itself plays a significant role in maintaining and promoting its prominence in the global industry—supporting and leading international representative and regulatory bodies such as the Roundtable on Sustainable Palm Oil (RSPO) and Council of Palm Oil Producing Countries (CPOPC), and funding industry bodies and events such as the biannual International Palm Oil Congress and Exhibition (PIPOC), alongside continued investment in research and palm oil technologies.

Yet what this collection illustrates is the varied forms that industry takes in Southeast Asia itself and also globally. Much as industry players and national governments might have sought to "copy" the Malaysian and Indonesian models, in each country, local socio-cultural and historical circumstances, as well as local human and non-human entities, (re)produce unique palm oil assemblages.

Palm oil exceptionalism

The individual chapters in this book also challenge ideas about the inevitability of the relationship between the different participants in the oil palm supply chain and other entities incorporated in the oil palm assemblage, in addition to ideas about the distribution of knowledge, power, and benefits within the industry. Again, it might be suggested that both critics and advocates of the palm oil industry have tended to propagandise a particular way of doing palm oil.

Aspects of this model are described in chapters by Varkkey and Ali, as well as Khor and Tamilwanan, in this volume. It involves a focus on the organisation of the industry as components of a "value chain" that links the different actors and entities involved in cultivation, harvesting, sale, and distribution in specific ways. This way of doing palm oil tends to equate with what is described by Khor and Tamilwanan (this volume) as the "corporate-centric" model; it is large corporate entities that control investment and know-how in the industry, and consequently, the same entities tend to enjoy an outsized share of power and profits. Underpinning this model are a range of ideas about the industry and the different components within it, which provide a rationale and organisational imperative, as the biophysical characteristics of the crop demand specific growing conditions and cultivation techniques. These, in turn, require specific forms of investment in land clearance and preparation, in planting materials and inputs, and in management and marketing. The characteristics of the crop itself are portrayed as having a very strong bearing on the consequent shape these relationships take. The requirements of the oil palm tree itself and the characteristics of the fruits produced shape relationships between land, soil, and hydrological systems, as well as the wider range of human individuals and entities in this assemblage, including growers, labourers, processing mills, and facilities.

A key issue here concerns processing deadlines and the fact that fruits have to be processed within 24 hours in milling plants, which are perceived as requiring large-scale financial investment and expertise to build and maintain. These, along with other features of the industry, play an important role in the logic and ideas of necessity, which provide both the explanation and justification for the organisation of the industry. In particular, it plays a significant role in suggesting the need for the involvement and leadership of large corporate entities. Within this framework, relationships between these entities and others are framed in terms of the need for know-how and the requirement to mobilise resources on a large scale to construct mills and develop the large-scale plantations and associated infrastructure needed to ensure these mills have timely access to the volumes of fresh fruit bunches (FFBs) necessary to run them profitably. This logic also frames relationships between other entities in the assemblage. The state, for example, may assume a role in facilitating the development of these companies through its capacity to issue land concessions, provide research investment, and in its sponsorship of outgrower schemes and other measures through which rural landscapes and communities are incorporated into the oil palm assemblage. The terms under which such schemes are implemented can have the further effect of defining relationships between the state and these communities.

They also shape relations between the large palm oil companies and these communities: the former as the source of resources, knowledge, and market access: and the latter as a source of both labour and of course, the precious FFBs which feed the mills. In the language of assemblage, oil palm assemblages are territorialised in this iteration in ways that tend to fix contingent relationships between entities in configurations that benefit and confer power upon corporate actors, and at the same time, reduce the power and benefit enjoyed by others. In this context, it could be argued that efforts of the oil palm lobby in some countries and corporate players to extend their reach via global palm oil organisations constitute efforts to extend that territorialisation, and to align other entities involved in the global palm oil "boom" in a corporate-centric palm oil model (something that might be worth noting in countries considering their relationships with the global industry and with the supposedly autonomous international organisations promoted by the core countries). The consequences of these assertions are—to some extent—to divorce discussions of palm oil from similar debates concerning other crops; to create a kind of oil palm exceptionalism which suggests that palm oil production is governed by different rules. For the industry's advocates, particularly those who embrace the "corporate-centric" model, making the industry a success and securing the benefits that will flow from this success means embracing features of the corporate model.

Assemblages of desire and necessity

Intrinsic to this iteration of the oil palm assemblage are particular ideas about the palm oil industry, which serve as a rationale for the ways in which the elements of this are territorialised. A particular combination of desire and necessity is proposed. The world needs edible oils in seemingly ever-increasing abundance, and both states and rural locations in the Global South need economic growth. The oil palm offers a means to meet these needs, which in turn generates new needs and desires for land, technology, and infrastructure that can only be met through the organisation of supply chains. In the context of the industry's historical evolution, features of this supply chain have been seen to include a need for knowledge and investment at almost every step in the process, which are presented as necessitating the involvement of large corporate players as leaders and investors, states as enablers, and rural populations as compliant providers of labour and land. In more corporate-centric approaches to the development of the crop, the result is the creation of an assemblage in which large corporate entities are defined as being both essential to the provision of the investment and know-how required to facilitate the (re)production of an oil palm assemblage and to meet the needs and desires of the multiple actors involved. As is demonstrated repeatedly in this volume, these ideas are extended as a rationale for the involvement of large corporations in almost every country in which the palm oil industry operates (see Fromm et al., this volume; O'Reilly et al., this volume; Varkkey & Ali, this volume; Pido et al., this volume; Córdoba et al., this volume).

In countries where this corporate model has prevailed to the greatest extent, the result is an industry that delivers significant control to large corporate entities. In their purest form, such bodies may hold substantial power over primary production through their control of inputs and contacts with smaller growers, in milling and processing through their involvement and ownership of processing facilities, and a key role in marketing palm oil products. Such companies thus enjoy a powerful position underpinned by ideas of desire and necessity that serve to suggest that there is no alternative.

It is unsurprising, given the perceived dominance of this model, that the ideas associated with this version of oil palm have come to delimit the ideational realm in which both advocates and critics of palm oil tend to work. For some of those who advocate for the development of palm oil, the Malaysian and/or Indonesian corporate model offers a blueprint for a successful industry that incorporates a rationale and justification for the development of similar models of corporate-dominated palm oil expansion elsewhere. Equally, however, even a brief sweep of the literature on palm oil sustainability suggests that those advocating measures to reduce the crop's adverse environmental impacts are largely focussing their efforts on addressing the sustainability of large-scale palm oil production. Indeed, while some might point to some success in the significant efforts that have been made to extend the principal palm oil sustainability measure to smallholders (the RSPO standard), this primarily involves the adaptation of measures devised for large-scale producers to smallholder production. Only time will tell if the efforts and investments that have been made by multiple entities—such as the German Organisation for International Development (GIZ) in Thailand—are likely to result in take-up on a sufficient level to justify the investments involved. However, if the history of rural intervention and indeed of palm oil development itself tells us anything, it is that one-size-fits-all solutions designed for large-scale commercial agriculture may not be the best solution for small-scale producers in another time and place. What the studies in this volume suggest is that far from representing the overriding and inevitable logic of the palm oil industry, this corporate-centric model is simply one of a range of ideas about the industry that are deployed in the ongoing struggles and contestations through which different national oil palm sectors are continually being (re)produced.

We referred at the outset to the possibly negative impacts on debates and analysis that may stem from assumptions that the way an industry operates in specific national jurisdictions apply universally. Certainly, it cannot be assumed that the "industry norms" in major producing countries apply to that industry everywhere. In the case of oil palm, however, the debate does sometimes appear to have been reduced to an almost two-dimensional battle in which trade-offs between conservation and production—proposed in historically situated contexts in two Southeast Asian countries—can be understood to represent the best fit for an evolving global industry. Allied to this has been a tendency for oil palm to be the subject of very intense debates about environmental impacts which are also framed in particular national and even habitat-specific contexts. These debates, allied with the intensive and specific nature of research into the oil palm industry and its portrayal in the media, have contributed to perceptions of the crops' apparently singular capacity to produce and destroy. This adds to the sense of a product and an industry that is set apart—one that, for those who have been involved in advocating for it, seems to defy the laws of economic gravity that affect other agricultural products and markets, while its environmental costs also seem to mark it out as a crop of uniquely destructive potential.

The work in this book clearly shows that there are wide national and, in some cases, even local variations in terms of who benefits from palm oil, as well as concerning who and what bears the brunt of the social and environmental costs associated with the crop. Depending on where in the world oil palm is being culti-vated, different land titling and ownership challenges emerge. States bring different ideological, political, and administrative interests to their response to the industry, and the crop is integrated into a variety of landscapes, including dry and wet rain-forests, mires and peatlands, as well as agricultural land and defunct plantations by multiple growers and industry participants. We also know that oil palm is not unique in its destructive potential—in the southern cone of South America, for example, the large-scale production of soybean has resulted in serious degradation in water quality and biodiversity in the region and profound socio-political harm (Saguier et al., 2021; Baraibar Norberg, 2019; Siegel, 2016; Arancibia, 2013). Research into soy in this region suggests that large-scale agricultural expansion/extraction linked to any crop in areas regarded as frontiers often throws up ten-sions between large agro-industrial players, smaller producers, and traditional land users. This perhaps suggests that the "palm oil problem" is not really a palm oil problem at all. Instead, it may rather be an expression of the problems associated with the growing desire of some actors to incorporate increased areas of land into large-scale intensive monocultures associated with accelerated processes of capital accumulation in areas which had previously been regarded as of only marginal importance for agriculture and economic development.

The chapters included here demonstrate the global oil palm industry's com-plexity and diversity. Even in the core-producing countries, the picture of the palm oil assemblage is more complex and nuanced than is sometimes assumed in policy debates. In some ways, it is more reminiscent of other crops, food sup-ply chains, and farming systems. There is, for example, growing evidence that oil palm smallholders in Indonesia are subject to similar cost price squeezes that impact smaller-scale producers of other crops and intensive monoculture com-modities. In addition, as the work by O'Reilly et al. (this volume) illustrates, even in Indonesia, communities and individuals have considerable scope to shape the industry in their specific localities. In effect, they make palm oil policy on the ground, utilising that country's dense and overlapping administrative and govern-ance arrangements to maximise the benefits they accrue from their involvement via endogenous responses to the industry. This is consistent with recent work by the Centre for International Forestry Research (CIFOR) (Andrianto et al., 2019; Jelsma et al., 2017), which has increasingly, if perhaps belatedly, recognised the important role and growing presence of the so-called independent smallholder sector in the development of the palm oil industry in Indonesia over the past two decades.

There is also ample evidence that, as with other food products, palm oil mar-kets are subject to extensive state interventions and subsidies. In Mexico (Pischke, this volume), Honduras (Fromm et al., this volume), and Colombia (Furumo, this volume), palm oil production was specifically targeted as a means of substituting for imports, a fact that also shaped state policy towards production in Thailand for many years. We also see many examples of interventions by governments in Malaysia and Indonesia in the form of subsidies and stock management to control palm oil prices. Indeed, the chapters in this volume demonstrate the ubiquitous presence of the state in shaping national palm oil assemblages, both through action (such as incentives, market regulation, and other measures) and inaction (such as selective overlooking of industry transgressions and failures to implement environmental and human rights laws). Far from being the case that the "corporate-centric" model of the industry represents the norm, the message these studies convey is that different ways of doing palm oil exist. Taking Africa as an example, Olayide and O'Reilly (this volume) highlight the artisanal nature of palm oil production and the existence of indigenous niche markets for hand-pressed palm oil in this region. Across the chapters, we show that while the crop's requirements remain the same and while the various stages in moving the product along a supply chain from land preparation and planting through to production, sales, marketing, and distribution may still need to be taken, the way that these steps are undertaken and by whom exhibits a substantial degree of national variation. Countries may share ideas, technologies, concerns, and expectations. However, the ways in which they make use of ideas, address concerns, and respond to expectations reflect the unique historical, political, social, and cultural circumstances of different nations, regions, and communities. This is reflected in the very different ways that the industry has been understood in different states and regions. For example, while the concept of the developmentalist state may more accurately inform the industry's evolution in Southeast Asia (Vu, 2007), explanations of (neo-)extractivism may better reflect the historical emergence of the industry in Latin America (Córdoba et al., 2017; McKay, 2017).

For this reason, the value of generalised conceptualisations and statements about the palm oil industry can only ever be limited. In some instances, academic analysis and public discourse concerning the industry associated with the crop and its costs and benefits—if they draw on reflections of the industry in only Malaysia, for example—tend to miss the point entirely. The subsequent shape of the oil palm assemblage there was not determined by the characteristics of the oil palm itself, nor does it simplistically constitute the necessary "blueprint" for global oil palm. Rather, the oil palm assemblage in that country exemplifies the fact that oil palm assemblages in different countries are always and inevitably the result of a historically unique series of events through which a range of actors and non-human entities are brought into a specific configuration of relationships, which in turn reflect the influence exerted by these actors and agencies in shaping the relations between elements of the assemblage. True, there are many examples of public and private texts and utterances that suggest otherwise and provide narratives in which Malaysia is depicted as founding a global industry, implying the merits of the Malaysian palm oil model. However, it may in fact be more a case of how Malaysia developed a crop that formed the basis of a Malaysian industry. While Malaysia's success was such that it encouraged the widespread adoption of the crop in different countries, it might be better to recognise that this resulted in the subsequent creation of a wide range of other national palm oil assemblages, each with characteristics that reflect both the universal constraints imposed by the crop itself, as well as diverse national circumstances and interests. The actual shape of these relationships in any state is never simply the outworking of the practices and arrangements demanded by the crop or the industry. They are only shaped by these things as they are also the embodiment of assemblage practices as they occur in specific times and spaces.

Informing ways forward

In the introduction, we outlined a series of questions which we hoped this collection would allow us to answer:

1 How do different governance arrangements enable and legitimise expansion?
2 How do power differentials affect oil palm governance?
3 How does governance enable the accumulation of wealth?
4 How is conflict governed and moderated in the oil palm sector?

What this collection shows above everything is that there are no simple, generalisable answers to these questions. As we have worked through these chapters as editors and co-authors, we have become ever more aware that such questions can only ever be addressed through detailed empirical analysis that is sensitive to the fact that in the case of palm oil, similar situations in different national, state-provincial, and local contexts involving similar activities in the production of an identical crop can be worked out in radically different ways, resulting in very different results. We believe that assemblage can serve as a useful tool in this process, alerting us to the need for a careful focus on the combination of relationships between humans and non-humans, the importance of the formal and informal dimensions of governance, the respective power of the biotic and the abiotic, the fundamentally political nature of economic and agricultural policy choices, as well as the fact that these relationships are contingent, hard to predict, and not particularly stable.

While we cannot hope to provide definitive answers to the questions we set in the introduction, we can at least refer to assemblage in indicating that in different countries, the answers to these questions involve similar challenges concerning the crop itself, the processes through which land, soil, inputs, and people produce it, and the working out of the relationships between them. This latter question, concerning relationships between entities in the palm oil assemblage, offers us our entry point into understanding questions of palm oil governance.

Taking our first question, it is clear that there is massive variation in how different national governance arrangements enable and legitimise expansion. On a global scale—and certainly as promoted in the core Southeast Asian countries—the productivity of the crop and its potential to generate additional revenues occupy a strong role in formal discourse. The idea that the oil palm offers a boon to those concerned with supporting the development of rural areas—in what are often regarded as low- to lower-middle-income countries—is widely held and persists. It is not a view that is solely restricted to national, provincial, and local elites. Evidence from Brazil (Córdoba et al., this volume), Honduras (Fromm et al., this volume), Thailand (Khor & Tamilwanan, this volume), Africa (Olayide & O'Reilly, this volume), as well as Indonesia (O'Reilly et al., this volume), and Malaysia (Varkkey & Ali, this volume) indicates that the view of the crop's commercial potential is also widely shared among smallholders. This is reflected in a certain amount of grassroots enthusiasm among some, whilst being equally regarded as a threat by communities who have traditional rights of ownership in areas targeted by the industry, as well as with reticence among farmers who simply do not particularly want to become palm oil growers. However, there is far less consensus as to what this means and what it legitimises.

In the chapter on Indonesia (O'Reilly et al., this volume) and Malaysia (Varkkey & Ali, this volume), we see how strong views about national interest and ambiguous land titling emboldened development states to invest significant resources in measures that resulted in the rapid expansion of an industry dominated by corporate players. By contrast, at some points in time, in the chapters on Brazil (Córdoba et al., this volume), as well as on Thailand (Khor & Tamilwanan, this volume) and Honduras (Fromm et al., this volume), we have seen that the impetus and much of the power over palm oil expansion was vested in smaller producers, through their capacity to exert political pressure (in the case of Thailand in particular), and through forms of collective action that enabled growers to exert control over the process of palm oil expansion and even processing. Critical to this was the role and position of the state in mediating relations between other parts of the assemblage, particularly the land, growers, milling companies, and markets. It is not so much a question of whether the state will intervene and assume a role in shaping the palm oil assemblage in each country as it is a question of how, to what extent, and in whose interests.

Likewise, in relation to our second question, it is entirely clear that pre-existing power differentials in different countries have a vast role in determining the shape of the industry. In addition, these differentials are in fact more influential than the characteristics of the crop itself or the demands of the oil palm supply chain in the configuration of oil palm assemblages. In general, states with more stable political systems, more adequate representations of rural peoples, and more transparent land allocation practices appear to have had less contentious palm oil transitions. Again, this is illustrated in the cases of Thailand (Khor & Tamilwanan, this volume), Mexico (Pischke, this volume), and Honduras (Fromm et al., this volume), where for the most part, the land use change for palm oil transitions has been associated with changes in agricultural land use rather than encroachments on non-agricultural land. By contrast, in areas where greater power differentials exist, where land titling is ambiguous, and where governments are subject to greater influence by commercial elites, palm oil adoption has been accompanied by much greater levels of conflict and even direct violence (Hazlewood et al., this volume).

In a similar vein, the third and fourth questions we raise can only ever be answered on a country-by-country basis. Crucially, however, in the case of conflict resolution and questions regarding the accumulation of wealth, it is not the crop itself, the landscapes in which it can be cultivated, or any intrinsic industry characteristic that determines how conflicts emerge and are dealt with, or ultimately who gets to benefit from the industry. The answers to these questions are to be found in understanding the relationships between different entities in oil palm assemblages in different countries. In understanding how oil palm assemblages are governed, questions of the economic and agronomic efficiency of the crop are nothing as compared to ideas concerning the political efficiency of the crop in delivering to those entities in the assemblage that enjoy power and influence. On the one hand, palm oil is unlikely to bring the kind of social development benefits promised to nations with poor governance. On the other hand, the crop is highly unlikely to fatally undermine institutions and practices in nations that possess more stable and inclusive governance systems.

This collection alerts us to the fact that oil palm is not a crop in and of itself and that there is a distinction to be made between "rural oil palm landscapes" and "rural landscapes with oil palm in them." In this context, we need to gain a deeper understanding of these landscapes and how the oil palm assemblages sit within them. Divested of its status as an industry apart, it becomes clear that questions around oil palm are in some ways, those facing agriculture more generally. Writing seven years apart, seminal papers by Rosegrant and Cline published in 2003 and by Godfray et al. in 2010 succinctly outline these challenges as being how to maintain yield increases in a context in which environmental and social constraints have changed (Godfray et al., 2010; Rosegrant & Cline, 2003). Both propose solutions that are similar to those that have dominated agricultural research since the dawn of the green revolution, if not before: further investment in a broad scientific research agenda, along with additional investment in relation to the provision of infrastructure and extension.

Despite the nuances that typify scientific discourse, these proposals are ones that are familiar to many of those involved in the palm oil industry, focussing on technical challenges and production issues. They are in line with current industry practices. Rosegrant and Cline (2003, p. 1918), for example, call for "investment in water harvesting technologies, crop breeding, and extension services, as well as good access to markets, credit, and supplies," while Godfray et al. (2010, p. 813) state that "existing technologies and best practices need to be spread by education and extension services, and market and finance mechanisms are required to protect farmers." Far from representing a break with previous practices or, as Godfray et al. (2010, p. 817) suggest, a "revolution in the social and natural sciences," both papers propose the incorporation of new issues into the "mainstream" of agricultural research, thus continuing long-established incrementalist traditions rather than critical analysis (Long & van der Ploeg, 1989). This reflects a broad tendency whereby rural and agricultural research and development advocates for further scientific research in line with existing practices in the face of changing global challenges.

This perspective is unsurprising given the historical narrative around scientific research and public policy related to agriculture. New machinery, agrochemicals, and extension strategies, as well as the breeding of new high-yielding varieties of a number of crops, undoubtedly contributed to boosting yields and securing development across the globe. Oil palm sits comfortably within these narrative fields as an outstanding example of a crop which, in the narrow and productivity-centric field of modern agronomy and agricultural policy, represents a success. In this respect, the story of palm oil is similar to that of agriculture more generally. With regard to its own objectives, there is no doubt that the oil palm revolution in Malaysia worked. In consequence, it has and continues to be presented as powerful evidence of the benefits that science and "big" agriculture can deliver to the modern world. The discourse around the crop also embodies many features of discourse common to responses to the problems of mainstream agriculture. The suggestion that more research, new technologies and further extension can overcome the problems of modern agriculture is intrinsic to RSPO and other palm oil sustainability initiatives.

Yet, as with other agricultural policies and programmes, the oil palm industry embodies a paradox. In much the same way as classical and neo-liberal economists have long railed against the global food supply, it is abundantly clear across almost all of the studies incorporated into this collection that palm oil cultivation involves the production of private goods in the form of oils and oleochemicals, which can also affect the provision of public goods such as food security, biodiversity, and climate regulation. It is also clear that the state plays a central role in the development of national palm oil assemblages and that, far from being the slave of market forces, domestic palm oil assemblages are capable of being shaped toward national priorities and preferences in terms of who grows, who processes, who sells, and who buys.

The chapters included in this volume confirm that the experience of palm oil supports broader critiques of approaches to agricultural development, which suggest an inevitable move towards intensification and bigger production units (see, for example, Hickey, 2009; De vries, 2007; Ferguson, 1994). Globally, the past two decades have seen re-engagement with smaller scale agriculture amongst academics, non-governmental organisations (NGOs), and policymakers. This has been driven by recognition that existing policies fail to recognise the multifunctional role agriculture serves in development and a movement to address this deficiency through research and policies that address this broader role (McIntyre et al., 2009). This re-engagement has been reflected in the Southeast Asian region (Wong & Lim, 2019; Timmer, 2008).

Amongst other things, this turn has contributed to a renewed recognition of the existence of alternative rural and agricultural development trajectories using low, as well as high technology. It has also contributed to an appreciation of the durability and crucial role of smallholders in global agriculture (Graeub et al., 2015; International Fund for Agricultural Development & United Nations Environment Programme, 2013; Maass Wolfenson, 2013) and a recognition that different ways of producing the same crops can result in a very different series of outcomes and benefits. Likewise, this has alerted commentators to the broader potential of small-scale production in food supply, economic development, and the support of local economic networks (Wiggins, 2009; Barrett, 2008). As we have seen, oil palm cultivation by smallholders enjoys an uneasy relationship with policy and research agendas at best, a feature shared with many agricultural activities (van der Ploeg, 2010; Collinson, 2001). The idea that the smallholder appears as a deviation from the gold standard of modern agriculture is far from being exclusively relevant to oil palm production. However, as in agriculture more generally, the studies in this book suggest that there is far more than simply one way to do palm oil and that different modes of production result in very different balances of positive and negative outcomes. Given the undoubted and enduring appeal of the golden crop, a recognition of the potential that exists to shape national, provincial and local oil palm assemblages around different priorities, needs, and demands alerts us to the possibility that political as well as economic concerns can shape the way the crop is cultivated, and to whose benefit.

Reference list

Andrianto, A., Fauzi, A., & Falatehan, A. (2019). The typologies and the sustainability in oil palm plantation controlled by independent smallholders in Central Kalimantan. In R. Kinseng, A. Dharmawan, D. Lubis, & A. Seminar (Eds.), *Rural socio-economic transformation: Agrarian, ecology, communication and community, development perspectives* (pp. 3–14). CRC Press.

Arancibia, F. (2013). Challenging the bioeconomy: The dynamics of collective action in Argentina. *Technology in Society, 35*(2), 79–92. https://doi.org/10.1016/j.techsoc.2013.01.008

Baraibar Norberg, M. (2019). *The political economy of agrarian change in Latin America: Argentina, Paraguay, and Uruguay.* Springer.

Barrett, C. B. (2008). Smallholder market participation: Concepts and evidence from eastern and southern Africa. *Food Policy, 33*(4), 299–317. https://doi.org/10.1016/j.foodpol.2007.10.005

Collinson, M. (2001). Institutional and professional obstacles to a more effective research process for smallholder agriculture. *Agricultural Systems, 69*(1–2), 27–36. https://doi.org/10.1016/s0308-521x(01)00016-6

Córdoba, D., Chiappe, M., Abrams, J., & Selfa, T. (2017). Fuelling social inclusion? Neo-extractivism, state-society relations and biofuel policies in Latin America's southern cone. *Development and Change, 49*(1), 63–88. https://doi.org/10.1111/dech.12362

Corley, R. H. V. (2009). How much palm oil do we need? *Environmental Science & Policy, 12*(2), 134–139. https://doi.org/10.1016/j.envsci.2008.10.011

Dauvergne, P. (2018). The global politics of the business of "sustainable" palm oil. *Global Environmental Politics, 18*(2), 34–52. https://doi.org/10.1162/glep_a_00455

De vries, P. (2007). Don't compromise your desire for development! A Lacanian/Deleuzian rethinking of the anti-politics machine. *Third World Quarterly, 28*(1), 25–43. https://doi.org/10.1080/01436590601081765

Ferguson, J. (1994). *The anti-politics machine: "Development," depoliticisation, and bureaucratic power in Lesotho.* University of Minnesota Press.

Fry, J., & Fitton, C. (2010). The importance of the global oils and fats supply and the role that palm oil plays in meeting the demand for oils and fats worldwide. *Journal of the American College of Nutrition, 29*(3), 245S252S. https://doi.org/10.1080/07315724.2010.10719841

Godfray, H. C. J., Beddington, J. R., Crute, I. R., Haddad, L., Lawrence, D., Muir, J. F., Pretty, J., Robinson, S., Thomas, S. M., & Toulmin, C. (2010). Food security: The challenge of feeding 9 billion people. *Science, 327*(5967), 812–818. https://www.science.org/doi/10.1126/science.1185383

Graeub, B. E., Chappell, M. J., Wittman, H., Ledermann, S., Kerr, R. B., & Gemmill-Herren, B. (2016). The state of family farms in the world. *World Development, 87*, 1–15. https://doi.org/10.1016/j.worlddev.2015.05.012

Haiven, M. (2022). *Palm oil: The grease of empire.* Pluto Press.

Hickey, S. (2009). The politics of protecting the poorest: Moving beyond the "anti-politics machine"? *Political Geography, 28*(8), 473–483. https://doi.org/10.1016/j.polgeo.2009.11.003

International Fund for Agricultural Development & United Nations Environment Programme. (2013). *Smallholders, food security, and the environment* (pp. 1–52). IFAD.

Jelsma, I., Schoneveld, G. C., Zoomers, A., & van Westen, A. C. M. (2017). Unpacking Indonesia's independent oil palm smallholders: An actor-disaggregated approach to identifying environmental and social performance challenges. *Land Use Policy, 69*, 281–297. https://doi.org/10.1016/j.landusepol.2017.08.012

Long, N., & van der Ploeg, J. D. (1989). Demythologising planned intervention: An actor perspective. *Sociologia Ruralis, 29*(3–4), 226–249. https://doi.org/10.1111/j.1467-9523.1989.tb00368.x

Maass Wolfenson, K. D. (2013). *Coping with the food and agriculture challenge: Smallholders' agenda* (pp. 1–47). FAO.

Manzo, K., Padfield, R., & Varkkey, H. (2019). Envisioning tropical environments: Representations of peatlands in Malaysian media. *Environment and Planning E: Nature and Space, 3*(3), 251484861988089. https://doi.org/10.1177/2514848619880895

McCarthy, J. (2010). Processes of inclusion and adverse incorporation: Oil palm and agrarian change in Sumatra, Indonesia. *The Journal of Peasant Studies, 37*(4), 821–850. https://doi.org/10.1080/03066150.2010.512460

McIntyre, B. D., Herren, H. R., Wakhungu, J., & Watson, R. T. (Eds.). (2009). *Agriculture at a crossroads* (pp. 1–590). International Assessment of Agricultural Knowledge, Science, and Technology for Development.

McKay, B. M. (2017). Agrarian extractivism in Bolivia. *World Development, 97*, 199–211. https://doi.org/10.1016/j.worlddev.2017.04.007

Meijaard, E., Brooks, T. M., Carlson, K. M., Slade, E. M., Garcia-Ulloa, J., Gaveau, D. L. A., Lee, J. S. H., Santika, T., Juffe-Bignoli, D., Struebig, M. J., Wich, S. A., Ancrenaz, M., Koh, L. P., Zamira, N., Abrams, J. F., Prins, H. H. T., Sendashonga, C. N., Murdiyarso, D., Furumo, P. R., & Macfarlane, N. (2020). The environmental impacts of palm oil in context. *Nature Plants, 6*(12), 1418–1426. https://doi.org/10.1038/s41477-020-00813-w

Naylor, R. L., Higgins, M. M., Edwards, R. B., & Falcon, W. P. (2019). Decentralisation and the environment: Assessing smallholder oil palm development in Indonesia. *Ambio*, *48*(10), 1195–1208. https://doi.org/10.1007/s13280-018-1135-7

Padfield, R., Varkkey, H., Manzo, K., & Ganesan, V. (2023). Time bomb or gold mine? Policy, sustainability and media representations of tropical peatlands in Malaysia. *Land Use Policy*, *131*, 106628. https://doi.org/10.1016/j.landusepol.2023.106628

Purnomo, H., Okarda, B., Dermawan, A., Ilham, Q. P., Pacheco, P., Nurfatriani, F., & Suhendang, E. (2020). Reconciling oil palm economic development and environmental conservation in Indonesia: A value chain dynamic approach. *Forest Policy and Economics*, *111*, 102089. https://doi.org/10.1016/j.forpol.2020.102089

Rosegrant, M. W., & Cline, S. (2003). Global food security: Challenges and policies. *Science*, *302*(5652), 1917–1919. https://doi.org/10.1126/science.1092958

Saguier, M., Gerlak, A. K., Villar, P. C., Baigún, C., Venturini, V., Lara, A., & dos Santos, M. A. (2021). Interdisciplinary research networks and science-policy-society interactions in the Uruguay River Basin. *Environmental Development*, *38*, 100601. https://doi.org/10.1016/j.envdev.2020.100601

Saleh, S., Bagja, B., Suhada, T. A., & Widyapratami, H. (2018). Intensification by smallholder farmers is key to achieving Indonesia's palm oil targets. World Resources Institute. https://www.wri.org/insights/intensification-smallholder-farmers-key-achieving-indonesias-palm-oil-targets

Siegel, K. M. (2016). Fulfilling promises of more substantive democracy? Post-neoliberalism and natural resource governance in South America. *Development and Change*, *47*(3), 495–516. https://doi.org/10.1111/dech.12234

Syahza, A., Irianti, M., Suwondo, & Nasrul, B. (2020). What's wrong with palm oil, why is it accused of damaging the environment? *Journal of Physics: Conference Series*, *1655*, 012134. https://doi.org/10.1088/1742-6596/1655/1/012134

Syahza, A., Rosnita, R., Suwondo, S., & Nasrul, B. (2013). Potential oil palm industry development in Riau. *International Research Journal of Business Studies*, *6*(2), 133–147. https://doi.org/10.21632/irjbs.6.2.133-147

Timmer, P. (2008). Agriculture and pro-poor growth: An Asian perspective. *SSRN Electronic Journal*, *5*. https://doi.org/10.2139/ssrn.1114155

van der Ploeg, J. D., Ye, J., & Schneider, S. (2010). Rural development reconsidered: Building on comparative perspectives from China, Brazil and the European Union. *Rivista Di Economia Agraria*, *65*(2), 163–190.

Vu, T. (2007). State formation and the origins of developmental states in South Korea and Indonesia. *Studies in Comparative International Development*, *41*(4), 27–56. https://doi.org/10.1007/bf02800470

Wiggins, S. (2009). *Can the smallholder model deliver poverty reduction and food security for a rapidly growing population in Africa* (pp. 1–20). FAO.

Wong, C., & Lim, G. (2019). A typology of agricultural production systems: Capability building trajectories of three Asian economies. *Asia Pacific Viewpoint*, *61*(1), 37–53. https://doi.org/10.1111/apv.12220

INDEX

Note: **Bold** page numbers refer to tables and *italic* page numbers refer to figures.

Printed in the United States
by Baker & Taylor Publisher Services